Public Branding and Marketing

Staci M. Zavattaro
Editor

Public Branding and Marketing

A Global Viewpoint

Springer

Editor
Staci M. Zavattaro
School of Public Administration
University of Central Florida
Orlando, FL, USA

ISBN 978-3-030-70504-6 ISBN 978-3-030-70505-3 (eBook)
https://doi.org/10.1007/978-3-030-70505-3

This Springer imprint is published by the registered company Springer Nature Switzerland AG
The registered company address is: Gewerbestrasse 11, 6330 Cham, Switzerland

To my parents

Contents

Author Biographies

Jan Boon (jan.boon@uantwerpen.be) is a senior postdoctoral researcher at the University of Antwerp and professor at the Antwerp Management School. His research interests include public sector innovation, change and reform, and the reputation and legitimacy of government organizations.

Afsan Chowdhury is a professional historian and media and development specialist. He has worked nationally and internationally for the UN and various agencies including UNICEF, BRAC, Panos South Asia, BBC, Ekushey Television, Real News Canada, and Bangladesh-Canada Community Services. Chowdhury received the Bangla Academy Award–2018 for his writing on the liberation war of Bangladesh. He was research associate at the York Centre for Asian Research, York University, Toronto (2007–2014) and visiting fellow at Oxford University (Centre for Forced Migration) in 2007. He became an Ashoka fellow for life in 1999. He was the Oak fellow on international human rights at Colby College, USA, in 2008. Chowdhury has edited, authored, and co-authored over a dozen of books on the history of Bangladesh, health and environmental issues, refugees, human rights, etc. Currently, he is teaching research methodology at the BRAC Institute of Governance and Development, BRAC University, Bangladesh.

Julia Cummings (julia.cummings@ucdenver.edu) has worked as a marketing director for CU Denver's School of Education & Human Development since 2007. She is also a PhD student in CU Denver's School of Public Affairs. Her research interests include public administration, public policy, marketing the value of public institutions, and voter engagement. She earned a BA in humanities *magna cum laude* from CU Boulder and an MBA with a specialization in marketing from CU Denver. Her previous work experience includes marketing and public affairs roles at the Smithsonian Institution, Audubon Society, and Denver Museum of Nature & Science.

Dag Yngve Dahle (dag.y.dahle@hiof.no) is assistant professor of organization and leadership at Østfold University College, Norway, and PhD candidate at Norwegian

University of Life Sciences, School of Economics and Business. His empirical and theoretical research interests include organizational branding, HRM, and employee voice within both the private and public sector.

Fred Gale (fred.gale@utas.edu.au) is professor of politics and international relations, School of Social Sciences, University of Tasmania, Australia. His research focuses on rethinking the discipline of political economy from a sustainability value perspective and working through the implications of trade, investment, finance, labor, and cultural practices for critical coupled human and natural systems (CCHANS). He has undertaken investigations into local, national, and global private governance systems, especially those using multistakeholder approaches to develop social and environmental standards. His sectoral expertise is in forestry and food governance and he has undertaken research from a political economy of sustainability perspective into deforestation, fisheries depletion, animal welfare, fair trade and organics.

Laura Ripoll González, PhD (l.ripollgonzalez@griffith.edu), is currently an adjunct researcher at Griffith's Centre for Social and Cultural Research, Griffith University, Australia. Laura completed her PhD at the University of Tasmania, Australia, and spent a period as a postdoc visiting scholar in the Department of Communications at Pompeu Fabra University, Barcelona. Laura has recently been awarded a European Union (EU)-funded Marie Curie Postdoc Fellowship at Erasmus University, Rotterdam. Laura's project, BRANDSUS, focuses on the study of inclusive governance arrangements for sustainable place development, responding to calls for more holistic models of regional development in the EU following the goals of the 2030 Agenda for Sustainable Development, including economic social and environmental sustainability.

Jenny Kosek (joksek@westalliswi.gov) is the communications strategist for the City of West Allis, Wisconsin. She has more than a decade of experience in writing, marketing, and branding. She also is an industry blogger, copywriter, and messaging strategist.

Ulrika Leijerholt, PhD (ulrika.leijerholt@umu.se), is a senior lecturer at Umeå School of Business, Economics and Statistics, Umeå University. As an experienced marketing practitioner, Dr. Leijerholt has worked in strategic and operative roles across multiple industries and within both the public and private sector. In her research, her focus is primarily on various aspects of public sector branding.

Vincent Mabillard (vincent.mabillard@ulb.be) holds a PhD in public administration from the University of Lausanne, Switzerland. He is currently working as assistant professor at the Solvay Brussels School of Economics & Management, Université Libre de Bruxelles (ULB), Belgium. His research interests include transparency in the public sector, place branding, public management and government communication.

Tim A. Mau (tmau@uoguleph.ca) is associate professor in the Department of Political Science at the University of Guelph. In addition to teaching undergraduate and graduate courses on public management and administration and leadership, he has provided leadership training and development for senior public sector officials in Canada (for the Municipal Property Assessment Corporation) and across Asia (for the Asian Productivity Organization). His current research interests are primarily in the area of political and public sector (administrative) leadership. He has published articles and book chapters on a wide range of topics, including the use of public sector leadership competencies, the Canadian model of public sector leadership, leadership in the Canadian military, scandal and corruption, French-English relations in business interest associations, and nationalist mobilization in Quebec and Scotland.

Ken McConnellogue (ken.mcconnellogue@cu.edu) has served as the University of Colorado's vice president for communication for nearly 15 years. His responsibilities include leading the overall branding and marketing of the university, and he has both scars and smiles to prove it. He earned a journalism degree (with honors) from CU Boulder.

Ellen Minkman (e.minkman@tudelft.nl) is a lecturer and researcher in the Faculty of Technology, Policy and Management at Delft University of Technology in the Netherlands. Her research concentrates on the governance of transitions and international cooperation in water management. Previously, she completed a PhD at Erasmus University Rotterdam, where she studied policy transfer with a focus on the use of Dutch ideas adaptive delta management in Bangladesh, Vietnam, and Indonesia.

Peter Nafzger (p.nafzger@mulierinstituut.nl) is a researcher at the Mulier Institute. He studied public administration and organisation science at Utrecht University. His interest focuses on organisational development, digitalisation and public health issues, and his current research monitors the progress of the national sport policy programs and corona-impact on the Dutch sport landscape.

Arjuman Naziz (arjuman@juniv.edu) is an assistant professor in the Department of Public Administration, Jahangirnagar University, Bangladesh. Her research focuses on media framing, public health, policy advocacy, and informal institutions. Prior to joining the university, Naziz worked at the Advocacy for Social Change, BRAC Bangladesh, where she was involved in designing and implementing program communication interventions, including national level policy advocacy. Naziz is a Commonwealth Scholar. She has earned her MSc in international development (public policy and management) from the University of Manchester.

Wawrzyniec Rudolf (wawrzyniec.rudolf@uni.lodz.pl) is a university professor in the Department of City and Regional Marketing at the Faculty of Management, University of Lodz, Poland. He teaches territorial marketing and the use of European funds in Poland. In his research activity, he is interested in territorial marketing and public management. He has published more than 60 articles on these topics.

Vidar Stevens (v.stevens@mulierinstituut.nl) is a researcher at the Mulier Institute in Utrecht, the Netherlands. His research focuses on collaborations, network governance, public branding and violence and integrity in sports.

M. Blair Thomas, PhD, MPH, MPA (michaelbthomas@valdosta.edu), is an assistant professor in the Department of Political Science at Valdosta State. His research interests broadly focus on how local and state governments engage in marketing and branding from a strategic management perspective.

Roanne van Voorst (r.s.vanvoorst@uva.nl) is a researcher (PhD) and writer. As an "anthropologist of the future," her core research focuses on future scenarios such as urbanization, robotification and new tech, and their impacts on society and humanity. Roanne is affiliated to the University of Amsterdam and is a board member of the Dutch Future Society.

Axelle Verdonck was a master's student of political communication at the University of Antwerp and is currently working as junior CRM consultant at C-Clear Partners.

Renaud Vuignier (renaud.vuignier@icloud.com) is currently working as Deputy Head of the Swiss Agency for Development and Cooperation China Office. He obtained his PhD from the University of Lausanne, Switzerland, where he is active as fellow researcher at the Swiss Graduate School of Public Administration (IDHEAP). His research interests focus on place branding, economic development, innovation policy and public management.

Arild Wæraas (arild.waeraas@vid.no) is professor of organization and leadership at VID Specialized University in Oslo, Norway. His research interests include reputation and branding, organizational identity, and the translation of ideas and practices.

Staci M. Zavattaro, PhD (staci.zavattaro@ucf.edu), is professor of public administration at the University of Central Florida. Her research focuses on public branding and marketing, administrative theory, and municipal management. She serves as editor-in-chief of *Administrative Theory & Praxis*.

Introduction

Staci M. Zavattaro

When I put in a proposal for this edited volume, things seemed almost simple in the world. I proposed this book on global public branding in early 2019, meant to be a dialogue between place branding scholars and practitioners. As authors began writing chapters, and as I sit and write this introduction in late 2020, a global pandemic hit and continues to rage. COVID-19 gripped the world starting in January 2020 (more formally at the end of 2019, given the "19" in the nomenclature), causing millions to fall ill and hundreds of thousands (as of this writing in December 2020) to die. I am based in the United States, and as such have been watching the federal government and my state of Florida bungle the response – to say the least. In October 2020, members of the White House staff indeed gave up, telling people they will no longer "fight" the virus, so we just have to live with it. Florida in December 2020 hit a million confirmed virus cases and climbing. The policy response in the United States has been truly problematic and compounded by our federal structure (Kettl 2020).

Germane to this volume on global public branding, the United States' reputation among allied countries plummeted in the wake of the country's failed response to the pandemic (Wike et al. 2020). Indeed, most respondents to the Pew Research Center's survey indicated the United States has handled the pandemic the worst globally. Eighty-three percent of respondents reported President Trump had not done the right thing regarding world affairs (Wike et al. 2020). To be blunt: these assessments of the country's reputation are bleak.

I realize the golden rule of writing is not to include material that will "date" your work. But it seems naïve and irresponsible to ignore a global pandemic wreaking social, economic, political, cultural, and moral havoc on the world. President Trump himself caught the highly contagious virus, took a helicopter to Walter Reed Medical Center, received experimental treatment not available to other Americans, then

S. M. Zavattaro (✉)
School of Public Administration, University of Central Florida, Orlando, FL, USA
e-mail: Staci.Zavattaro@ucf.edu

S. M. Zavattaro (ed.), *Public Branding and Marketing*,
https://doi.org/10.1007/978-3-030-70505-3_1

came back to the White House to tweet about not letting the virus control your life. Needless to say, such brashness was met with much resistance from people who demanded empathy from a leader who has never shown any. While there is much to learn and write about the pandemic, this book is focused on global branding, and thanks to the Trump administration, the United States' brand is hurting (Wike et al. 2020; Zavattaro and Fay 2019).

And while I write from experience being in the United States and watching firsthand its reputation plummet, this problem is not contained to only this country. As the World Health Organization declared a global pandemic, much of the global economy came to a screeching halt. Planes were grounded. Hotels closed. Tourism basically stopped to areas needing it (Vegnuti 2020). Shops shuttered their doors, possibly for good. People became disenchanted with government responses. Consumers learned firsthand the tightly coupled system that is the global supply chain, as goods such as toilet paper, cleaning supplies, and even some food items were and remain in short supply (Rossolatos 2020).

The virus seemed in a way to take on a life of its own, creating its own narrative and brand (Rossolatos 2020). As Rossolatos (2020) describes, the narrative about and of the virus seemed to transform from a purely medical one to one rooted in terror and destruction, which then created more war-like narratives about fighting a common enemy. Additionally, when the virus became branded as the "China virus" that created another layer of problems including a rise in xenophobia and violence against Asian communities (Roberto et al. 2020). From a branding standpoint, Rossolatos (2020) notes the power of narratives and storytelling expands the power of place brands beyond a specific geographic location to a movable thing.

Public brands, reputations, and identities are crucial for success on myriad levels, not just an economic one (Eshuis and Klijn 2012). But what is a brand anyway? We usually associate the term, and its cousins marketing and public relations, with practices from the corporate world. But as New Public Management gripped the public sector, importing business-based practices in the name of efficiency, it was natural this one followed as well. Eshuis and Klijn (2012, p. 6) define a brand as "a symbolic construct that consists of a name, term, sign, symbol or design, or combination of these, intended to identify a phenomenon and differentiate it from similar phenomena by adding a particular meaning to it." Branding, then, is the active process of distilling, refining, and communicating a brand identity (Anholt 2010). Place branding is the application of this logic to geographic entities (Eshuis and Edwards 2013), but the interesting realization is that without trying, many places already have a brand image in the minds of consumers and people (Anholt 2010).

It might sound simple enough – import another concept rooted in corporate ideals into the public sector and see results. The complication comes when myriad stakeholders have a different vision for what constitutes a successful public brand (Hanna and Rowley 2011). As Hanna and Rowley (2011) argue, a branding process ideally begins with identifying relevant stakeholder groups to come together to influence the brand creation and vision. Once this is done, the brand and its associated communication elements are then evaluated and changed based on feedback and how users perceive the brand – the ultimate marker of success. It should be obvious this is challenging given all those stakeholder groups have unique visions of how to measure and evaluate success (Kavaratzis and Kalandides 2015; Van Ham 2008).

I am sure some readers are asking the question: Why do public entities want to brand given there is little competition for resources? Wæraas et al. (2015, p. 1290) answer this question by arguing: "Municipalities compete with other municipalities with respect to reputation, human capital, financial resources and organizational culture, as well as recruitment and employer policies". While marketing and branding in the corporate world might focus solely on increasing a financial bottom line, in the public sector this concept manifests in terms of taxes, social infrastructure, and place loyalty (Butler et al. 2007). And when we see branding as a key governance strategy, rather than an add on, it shifts the purpose of the task from one of simply "keeping up with the Joneses" to that of gaining advantage in key social, cultural, political, and economic arenas (Eshuis and Klijn 2012; Go and Trunfio 2012).

If we come to realize places and public entities are branding – not debating about *if* they should – we can move on to answering additional questions about how to know if a brand strategy is working. Again, this becomes complex given the various views of success (Zenker and Martin 2011). The easiest measures of success include economic ones because managers can see changes from year to year – tax revenues are going up or down, hotel stays are going up or down, and receipts in local businesses are going up or down. In addition to this logical approach, Teodoro and An (2018) introduce citizen-based brand equity, again borrowing a term from corporate branding referring to someone's proclivity to buy a branded product versus a non-branded alternative. They borrow the concept and test it on people exposed to US federal government brands and reputational materials, noting how organizations can focus on building emotional ties to increase brand equity.

What this short introduction highlights is the realization that scholarship in public administration and management specifically related to public branding is nascent yet growing – and this volume aims to expand that conversation by taking a distinctly global view of how public entities throughout the world use and understand branding practices to gain some kind of advantage, economic or otherwise. In this field, most of the research into public branding and marketing comes from European scholars – though there is a small group of American-based and globally based scholars attempting to shift this trend. To be clear, I am specifically referring to public management scholars studying place branding, not, for instance, corporate marketing scholars who study branding in public entities or tourism scholars studying destinations. Coming from a public administration and management lens brings a different knowledge background and incorporates administrative theory and practice into the studies. My hope is the scholars in this book begin to expand this conversation and give readers additional insights into branding in the public sector.

The first part of this book contains chapters more macro in nature, focusing on theories or ideas related to place branding. The second part of this book focuses on internal branding efforts in public organizations, perhaps the meso-level of place branding implementation. Finally, the third part contains interesting case studies of how places are using branding strategies for particular ends. These chapters highlight the on-the-ground realities of implementing and evaluating branding strategies. There are still a couple of chapters from practitioners, though most pulled out when COVID-19 hit, focusing on organizational crisis communication.

References

Anholt, S. (2010). Defintions of place branding: Working towards a resolution. *Place Branding and Public Diplomacy, 6*, 1–10.

Butler, P., Collins, N., & Fellenz, M. R. (2007). Theory-building in political marketing. *Journal of Political Marketing, 6*(2–3), 91–107.

Eshuis, J. & Klijn, E. (2012). Branding in governance and public management. Routledge.

Eshuis, J., & Edwards, A. (2013). Branding the City: The democratic legitimacy of a new mode of governance. *Urban Studies, 50*(5), 1066–1082.

Go, F., & Trunfio, M. (2012). A paradigm shift from tourism destination management to democratic governance of place branding. *Journal of Travel and Tourism Research, S1*, 4–17.

Hanna, S., & Rowley, J. (2011). Towards a strategic place brand-management model. *Journal of Marketing Management, 27*(5–6), 458–476.

Kavaratzis, M., & Kalandides, A. (2015). Rethinking the place brand: The interactive formation of place brands and the role of participatory place branding. *Environment and Planning A, 47*, 1368–1382.

Kettl, D. (2020). States divided: The implications of American federalism for COVID-19. *Public Administration Review, 80*(4), 595–602.

Roberto, K. J., Johnson, A. F., & Rauhaus, B. M. (2020). Stigmatization and prejudice during the COVID-19 pandemic. *Administrative Theory & Praxis, 42*(3), 364–378.

Rossolatos, G. (2020). A brand storytelling approach to COVID-19's terrorealization: Cartographing the narrative space of a global pandemic. *Journal of Destination Marketing & Management*. https://doi.org/10.1016/j.jdmm.2020.100484.

Teodoro, M. P., & An, S. (2018). Citizen-based brand equity: A model and experimental evaluation. *Journal of Public Administration Research and Theory, 28*(3), 321–338.

Van Ham, P. (2008). Place branding: The state of the art. *The Annals of the American Academy of Political and Social Science, 616*(1), 126–149.

Vegnuti, R. (2020). Cinque Terre, Italy – A case of place branding: From opportunity to problem for tourism. *Worldwide Hospitality and Tourism Themes, 12*(4), 471–483.

Wæraas, A., Bjørnå, H., & Moldenæs, T. (2015). Place, organization, democracy: Three strategies for municipal branding. *Public Management Review, 17*(9), 1282–1304.

Wike, R., Fetterolf, J., & Mordecai, M. (2020). U.S. image plummets internationally as most say country has handled coronavirus badly. Retrieved from: https://www.pewresearch.org/global/2020/09/15/us-image-plummets-internationally-as-most-say-country-has-handled-coronavirus-badly/

Zavattaro, S. M. & Fay, D. L. (2019). Brand USA: A natural quasi-experiment evaluating the success of a national marketing campaign. *Tourism Management, 70*, 42–48.

Zenker, S. & Martin, N. (2011). Measuring success in place marketing and branding. *Place Branding and Public Diplomacy, 7*, 32–41.

Part I
A Macro View of Place Branding

As you read this, think about a place you like to vacation. Maybe you prefer sunshine and sandy beaches. Maybe you are a mountain person bent on skiing and drinking hot chocolate in the lodge. Perhaps you seek arts and culture so prefer destinations with museums and galleries. Or maybe you want a place with quiet and solitude so seek a remote destination where you can unplug and escape.

Now repeat this same experiment again with products. When you shop down the laundry detergent aisle, what brand do you choose and why? Maybe you need something without harsh chemicals, or perhaps there is a brand your parents chose so you simply carry on that tradition. You can do the same thought experiment with soap, clothing, wine, etc. The list is literally endless.

Inherent in these scenarios is economic competition. As Schwartz (2004) reminds us, as more choices become available, the harder it is to actually make a decision. Brands can serve as those cognitive shortcuts helping aid our decision process. We choose something because we know it works. It is reliable. It is what our parents did. It makes us happy. In a world fraught with decisions, cognitive shortcuts are not nice to have – they are necessary (Schwartz 2004).

There is a rich literature on destination branding and product branding – the two styles used above. But what needs more attention is why places would want a brand apart from being grand destinations or well-known products or organizations. Why would a city in the middle of Iowa or Florida want a brand identity? What would it do for the place? Why might a small museum on the Blues Trail in Mississippi seek its own brand identity? Why should a public hospital or university brand and market its assets? Traditionally people think there is no competition in the public sector, but as market logic rooted in business practices became prevalent, places were forced to compete (Eshuis and Klijn 2012; see also Dahle in this volume).

Typically, place branding strategies have an economic focus, and Papadopoulos (2004) outlines four of them: improve a place's exports, insulate domestic businesses from foreign competition, encourage development, and "generally position the place for advantage domestically and internationally in economic, political, and social terms" (p. 37). Places can view themselves – for good or ill – as products worthy of gaining a competitive advantage, so engaging in branding and marketing is one way to achieve those ends. There is coming realization, though, that place brands can do more than deliver economic value but also can contribute to creating positive public values as well (Herezniak and Anders-Morawska 2020).

Vuignier (2017) undertook an extensive literature review to understand the evolution of place branding research. By analyzing 1,172 articles in 98 journals between 1976 and 2016, he found several patterns of note: much of the work comes from private marketing journals, with public management journals coming second, most articles are descriptive, and there is a large focus on image and identity. Overall, Vuignier (2017) concludes the field lacks conceptual clarity, relies heavily on best practices when more critique is needed, disconnects from the political aspects of places, and lacks empirical evidence to substantiate many hypotheses. Put simply: more work, especially within public management, needs to be done to understand the emergence of the branding phenomenon (van Ham 2002).

In 2018, I posed some big questions for place branding via a public management lens, touching on many of the themes Vuignier (2017) identified. In that article (Zavattaro 2018), I asked questions such as: why even brand in the first place? How do people perceive the brand, and how do brand creators know? Who owns the brand anyway? How can places measure success, especially when brands are more about economic attraction? Chapters in this part, and throughout the book, touch on some of these questions by explaining the challenges public entities face when engaging in brands, the reasons to engage stakeholders meaningfully, and the problems faced when trying to access data to evaluate brand success.

References

Eshuis, J., & Klijn, E. (2012). *Branding in governance and public management.* London: Routledge.

Herezniak, M., & Anders-Morawska, J. (2020). Public value-drive place branding. The way forward? *Place Branding and Public Diplomacy*, https://doi.org/10.1057/s41254-020-00185-0

Schwartz, B. (2004). *The paradox of choice: Why more is less.* Harper Collins.

Vuignier, R. (2017). Place branding & place marketing 1976–2016: A multidisciplinary literature review. *International Review of Public and Nonprofit Marketing, 14*, 477–473.

Economic Coopetition and the Absence of Comprehensive Data: Why Scholars and Practitioners Should Work Hand in Hand

Vincent Mabillard and Renaud Vuignier

Introduction

Policies designed to attract and retain residents, tourists, and companies are of vital importance for public sector organizations (Poirot and Gérardin 2010). The added value of all these stakeholders often proves critical to a strong economy and the global quality of life in a given area. Taking this assessment into account, place branding strategies are increasingly being shaped by policymakers in numerous regions and countries. These strategies remain vague and encompass several aspects like marketing, branding, positioning, repositioning, and regeneration of places at all levels – national, regional, and local (Lucarelli and Berg 2011). Such vagueness has even led to superficial use of the concept and conceptual confusion (Kumar and Panda 2019), and debates have been lively among the research community. Consequently, contributions about place branding and marketing have flourished in the last few years, including comprehensive literature reviews (e.g., Vuignier 2017).

In this regard, the present chapter focuses on one specific aspect: the issue of exogenous economic attractiveness. Often regarded as the ultimate goal of place branding, attractiveness policies aim to turn theory and formulation into outputs (e.g., new firms and jobs) and hopefully desired outcomes (e.g., improved quality of life, better place image). In the same vein, Bose et al. (2016), p. 621) argue that "investment is one of the major reasons for place branding." As a result, investors and companies, residents, tourists, and talents have become important target groups for evolving regions and cities. Place branding strategies have therefore been

V. Mabillard (✉)
Solvay Brussels School of Economics and Management, Université Libre de Bruxelles (ULB), Brussels, Belgium
e-mail: vincent.mabillard@ulb.be

R. Vuignier
Swiss Graduate School of Public Administration (IDHEAP), University of Lausanne, Lausanne, Switzerland

S. M. Zavattaro (ed.), *Public Branding and Marketing*,
https://doi.org/10.1007/978-3-030-70505-3_2

adopted by many regions and municipalities to attract those groups. This of course includes businesses since the construction of place image is a central criterion in the decision process of companies to establish their facilities (greenfield investment) or invest in existing facilities (brownfield investment) in a potential host place (Metaxas 2010).

From a global perspective, it is easy to understand how and why attractiveness policies regarding companies are central for regional public authorities. The settlement of new firms can indeed create jobs, revitalize the economic potential of the area, establish attractive clusters for other companies, or even position a region as a leader in a specific field of activity, which enhances the place reputation. As explained below, trade-offs and priorities given to certain segments of the economy over others make this image far more complex than initially thought. Moreover, the reconstruction of places to better match the target groups' desires and needs (Arku and Oosterbaan 2015) have led to situations in which coopetition increasingly goes together with competition. Although factors of attractiveness have been widely studied, contributions about effects in terms of outputs and outcomes have remained quite rare due to the complexity of such an approach.

This chapter aims precisely at exploring one major obstacle that impedes a more thorough analysis of attractiveness policies' concrete effects: the opacity surrounding economic data. This lack of transparency results mainly from reputation issues, the context of coopetition, and their sensitive nature. While data about residents and tourists abound in the case of Switzerland – they are regularly published by agencies in charge of demographic- and tourism-related data (FSO 2020) – precise information about businesses and investment often remains unavailable. This issue negatively impacts scholars who cannot conduct detailed and systematic research. It is also partly detrimental to practitioners, who cannot always identify best practices and make a fully informed evaluation of the situation, although they also have legitimate reasons (e.g., trade secrets) not to disclose all information. At the same time, this joint lack of knowledge calls for a deeper collaboration between scholars and practitioners, as it may be advised by organizations active in economic development and subsequently reported here.

Such implications will be presented and further explained in the main part of the chapter, which is structured as follows. The first part capitalizes on the existing literature to better assess the relationship between place branding and economic attractiveness and the identification of companies as target groups. The second part specifies the context of the study, presenting the peculiarities of the Swiss case and the prevailing context of coopetition. The third part focuses on the (non-)availability of data, insisting on transparency issues. In the fourth part, implications for both research and practice are extensively presented, along with reasons explaining why all data are not publicly disclosed. Finally, the fifth part discusses possible solutions to overcome the opacity/confidentiality problems and to improve research on the effects of attractiveness policies.

Place Branding, Economic Attractiveness, and Companies as a Target Group

Places are often subject to marketing strategies due to the strong willingness of policymakers to promote and to value them, with a clear objective of attracting and retaining diverse target groups. As such, places have often launched a branding strategy aiming to develop a proper identity and to better manage their reputation on the long term. To do so, they have mainly used marketing tools, accompanied by a strong narrative, sometimes regarded as "magical" (Miles 2013). This typically refers to the strong beliefs in the power of strategic objectives regarding place attractiveness and turns the notion of "image" into the concept of "brand image," resulting in great enthusiasm about the creation of logos, mottos, and other visual tools not only to give substance to a sophisticated strategy but rather to generate an attractive place identity (Govers 2013). The performative language typical from this kind of processes, underpinned by other projects, tends to reinform this belief that branding tools and strategies are the best way to attract target groups. Moreover, the confusion between the brand itself and the global place branding approach makes it difficult to clearly identify the role and impact of the brand on the attractiveness of a place. This confusion is reinforced by the tendency to consider cities as brands themselves, since they convey a certain image in people's minds, and to affirm that "every place is a brand" (Boisen 2015).

Although every place is obviously related to a certain image, places are not intrinsically brands. They become brands through a strategic marketing process, supported by a brand effect affecting target groups at various levels of intensity. As proposed by Vuignier (2018), the literature on the matter enables us to identify four aspects related to the branding process from a strategic point of view: (1) places should have recognized limits about what is/is not the place, (2) specific attributes (3) commonly acknowledged by both internal and external actors, and (4) they should have an organizational structure to design a dedicated strategy. Clear delimitations may go beyond institutional borders, as shown by the example of cities' brands applied to a broader region; it should be added here that there is a frequent mismatch between these institutional borders and identity, cultural, and functional areas.

Attributes refer to services delivered by places; without those services, the brand would lose its meaning and substance (Kapferer 2013). In addition, the literature shows that attributes considered as assets of the place must be perceived as such by the residents and external actors. The willingness of people to visit or invest in a given area will heavily depend on this factor. Place audit, calling for a better understanding of the strengths and weaknesses of a place, is therefore desirable and highlighted in several contributions (see Chamard et al. 2014). Finally, politico-administrative capacity to support a brand and to implement a strategy remains an essential factor. The creation of a network of stakeholders or the establishment of a specific agency is indeed a central condition to develop and to strengthen the place brand.

These preliminary remarks show that the brand issue has become particularly salient. Therefore, it has triggered multiple initiatives around the world, ranging from cities (e.g., Amsterdam or Reno, NV; see Zavattaro 2014) to countries (e.g., Ghana or the United Kingdom; see Dinnie 2015).

Place brands have several functions including distinctiveness, orientation and information, trust or identity (Vuignier 2018). They refer to the capacity for places to develop a sense of distinctiveness and to trigger positive associations in people's minds, whether they are internal stakeholders or external target groups. They also aim at creating a proper identity, characterized by a strong symbolic component and the motivation to reach a collective construct (Chamard et al. 2014). Keeping these functions in mind, the ultimate goal of place brands is to attract and retain target groups. Depending on the strategy adopted, these groups can be targeted separately or collectively, especially in the case of umbrella brands. Public managers mostly focus on differentiation, an approach justified by the needs to capitalize on their strengths and to position themselves vis-à-vis other places. The tension between specialization and the necessity to maintain a global perspective (to ensure recognition by both internal and external actors) has often led to vague and paradoxically similar promotion communication and close – sometimes almost standardized – brand strategies (Andersson 2014).

As mentioned just above, the main objective of these strategies relates to the capacity to attract and to retain target groups – current and future residents, talents, tourists, companies, and investors. According to different theoretical perspectives on country attractiveness, places can be regarded alternatively as a market to invest (business approach), as a site to visit (tourism approach), or as a residence to live (migration approach), depending on the concerned parties (Lee 2016). Fewer contributions have been written on the "retention" side of place branding. However, some authors have suggested to take a broader perspective, focusing on the different processes of place branding and taking into account its multidimensionality, including retention. Others have proposed measurement tools of place marketing costs, including those associated with citizen retention in a given place (Zenker and Martin 2011). More recent articles have focused on the development of smart cities, conceived as (potentially) common places for residents and tourists. The analysis of the determinants of such cities indicates that high scores on cultural interaction are of particular importance to their global attractiveness, while economic strength is especially appealing to residents (Romao et al. 2018).

Although all phases and groups are relevant to the field of place branding, this chapter focuses on the existing structures to attract companies and investments in the case of Switzerland (i.e., exogenous economic promotion). The country provides a great example of multi-stakeholder networks, involved at several institutional levels. As such, there are simultaneously competition between the federated states in Switzerland (hereafter: the cantons), which have different regulations, and cooperation since most of them have joined their forces in supra-regional agencies, which possess their own brand. It is then particularly interesting to assess the efficiency of such an institutional system and see how it performs in a globalized world, given the recent tendency to favor more integrated approaches. Here, the evolution from a strong decentralized institutional landscape to the creation of supra-cantonal organizations seems to be a response but, at the same time, raises challenges in terms of effective collaboration and efficiency. In the case of economic development, such assessments are typically based on a list of criteria related to attractiveness for businesses (see Snieska et al. 2019).

Data regarding these factors have also been compiled in the Swiss case, but they are unfortunately either provided through confidentiality agreements or not provided at all, seriously limiting the potential for any advanced analysis. Somehow paradoxically, this difficulty – partially shared by decision-makers – may pave the way to the construction of better assessment tools in a joint effort between scholars and practitioners. Before we comment the related implications of this opportunity for a deeper theory-practice exchange, we briefly describe the structure of the Swiss institutional system in the next section.

Background: Economic Coopetition and Cantonal Competitiveness

The federal structure of the Swiss institutional system attributes diverse tasks and degrees of autonomy to the parties involved in economic promotion. At the national level, government has promoted investment and the establishment of foreign companies in the country since 1996. Under the Federal Act on the Promotion of Switzerland as a Business Location,[1] it has encouraged the location of firms through a program called "Switzerland. Trade & Investment Promotion." Within this framework, operational responsibility is assumed by a private association, Switzerland Global Enterprise (S-GE), which has concluded a service agreement with the State Secretariat of Economic Affairs (SECO). In turn, the SECO develops strategies devoted to location promotion and, in the context of economic policy legislation, acts as the competent administrative body to represent the country's interests as a location for business. Adopting a liberal approach, it intends to offer a competitive framework to companies rather than direct support. At the same time, cantons oversee attracting specific private corporations to settle in their respective jurisdiction. This policy area is thus illustrative of the implementation of a joint task between the cantons and the national government. This division of labor is supplemented by investment promotion agencies at the supra-cantonal level, as presented in Table 1, created autonomously by the cantons. These agencies work together with cantonal economic promotion agencies, guiding companies to launch their activities in the targeted region. They have established strong networks of representatives abroad, with the aim of identifying companies that could benefit from the business

[1] *Federal Act on the Promotion of Switzerland as a Business Location* SR 194.2 from October 5, 2007 (last updated March 10, 2020). https://www.admin.ch/opc/de/classified-compilation/20070050/index.html

Table 1 Main actors and tasks regarding economic promotion in Switzerland

Organization	Actors	Promotion tasks
Steering group: Promotion of the country's image	Swiss confederation, cantons (mostly represented by areas), S-GE (no voting rights)	Exchanges between actors, strategic orientation, and definition of a strategy plan to find new markets
Switzerland global Enterprise (S-GE)	Member companies-based association	Promotion of Switzerland as a business location, presentation and diffusion of information regarding Switzerland as a business location, communication with the cantons about cantons and prospects
Swiss business hubs	Managed by S-GE and administratively linked to embassies and consulates	Contacts with businesses abroad interested in investing or in locating a company in Switzerland
Economic regions	Greater Zurich area (GZA)	International promotion and marketing for the cantons regrouped in supra-cantonal organizations, support for the settlement of foreign companies and for the establishment of newly created businesses
	BaselArea.Swiss	
	Greater Geneva Bern area (GGBa)	
	St.GallenBodenseeArea (SGBA)	
Intermediaries	Private companies, e.g., consulting firms	Consulting activities for investment at the international level, especially regarding aspects of companies' settlement not directly related to public authorities

Based on Von Stokar et al. (2014)

environment in Switzerland. They were created recently; for instance, the Greater Geneva Bern area (GGBa), also known as Invest Western Switzerland, was founded in 2010.

The active work of these agencies, supported financially by their member cantons, creates a special situation in which cantons still compete to attract companies and investment (competition) while joining their forces in supra-cantonal organizations (cooperation). How do the cantons deal with this situation? When S-GE is solicited by investors, the information is then shared with the supra-cantonal organizations and the cantons. Although the latter have committed to join supra-regional bodies, they are still free to send their own file to convince potentially interested companies to locate on their soil. During the whole process, supra-cantonal agencies commit to remain neutral and thus not to favor one place over another. Cantons benefit from those agencies since they strongly promote themselves as part of the global "Switzerland" brand, which generally holds a positive image, and through competitive clusters at the regional level. Both aspects have proved crucial for the cantonal members, which have, for instance, attracted almost 80 companies in the GGBa in 2019. The goal of fostering more cooperation rather than competition stems from the idea that in a small territory like Switzerland, the location of a private firm in a certain canton will probably benefit the whole region regardless of the cantonal borders. For instance, employees in newly established companies may decide to live and to pay their taxes in neighboring areas, while the jobs created may benefit the inhabitants of other cantons. More generally, the success of a canton in attracting companies may well have spillover effects, which contribute to the development of trans-cantonal clusters. This rationale has supported the daily operation of supra-cantonal agencies and has sustained a healthy system of coopetition so far.

Despite these coordination efforts, established through a bottom-up approach, competition remains a reality in the current work of economic developers. In this sense, cantonal economic development agencies promote the settlement of foreign companies in their respective cantons, as illustrated by the example of Vaud, where the promotion agency (DEV) has long operated as a nonprofit organization under the mandate of the Department of Economy (Sandoz 2019). The DEV has played an intermediary role to facilitate contacts between local institutions and foreign firms.[2] However, Vaud cannot prospect independently in areas where the GGBa is active.

At the same time, cantons can also count on a network of prospectors deployed abroad. It should be added that critical works have pointed to potential coordination problems in addition to this dynamic of coopetition, as well as governance issues within the supra-cantonal agencies (Monnier 2015). This results mainly from overlapping institutional layers, the multiplication of actors, and the complex structures of these supra-cantonal agencies. In sum, cooperation and competition characterize the behavior of cantons when it comes to attractiveness policies. This situation derives from the political and institutional system of the country, which is based on multi-level federalism and a relatively weak center (Vatter 2018). Consequently, subnational units can use the economic, legal, and political flexibility of the system to increase their attractiveness vis-à-vis other areas.

Assessing Attractiveness Policies: The Issue of Data Availability

The scope of this chapter is limited to the attractiveness of foreign companies at the cantonal level. Of course, such an approach is restrictive in nature but fully belongs to broader discussions related to place branding. Indeed, attractiveness can be addressed as the ultimate goal of place branding (Mabillard and Vuignier 2017). In the Swiss case, external branding strategies focus mainly on the attraction of companies rather than individuals, and the creation of jobs is strongly highlighted in reports published by organizations such as the DEV or supra-cantonal agencies. The attribution of public funds is therefore related to the number of jobs potentially created and benefiting the local labor market (Sandoz 2019). Unfortunately, only few studies – which are publicly available or have been published – have attempted to empirically link the issue of attractiveness with the establishment of foreign companies at the subnational level (e.g., Nilles and Conde 2015). It should be added here

[2] On March 12, 2020, the DEV merged with Innovaud to promote both foreign investment and innovation. The new entity is called Innovaud. https://www.innovaud.ch/en/news-impact/news/2020/the-merger-of-innovaud-and-economic-development-canton-of-vaud-dev-officially-approved

that even at the national level, the literature emphasizes that accurate assessments are difficult to make. Regarding, for instance, branding, marketing, and investment, Papadopoulos et al. (2016: 624) argue that causal mechanisms remain largely unknown:

> Many different factors affect investor decisions and much activity is observed in the nation branding field, but we do not know which of these factors are more important than others; we do not know which place marketing approaches work better in what contexts; and we seem to be not anywhere near to a reasonably comprehensive understanding of the intersection between the well-established (even if in some ways lacking) literature on FDI and the much newer field of nation branding, which can readily be said to still be in its infancy.

The recency of place branding as a research field may explain the difficulty to make solid impact assessments. Among the rare academic studies addressing the issue of attractiveness and the settlement of foreign companies, Brülhart et al. (2012) show how high corporate taxes are a deterrent to company location but that the effect of this factor is weaker in sectors where clusters are more strongly established. In other words, location choices are less sensitive or impacted by taxes in more agglomerated sectors. Insisting on the central role of spatially concentrated clusters, this study focuses on the number of new firms (start-ups) in the period 1999–2002, the effect of the local corporate tax burden, and agglomerations in Switzerland ($n = 213$ largest Swiss municipalities).

While it presents interesting results regarding the impact of taxation and clusters, this contribution does not provide any precise information about the attractiveness of places for foreign companies. This shortcoming, also present in any attempt to measure such an effect, is mainly due to the lack of data availability regarding the establishment of foreign firms in the cantons and municipalities. Analyses conducted annually by two banks (Credit Suisse and UBS) highlight the challenges faced by Swiss cantons and subregions in terms of attractiveness and create a ranking based on their respective strengths and weaknesses. However, they do not publish any figures about the outcomes of such attractiveness (number of companies established and jobs created). At the same time, the Federal Statistical Office (FSO) provides data about the total number of foreign companies established in the cantons, without giving details about the global evolution of settlements. Moreover, most statistics remain confidential and cannot be fully exploited for research purposes, as presented in Table 2.

As detailed earlier, cantons have the legal competence to set their own foreign investment and company location attraction policies. As such, they can also communicate freely about the number of foreign firms established on their soil. While the GGBa provides aggregated data for its members, some of the cantons publish detailed figures. This is, for instance, the case of the canton of Vaud, which has published reports about the new settlements for the period 2015–2018 (SPEI 2019), and of the canton of Geneva, where the authorities have presented the decrease of new settlements of companies supported by their cantonal development agency and/or the GGBa for the period 2014–2019 (Canton of Geneva 2020). These two examples show that data are available but are shared individually. Unfortunately, this lack of coordination is supplemented by missing data since the other cantons have

Table 2 Freely available, confidential, and uncommunicated data about companies and attractiveness factors at the cantonal level

Category	Data	Available years	Availability and use	Source	Recent reference
Companies	Total number of foreign companies	2014–2017	Partly confidential – Aggregated treatment of the data	FSO (2018)	Portraits of companies' groups in Switzerland (2018)
	Establishment of foreign companies	–	Partly available – Unexploitable for research	VDK/CDEP, S-GE, SECO, cantons, supra-regional organizations	For example, Greater Zurich Area Annual Report (GZA 2018)
Factors of attractiveness	Locational quality Indicator (LQI)	2013–2018	Confidential – Aggregated treatment of the data	CS (2018)	LQI 2018
	Cantonal competitiveness Indicator (CCI)	2018	Partly available – Unexploitable for research	UBS (2018)	CCI 2018
	Cantonal indicators (population, economy, quality of life, infrastructure)	1979 to 2018 (misc.)	Freely available – Fully exploitable for research	FSO (2019)	Portraits of the cantons (2019)

Acronyms of organizations
FSO Federal Statistical Office (Switzerland), *VDK/CDEP* Conference of the Directors of public economy cantonal departments (Switzerland), *S-GE* Switzerland Global Enterprise, *SECO* Swiss State Secretariat for Economic Affairs, *CS* Credit Suisse Group AG (investment bank and financial services company), *UBS* UBS Group AG (investment bank and financial services company)

decided not to publish any detailed information about their past and current situation. To the extent of our knowledge, no study has investigated the explanations for this situation. We argue here that although data about the location of foreign companies are public in nature, their sensitive character seems to erode transparency, thus diminishing the accountability of those cantons (Mabillard and Zumofen 2017) regarding attractiveness policies. The sensitivity of the data mostly refers to business secrecy and the situation of coopetition, in which cantons still compete to attract companies.

Moreover, there are discrepancies regarding the economic situation of the cantons: while Jura contributes to 3% of Western Switzerland's GDP, Geneva and Vaud account for 30% and 34%, respectively (BCF et al. 2018). This may prevent less successful cantons from publishing (low) figures to avoid criticism and to display other assets of their policies. The canton of Valais, where an almost all-encompassing umbrella brand has been created (Vuignier 2017), insists, for instance, on its attractiveness for small- and medium-sized enterprises (SMEs) and the development of the local economy. All in all, the absence of a systematic communication of the data makes it extremely difficult for practitioners and scholars alike to conduct high-quality studies or to develop benchmarks. While this challenge is predominant in the private and the third sectors, often characterized by high opacity levels of their operations, the issue is also salient in the public sector due to the risk-averse attitude of public servants and a long-standing culture of administrative secrecy. In Switzerland, legislation on administrative transparency even remains inexistent in certain cantons (Keuffer and Mabillard 2019).

More generally, this situation points to the fact that most data are available but not disclosed. In an era of big data, which is increasingly regarded as a "gold mine for business intelligence" in the current economy (Chen et al. 2012), such a lack of data sharing may prevent economic development agencies from making better-informed decisions. Indeed, opportunities offered by data appear immense (e.g., the development of artificial intelligence), although privacy issues remain no less significant and sensitive (e.g., mass surveillance). Scholars advocating for open innovation argue, for example, that the Internet – through data sharing – is transforming our collective intelligence (Nielsen 2020). Applied to public administration, information and communications technologies can improve (data) exchanges among stakeholders and therefore effectiveness of policy decision-making processes (Schmidthuber and Hilgers 2018).

Interestingly, analyses of evidence-based policymaking have shown that behind the challenges of data availability, access, and management stands a crucial issue: the lack of policy analytical capacity affects the ability of policymakers to engage in prioritizing evidence-based decision-making (Howlett 2009). Given the growing knowledge and know-how gathered by the variety of stakeholders concerned directly or indirectly by place attractiveness (public bodies, statistical offices, banks, and researchers), this could be solved by killing several birds with one stone: a more integrated and systematic communication between the different stakeholders would then benefit the economic development agencies. The latter may then seize the opportunities offered by a mutual analysis of available data. Acknowledging these trends and our observations, the next section discusses what data access implies at the micro level for both scholars and practitioners active in the field of attractiveness.

Implications for Scholars and Practitioners

Researchers are often confronted with the absence of data, mainly resulting from non-responses to certain items, unusable responses, or unreturned questionnaires in the example of surveys (Brick and Kalton 1996). Many papers in the methodology literature have discussed the issue and more particularly the way they can be dealt with technically. In this sense, most contributions typically propose solutions such as maximum likelihood estimation, weighting, and imputation to properly handle missing data in statistical analysis (Kleinke et al. 2020). However, to the extent of our knowledge, few contributions have focused on the incidence of limited datasets due to partial/confidential or non-communication of information. Our case study belongs to this category and raises several challenges in terms of research and practice.

First, implications for scholars point to the impossibility to conduct a systematic study, based on comparative and reliable data. This problem is reinforced by the fact that data aggregated at the meso or macro level undermine any effort to carry out a detailed and precise analysis. Second, confidentiality of the data prevents researchers from publishing descriptive statistics, the agreement with providers stipulating that individual information must not be disclosed. Usually displayed in the findings section of most academic works, this source of information remains unavailable to the reader under the terms of the contract signed between the researcher and the data provider.

Third, this aggregated form and the fact that data are only available at the regional scale strongly limit the opportunities for statistical inference. Indeed, technical issues quickly arise with regression, for instance, where a common rule of thumb states that there should be at least ten observations per variable included in the analysis model. However, data gathered at the regional level barely fit with the institutional structure of most states, in which the number of regions remains rather limited. As a result, small samples are often not suitable for technical inference statistics. Finally, the absence or partial character of the data presented in this particular case does not enable any comparative analysis with other contexts, which always constitutes a strong limitation to the explaining scope of any study, especially those with a deep exploratory character.

Applied to the case of attractiveness for companies in Switzerland, these empirical problems concern more particularly the lack of systematic data. As mentioned earlier, the absence of communication by cantons about the establishment of new foreign firms on their soil does not allow researchers to take full advantage of this variable. Consequently, it would be extremely difficult to build an analysis model using this factor – the establishment of new companies in a given area – as the dependent variable, while independent variables such as attractiveness factors remain fragmented. For example, although the FSO (2019) provides regularly updated figures about cantonal characteristics, individual data stemming from the CS LQI (see abbreviations above) are confidential. Unfortunately, information from the UBS CCI are not exploitable due to the unwillingness of the company to give access to their data. This is particularly unfortunate since their data are aggregated at the subregional level and would enable scholars to conduct a deeper analysis, probably using multiple linear regression.

As a result, research remains limited to correlation analysis (Mabillard and Vuignier 2020), showing, for instance, that, surprisingly, quality of life is not significantly related to the attraction of new foreign firms. On the contrary, cantonal wealth and English as main language are positively and significantly associated with the number of foreign companies settled in a canton. It should be noted that such correlation analysis remains bidirectional; the reasoning may therefore be reversed and limits the analytical power of the findings. Moreover, it seems particularly complicated to envision comparative work and to extend it to other countries at this stage.

At the same time, politicians rely mostly on economic development agencies to attract investment and companies to their jurisdiction. As presented above, economic regions play a major role in Switzerland, where supra-cantonal organizations actively look for new opportunities and support the settlement of foreign companies and the establishment of newly created businesses in a given area. As such, the absence of systematic data may be detrimental to their work because evaluation tools remain incomplete. Along the same lines, it is difficult to conduct impact assessments, to create benchmarking tools, or to identify best practices.

In addition, the absence of comprehensive and up-to-date datasets makes it virtually impossible for scholars to provide expertise to decision-makers, which reveals a serious impact on practice. Indeed, how can attractiveness factors and policies be compared in the absence of publicly available indicators, and how can they be rated? Finally, the nonexistence of scientific research and the lack of evidence tend to reinforce studies carried out by external stakeholders (banks, consulting firms), which insist on a narrow set of indicators, not reflecting the complexity of the issue at stake. In this sense, cantons sometimes suffer from a negative image, which is partly related to the existing situation. A more detailed view at the big picture and a more nuanced approach to the question of attractiveness, based on accessible and high-quality datasets, would potentially improve the understanding of the situation, taking into account the contextual specificities of each canton. This would in turn facilitate smooth external communication for those often impacted by the above-mentioned unfavorable studies.

The issue of the absence of systematic data and corresponding analysis sheds light on three main challenges for practitioners working in economic development agencies. Those have been extensively discussed during our interview of the Director-General of the GGBa.[3]

Firstly, the mandate to attract foreign companies encompasses many activities that cannot be translated into the single number of new company settlements. The narrow mandate hides in fact a wide-ranging spectrum of activities. As a result, regional organizations often fulfill tasks that are not directly reported in annual reports, such as exports, imports, and scientific exchanges/partnerships. Such exchanges can be regarded as the "engagement time before getting married"; they are important yet barely considered and should be better considered when looking at the figures. Furthermore, data often refer to visiting delegations and new settlements, but only globally (aggregated data), and do not take into consideration the number of firms that may have left the area in the meantime. Jobs created are only measured through projections (in the next 3 years). Based on these approximations, no information is systematically disclosed in detail, leaving room for cantons belonging to the GGBa to freely communicate about the numbers of companies

[3] Interview with the Director-General of the GGBa, February 4, 2020.

established in their jurisdiction. Although the principles of competition and cooperation are in this way ensured (coopetition), the lack of information echoes the difficulties mentioned earlier and may well refrain the full potential of the activities carried out.

Secondly, the attraction of new companies cannot be isolated from activities led by other stakeholders. Those actors typically set the economic framework or lead the innovation policy. In this regard, we observe a shift in the political priority regarding attractiveness: from attracting mainly new companies from abroad, cantons focus now on the existing ones (retention) and want to develop clusters based on local know-how (see Canton of Geneva 2020). Although the mandate of the GGBa remains dedicated to exogenous economic development, this shift impacts its work, requiring focusing more on the regional added value of foreign companies attracted rather than their quantity. The Director-General of the GGBa turns the attention to the complexity of processes related to attracting companies, including the multiple channels available for companies in their decision to locate their activities in a certain place. The GGBa is only one agency among many actors. In this regard, statistics falling into the "number of companies established in a canton" category will not reflect (or very indirectly) the impact of attractiveness policy measures implemented or showcased by this organization. Moreover, contextual economic variables have a strong influence as well. For example, sanctions imposed on Russian companies by the European Union (of which Switzerland is not a member) may also explain the relocation of companies in Switzerland, regardless of economic development agencies' efforts to attract investment in their region.

Thirdly, collecting, understanding, and analyzing data related to attractiveness remains a daunting task that requires coordination and capacity at all levels. The coordination between all federal and cantonal bodies involved in attractiveness and innovation policies currently remains at a very early stage, with no integrated approach yet. In contrast, at the cantonal level, one can notice the example of BaselArea.swiss. This supra-cantonal agency favors an integrated approach since it works as a one-stop hub for both economic development and innovation. Regarding capacity, these observations give credit to Howlett (2009): in the Swiss case, there is predominantly a lack of policy analysis capacity in all government levels to offer a comprehensive overview and to draw strategic priorities.

These three challenges call for deeper exchanges between practitioners and scholars. Short practical steps, such as giving access to existing data, favoring open exchanges, and data analysis capacity-building, can emerge as attractive easy-to-implement solutions. During his interview, the Director-General of the GGBa suggested that measurement tools, necessary to assess the organization's efficiency and identify the strengths and weaknesses of the current policy approach, may be essential. The creation of such a tool would be a great opportunity for a collaborative solution, based on solid evidence and thorough analysis. Performance evaluation by practitioners may benefit from academic expertise, as shown, for instance, by the strong influence ("effet résumé" or summary construct) of the "Switzerland brand" on companies' decisions to settle in the country (Vuignier 2018).

In turn, knowledge of the field and experience of development agencies' staff and decision-makers could certainly provide critical information to scholars in order to build a better assessment tool. All told, this ambitious project would require openness regarding data provision from all potential partners, including cantons, regional agencies, other stakeholders (banks and consulting firms), and private companies themselves. It would be possible to gather data from existing surveys and studies and to capitalize on feedbacks collected among firms established in the cantons, provided that some kind of confidentiality is ensured. These transparency efforts concern academia as well, since methodological aspects should be discussed openly, and results should be made available to all partners. Hence, this case study lays, potentially, the foundations of greater exchanges between theory and practice, based on mutual trust and common interests.

Conclusion

This chapter has presented the main pitfalls raised by the absence of systematic communication about exogenous economic promotion. While the partial disclosure of data prevents researchers from conducting detailed analyses, practitioners often suffer from incomplete information to assess their attractiveness policies' efficiency. In the Swiss case, data regarding the settlement of foreign companies in a canton are typically communicated by supra-cantonal agencies in an aggregated manner, and cantons do not provide figures regularly. Moreover, data about regional and subregional attractiveness, compiled by private sector organizations, such as banks, often remain confidential or not disclosed at all. This is supplemented by shortcomings in terms of policy analysis capacity and the global underestimated potential for stronger data-driven policies. Improvements in this domain would certainly lead to design strategic priorities based on more solid evidence and subsequently help conduct better evaluations of such policies. We argue here that a more accurate assessment tool should be developed through a collaborative work between all stakeholders involved in economic development. A more sustained dialogue between all parties (governments, private companies, banks, and researchers) would potentially lead to the creation of a promising "task force" to address this issue.

However, a better exchange between practitioners and scholars would certainly be insufficient since data about companies' decisions to locate remain in the hands of companies themselves. Unfortunately, previous studies have shown that firms are often reluctant to provide such information, either for economic reasons (business or trade secrets) or because they do not see the added value of such a disclosure (what would the company earn from this effort?). Surveys conducted with private corporations and data sharing would certainly help understand their motivations, including a potential effect of supra-cantonal and cantonal development agencies in the process. In turn, this would allow practitioners and scholars alike to make a better assessment of attractiveness policies at different institutional levels. Directors of several economic development agencies have already mentioned that some companies have often located their activities or invested in certain jurisdictions for contextual reasons. Penalized by sanctions from the United States and the EU in 2014, in the wake of the Crimea affair, certain Russian companies have, for instance, decided to settle in Switzerland (Rossé 2018). Other factors include the importance of private networks, as illustrated by the case of Chinese companies choosing to invest or to locate their activities in Switzerland (Kessler et al. 2014).

Another obstacle to data sharing relates to the unwillingness of several stakeholders to make their findings publicly available, although their studies may have been mandated by public bodies. If published, data are most of the time aggregated and do not provide any detailed results, impeding once again comprehensive analyses of attractiveness policies. Apart from this lack of transparency, such studies often remain very specific and do not embrace the complexity of attractiveness; they may also create tensions between residents' desires and consultants' recommendations (Kalandides 2011). In a way, they mirror the growing involvement of consulting firms in public sector organizations (Steiner et al. 2018), which is sometimes biased by the tendency of consultants to ensure a nice fit between their area of expertise and the issue at stake (Ashraf and Uddin 2013). Nevertheless, there is an increasing call for more collaborative processes, especially in place branding, to gather all stakeholders (including consultants) and to prefer a more participatory approach, which may lead to the adoption of a more constructive practice (Kavaratzis 2012). This perspective would stimulate partnerships and encourage more sustained exchanges between theory and practice. Hopefully, this would then incite all parties to share more information and make a better assessment of current policies.

In conclusion, it is often acknowledged that (big) data is the gold of an increasingly digitalized economy (Chen et al. 2012). This influences innovation and, more interestingly here, on the way collaboration between economic actors can be enhanced. In this regard, our analysis adds a new element to the global reflection on collaborative networks: it suggests that broader data exchanges would improve both scholars' and practitioners' endeavors, through, for example, a better toolbox to assess the efficiency of economic development agencies. However, it must be reminded that apart from measurement issues, secrecy is still deeply rooted in the daily operation of many organizations, in all sectors. As explained by O'Toole and Bennis (2009), transparency is an attitude that should be anchored in any organization, either public, private, or not-for-profit. The creation of a "culture of candor" would reinforce evidence-based policies and help make better decisions. Rapid innovation and technological change call for openness measures before organizations will have transparency forced upon them. In this vein, accurate assessment tools would prevent public bodies from proceeding by trial and error and exposing them to management failures. Promotion of the agencies' actions could also be rewarded on more rational grounds, going beyond crude numbers and aggregated data. This could be achieved through a better understanding of the specificities of places, the activities carried out by economic development agencies, and the successes recorded along the way.

References

Andersson, I. (2014). Beyond "Guggenheiming": From flagship buildings to flagship space in Sweden. *Norwegian Journal of Geography, 68*(4), 228–237.

Arku, G., & Oosterbaan, C. (2015). Evidence of inter-territorial collaborative economic development strategies in Ontario, Canada. *GeoJournal, 80*(3), 361–374.

Ashraf, M. J., & Uddin, S. (2013). A consulting giant, a disgruntled client: A 'failed' attempt to change management controls in a public sector organization. *Financial Accountability & Management, 29*(2), 186–205.

BCF, BCGE, BCJ, BCN, BCVS, & BCV. (2018). La Suisse romande et la stratégie énergétique 2050 : Rapport sur le produit intérieur brut (PIB) romand par les six banques cantonales de Suisse romande. Retrieved, https://www.bcj.ch/fr/La-Banque/Actualites/PIB-Romand-2018.html.

Boisen, M. (2015). Place branding and nonstandard regionalization in Europe. In S. Zenker & B. P. Jacobsen (Eds.), *Inter-regional place branding: Best practices, challenges and solutions* (pp. 13–23). Cham, Switzerland: Springer.

Bose, S., Roy, S. K., & Tiwari, A. K. (2016). Measuring customer-based place brand equity (CBPBE): An investment attractiveness perspective. *Journal of Strategic Marketing, 24*(7), 617–634.

Brick, J. M., & Kalton, G. (1996). Handling missing data in survey research. *Statistical Methods in Medical Research, 5*(3), 215–238.

Brülhart, M., Jametti, M., & Schmidheiny, K. (2012). Do agglomeration economies reduce the sensitivity of firm location to tax differentials? *The Economic Journal, 122*(563), 1069–1093.

Canton of Geneva. (2020). *Genève attire toujours moins d'entreprises étrangères.* Press release, March 10, 2020. Retrieved, https://www.ge.ch/document/geneve-attire-toujours-moins-entreprises-etrangeres.

Chamard, C., Gayet, J., Alaux, C., Gollain, V., & Boisvert, Y. (2014). *Le marketing territorial: comment développer l'attractivité et l'hospitalité des territoires ?* Louvain-la-Neuve, Belgium: De Boeck.

Chen, R. H., Chiang, L., & Storey, V. C. (2012). Business intelligence and analytics: From big data to big impact. *MIS Quarterly, 36*(4), 1165–1188.

Credit Suisse. (2018). Locational quality Indicator 2019. Retrieved, https://www.credit-suisse.com/media/assets/private-banking/docs/ch/unternehmen/unternehmen-unternehmer/publikationen/standortqualitaet-2025-en.pdf.

Dinnie, K. (2015). *Nation branding: Concepts, issues, practice* (2nd ed.). New York: Routledge.

Federal Statistical Office (FSO). (2020). *Swiss tourism statistics 2018.* Retrieved, https://www.bfs.admin.ch/bfs/en/home/statistics/tourism.assetdetail.11507548.html.

FSO. (2018). *Portrait of corporate groups in Switzerland 2014–2017.* Retrieved, https://www.swissstats.bfs.admin.ch/collection/ch.admin.bfs.swissstat.fr.issue18061845170005.

FSO. (2019). *Cantonal portraits 2019.* Retrieved, https://www.bfs.admin.ch/bfs/en/home/statistics/regional-statistics/regional-portraits-key-figures/cantons.assetdetail.7866378.html.

Govers, R. (2013). Why place branding is not about logos and slogans. *Place Branding and Public Diplomacy, 9*(2), 71–75.

Greater Zurich Area (GZA). (2018). *Annual Report 2018.* Retrieved, https://www.greaterzurichaarea.com/sites/default/files/media/files/gza_jahresbericht_2018_upload_web.pdf.

Howlett, M. (2009). Policy analytical capacity and evidence-based policy-making: Lessons from Canada. *Canadian Public Administration, 52*(2), 153–175.

Kalandides, A. (2011). City marketing for Bogotá: A case study in integrated place branding. *Journal of Place Management & Development, 4*(3), 282–291.

Kapferer, J.-N. (2013). *Ré-inventer les marques: La fin des marques telles que nous les connaissons.* Paris, France: Editions Eyrolles.

Kavaratzis, M. (2012). From "necessary evil" to necessity: Stakeholders' involvement in place branding. *Journal of Place Management & Development, 5*(1), 7–19.

Kessler, E., Prandini, M., & Wu, J. (2014). Chinese companies in Switzerland. *Central European Business Review, 3*(3), 23–30.

Keuffer, N., & Mabillard, V. (2019). Administrative openness and diversity in Swiss municipalities: How does local autonomy influence transparency practices? *International Review of Administrative Sciences*. https://doi.org/10.1177/0020852318823278.

Kleinke, K., Reinecke, J., Salfrán, D., & Spiess, M. (2020). *Applied multiple imputation*. Cham, Switzerland: Springer.

Kumar, N., & Panda, N. K. (2019). Place branding and place marketing: A contemporary analysis of the literature and usage of terminology. *International Review on Public and Nonprofit Marketing*. https://doi.org/10.1007/s12208-019-00230-6.

Lee, K.-H. (2016). The conceptualization of country attractiveness: A review of research. *International Review of Administrative Sciences, 82*(4), 807–826.

Lucarelli, A., & Berg, P. O. (2011). City branding: A state-of-the-art review of the research domain. *Journal of Place Management and Development, 4*(1), 9–27.

Mabillard, V., & Vuignier, R. (2017). The less transparent, the more attractive? A critical perspective on transparency and place branding. *Place Branding and Public Diplomacy, 13*(4), 348–359.

Mabillard, V., & Zumofen, R. (2017). The complex relationship between transparency and accountability: A synthesis and contribution to existing frameworks. *Public Policy and Administration, 32*(2), 110–129.

Mabillard, V., & Vuignier, R. (2020). Exploring the relationship between transparency, attractiveness factors, and the location of foreign companies: what matters most? *Place Branding and Public Diplomacy*. https://doi.org/10.1057/s41254-020-00171-6.

Metaxas, T. (2010). Place marketing, place branding and foreign direct investments: Defining their relationship in the frame of local economic development process. *Place Branding and Public Diplomacy, 6*(3), 228–243.

Miles, C. (2013). Persuasion, marketing communication, and the metaphor of magic. *European Journal of Marketing, 47*(11/12), 2002–2019.

Monnier, P. D. (2015). *Promotion économique de la Suisse occidentale : radiographie sans complaisance*. Geneva, Switzerland: Slatkine.

Nielsen, M. (2020). *Reinventing discovery: The new era of networked science*. Princeton, MA: Princeton University Press.

Nilles, D., & Conde, D. (2015). *La Genève internationale : Impact sur l'économie genevoise et régionale*. Lausanne, Switzerland: Université de Lausanne HEC CREA.

O'Toole, J., & Bennis, W. (2009). A culture of candor. *Harvard Business Review, 87*(6), 54–61.

Papadopoulos, N., Hamzaoui-Essoussi, L., & El Banna, A. (2016). Nation branding for foreign direct investment: An integrative review and directions for research and strategy. *Journal of Product & Brand Management, 25*(7), 615–628.

Poirot, J., & Gérardin, H. (2010). L'attractivité des territoires: un concept multidimensionnel. *Mondes En Développement, 149*(1), 27–41.

Romao, J., Kourtit, K., Neuts, B., & Nijkamp, P. (2018). The smart city as a common place for tourists and residents: A structural analysis of the determinants of urban attractiveness. *Cities, 78*, 67–75.

Rossé, N. (2018). De plus en plus de firmes russes en Suisse pour éviter les sanctions. *Le Temps*, August 24, 2018.

Sandoz, L. (2019). *Mobilities of the highly skilled towards Switzerland*. Cham, Switzerland: Springer.

Schmidthuber, L., & Hilgers, D. (2018). Unleashing innovation beyond organizational boundaries: Exploring Citizensourcing projects. *International Journal of Public Administration, 41*(4), 268–283.

Service for the Promotion of the Economy & Innovation (SPEI). (2019). *Annual Report 2018*. Retrieved, http://www.vd.ch/fileadmin/user_upload/organisation/dec/selt/fichiers_pdf/Rapport_annuel_2018.pdf.

Snieska, V., Zykiene, I., & Burksaitiene, D. (2019). Evaluation of location's attractiveness for business growth in smart development. *Economic Research, 32*(1), 925–946.
Steiner, R., Kaiser, C., & Reichmuth, L. (2018). Consulting for the public sector in Europe. In E. Ongaro & S. Van Thiel (Eds.), *The Palgrave handbook of public administration and management in Europe* (pp. 475–496). London, UK: Palgrave Macmillan.
UBS. (2018). Cantonal Competitiveness Indicator 2018. Retrieved, https://www.iszl.ch/uploaded/2018-19/DIA/UBS_cantonal_competitiveness_indicator_en_1383259.pdf.
Vatter, A. (2018). *Swiss federalism: The transformation of a federal model*. Abingdon, UK: Routledge.
Von Stokar, T.,Vettori, A., Zandonella, R., Scherer, R., Zumbusch, K., & Schoenenberger, A. (2014) *Standortpromotion Schweiz*. Evaluation report for the Swiss Secretariat of Economic Affairs (SECO).
Vuignier, R. (2017). Place branding & place marketing 1976–2016: A multidisciplinary literature review. *International Review on Public and Nonprofit Marketing, 14*(4), 447–473.
Vuignier, R. (2018). *Attractivité des territoires et place branding: Etude exploratoire de la sensibilité des décideurs d'entreprise à la marque territoriale*. PhD thesis, University of Lausanne, Lausanne, Switzerland.
Zavattaro, S. M. (2014). *Place branding through phases of the image*. New York, NY: Palgrave Macmillan.
Zenker, S., & Martin, N. (2011). Measuring success in place marketing and branding. *Place Branding and Public Diplomacy, 7*(1), 32–41.

Theorizing “Value” in Sustainable Urban Branding Strategies

Laura Ripoll González and Fred Gale

The twenty-first century has been officially dubbed as the “century of cities” (Carrillo et al. 2014). With the advent of globalization, the concept of the Anthropocene grasps the idea that it is human rather than natural impacts now shaping Earth’s fundamental geological features, most notable in the worldwide spread of pollutants and microplastics, the melting of glaciers, climate change, but also perhaps global pandemics. Sometimes characterized as an *ideology* (Yigitcanlar and Kamruzzaman 2018) like liberalism, socialism, nationalism, feminism, and ecologism, the concept of sustainability and associated ideas have resulted in a new conception of the ends of political economic activity in urban areas necessitating emerging, and new combinations of, available governance means and associated resources.

However, while cities are increasingly engaging in planning and branding for sustainability, to date, there has been insufficient attention or scrutiny applied to what these new ends and means are and might need to look like (for notable exceptions, see Anttiroiko 2014; Eshuis and Edwards 2013; Lucarelli 2018; Sevin 2011; Zavattaro 2018). In particular, little attention has been devoted to reflecting on how sustainability is calling into question our understanding of one of the English language’s most slippery concepts, “value.”

In its origins, city branding resulted from the application of mainstream marketing and branding concepts to cities. Focusing on commoditizing places as products, marketing has long been used as “a strategy to provide cities with an image, a

L. R. González (✉)
Department of Public Administration, Erasmus School of Social and Behavioural Sciences, Erasmus University, Rotterdam, Netherlands
e-mail: ripollgonzalez@essb.eur.nl; l.ripollgonzalez@griffith.edu.au

F. Gale
Politics and International Relations, School of Social Sciences, University of Tasmania, Hobart, TAS, Australia
e-mail: fred.gale@utas.edu.au

S. M. Zavattaro (ed.), *Public Branding and Marketing*,
https://doi.org/10.1007/978-3-030-70505-3_3

cultural significance, which will, ideally, function as a source of added symbolic and thus *economic value*" (Mommaas 2002, p. 34, emphasis added).

This understanding implies city branding is connected "primarily with the economically inspired [process] to position cities" (Kerr 2006). In focusing on economic value, the ends of city marketing and branding are little different from the ends of nation branding or corporate branding or, indeed, of development as modernization itself. At the broadest possible level, and commencing with the work of Adam Smith in *The Wealth of Nations* (1937), the goal of modern life has been to ensure that there is more exchange value taking place at each level – individual, corporation, city, region, state, and world – than there had been before. The end of a modern life well-lived – the good life – is a life of the highest quality (Chan and Marafa 2014) manifested through economic accumulation and unbridled consumerism as represented at the level of the individual by assets (house, car, savings) and endless new *experiences*, businesses (offices, plant, profits, brands), and nations (human capital, manufacturing capital, financial capital, and growing gross domestic product). Applied to the city, the end goal of traditional marketing and branding approaches has been to grow the economic wealth of the city as measured by an influx of tourists, students, and/or investors. As a consequence, through time, city branding has become increasingly entangled with economically motivated spatial planning (Eshuis et al. 2014; Oliveira 2015).

Marketing practitioners informed initial developments in urban branding via their everyday practices. Private marketing organizations and government agencies (Zavattaro et al. 2015) contributed to the development of local urban development policies to enhance certain positive attributes of place with the objective of increasing their competitiveness in the global marketplace (Boisen et al. 2018). Notably, however, scholars from a range of fields, including human geography, sociology, and political sciences, began to recognize cities differ from consumer products. Cities are socially constructed and dynamic entities with complex layers of hard and soft resources with complicated structuring of spaces (Giovanardi 2012), and their brands do not belong to one but many stakeholders (Kavaratzis and Kalandides 2015).

For these reasons, city marketing understood as the application of traditional marketing tools and concepts for products and services to cities has proven challenging. Brands are complex identities that exist in the minds of consumers (De Chernatony and D'All Olmo Reilly 1998). In addition, the emergence of the service-dominant logic of marketing (Merz and Vargo 2009; Warnaby 2009) and the paradigm of brand co-creation (Kavaratzis and Hatch 2013) led to the realization that to achieve compelling city brands, a range of stakeholders, including local governments and other quasi- and non-market stakeholders, needed to be involved in the branding process.

However, in practice, what seems to be occurring is that the "ends" of city branding have remained largely unchanged – the focus is still on generating more exchange value through enhanced market activity. Instead, what has changed is the means by which that might be done, from top-down, consultancy-driven marketing campaigns to various forms of "inclusive" or "participatory" place branding processes (Kavaratzis et al. 2018) leading to the commodification of everything (Gilbert 2008).

The domination of the exchange value mindset in the city branding literature, we argue, is hindering the development of place branding as a tool to achieve sustainability. Building on the concept of sustainable development but recasting it as its own noun by displacing "development," sustainability is identifying a different end beyond mere increased exchange value as the objective of political economic action. Unlike other ideologies, however, the end of sustainability is not to maximize exchange value (liberalism), labor value (socialism), use value (nationalism), or environmental (function) value (ecologism) but to seek to bring these various forms of value into balance in each and every decision (Gale 2018). Such a conception of the end goal of sustainability constitutes a deep challenge to the entire discourse of city branding, given branding by definition appears aimed at generating exchange value by increasing a city's competitive advantage through highlighting its unique attributes (Chan and Marafa 2014). Competitive advantage is realized by increased equity, that is, higher added value of the place's goods or services in the eyes of consumers compared to many others performing similar functions and sharing the same characteristics (Jaffe and Nebenzahl 2006).

The purpose of this chapter is to subject the concept of "value" in the city branding literature to a forensic analysis to determine the degree to which urban branding premises and practices can indeed contribute to creating the "sustainable city." (Rees, 1997) The next section introduces and reflects on some of the key concepts in value theory in the philosophical, psychological, and political economy literatures, notably differences between final values and instrumental values, held values and assigned values, and use, labor, and exchange value. Using the distinction between final values (ends) and instrumental values (means), we then develop a simple two-by-two matrix to classify and clarify four different usages that manifest as models in the city branding literature: the growth machine, public value, co-option, and participatory place branding models. Of these, we argue in our discussion and conclusion that it is the participatory place branding approach that is best fitted to deliver sustainability. However, to deliver on that potential, theorists and practitioners of participatory place branding need to be more specific about the exchange and non-exchange values it wishes to bring into greater balance and how participatory processes should be structured to ensure the appropriate degree of deliberation occurs.

Value Theory: Some Conceptual Distinctions

Value theory is a complex, transdisciplinary field with a bewildering variety of ways of conceptualizing the value concept. Philosophical reflections on the meaning of value are useful in drawing distinctions between final end values, which can be considered as referencing things with intrinsic value, and means toward their attainment that are considered instrumental values (Hirose and Olson 2015). For instance, for some people, a university education can be considered an end in itself, and the more educated an individual and society is, the better. Education from this perspective is a good in itself. For many others, however, a university education is not a good in itself but a means to another good – a better-paying job or a more productive economy. The distinction matters from a policy perspective because if other, better ways emerge for an individual to get a well-paying job or a government of making society more productive, then economic instrumentalists will quickly abandon their support for universal higher education and back different policies that better achieve their final objective.

A second useful distinction in value theory derives from the psychological literature that distinguishes between held and assigned values (Rokeach 1973). The number of assigned values appears almost infinite as these refer to the value that individuals place on (assign) to things. From the moment one wakes up to going back to sleep, one is constantly assigning value to things in terms of making choices – a process of quotidian valuation based on currently held desires, likes, opinions, norms, and preferences.

In both word and deed, people continuously engage in assigning value to people, places, and things. Values evidently come into play in important democratic processes such as voting in elections; one is more likely to vote for a politician whom one perceives as sharing one's perspective on critical issues such as national security, abortion, guns, equality, health care, and social insurance. However, values are also relevant to much more mundane behaviors, too. Many people are today choosing public transportation over automobiles and electric vehicles over internal combustion engines to limit their own personal carbon footprints.

One standard way to assess what people value is to look at how they spend their money. In neoclassical economics, spending money reveals a person's preferences, the assumption being made that they are rational individuals who maximize their utility by distributing their expenditure across the optimal basket of goods. Another approach is the opinion survey. Such surveys, appropriately constructed, assess the values that people (say city residents) have with regard to the city's current economic, social, and environmental configuration and the analyzed responses used to determine the public values the city council should seek to validate in city planning and branding. In responding to questions, those completing the survey are asked to express the values they assign to things such as a supermarket development, a cultural event, or a local wetland.

The existence of assigned values raises the important psychological question as to where these come from. Are there many held values that result in there being so many assigned values or, conversely, can a relatively small number of held values generate many assigned values? Extensive research by Rokeach (1973), Feldman (2003), and Schwartz (2012) has substantiated the latter view that there is, in fact, a relatively small set of held values that, according to Shalom Schwartz and colleagues, serve an evolutionary function in ensuring individual, group, and intergroup survival (Schwartz and Rubel 2005). Schwartz and Rubel (2005) have shown there are only a relatively small number of ten fundamental held values that that can be grouped along two dimensions: an "openness to change versus conservation" dimension and a "self-enhancement versus self-transcendent" dimension. For example, individuals who prioritize the conservation values of security, conformity, and tradition often score high on measures of right-wing authoritarianism; in contrast, those scoring high on the self-transcendent values of universalism and benevolence are statistically more likely to preference social equity and nature (Cohrs et al. 2012).

Broadly, the extensive literature on personal values supports the proposition that while our assigned values are clearly informed by several features of our social lives including norms, knowledge, family upbringing, and peer groups, they also derive in no small measure from the specific way we rank the ten fundamental held values in Schwartz's (2012) values circumplex. The existence of these fundamental personal values coupled with their exploitation by politicians helps explain why society is more polarized today than ever before. Political elites, advised by professional pollsters with their deep knowledge of personal values, have worked out how best to trigger them by highlighting certain political phenomena. Those on the political right seeking to mobilize their base focus on threats to security to trigger conservation values linked to conformity and tradition. Meanwhile, those on the left highlight deep social inequalities or threats to the natural environment to trigger self-transcendence universal and benevolence values (Common Cause 2012).

In addition to the distinctions between final/instrumental values and assigned/held values, a third body of political economy literature distinguishes among those values implicated in the production and consumption of goods and services (Dobb 1973). This literature contrasts the use value of things with their labor value and exchange value. The distinction itself traced back to political economist Adam Smith's *The Wealth of Nations* (1776). A common illustration of an important distinction between them can be found in Smith's water-diamond paradox, the oddity being that whereas water has enormous use value being essential for direct and indirect human survival, it usually has little exchange value: people do not often queue up to buy it. This stands in stark contrast to apparently completely useless diamonds as a person can survive perfectly well without them and which nonetheless have enormous exchange value.

In addition to use and exchange value, Smith identified a third type of political economic value: labor value. Indeed, he even argued that in the rude state of nature before markets and capital had become significant inputs into the production process, goods had exchanged based on the relative labor time required to procure them. He used the beaver-deer parable to illustrate the point. If it is twice as difficult (takes twice as much time) to bag a beaver as a deer, then according to Smith, and many later classical political economists, if a beaver hunter met a deer hunter and they were interested in exchanging their prey, they would do so on the basis of the embodied labor time involved in procuring them. Two deer would be exchanged for a beaver given that it required twice as much time to capture the beaver than the deer.

The history of political economy highlights how this relatively rich, threefold conception of economic value was slowly dispensed with over the course of the nineteenth century as some in the discipline sought to make it more scientific and capitalistic (Gale 2018). By the 1870s, a new school of political economists emerged known as Marginalists that began to narrow down the discipline to focus virtually exclusively on the exchange of goods and services in the marketplace (Steedman 2003). During the subsequent 20 years, concepts of use value and labor value were deeply criticized and dispensed with to enable the focus to be placed exclusively on exchange value, now understood as arising from the interplay of supply and demand on the market as synthesized by Alfred Marshall (1890).

Today, the discipline of neoclassical economics focuses almost exclusively on conceptualizing value this way, employing many ingenious and sophisticated methods to translate things that are not traded on markets, such as housework, wetlands, volunteering, and biodiversity, by imputing market values to them through such techniques as cost-benefit analysis, hedonic pricing, and contingent valuation (Common and Stagl 2005). So hegemonic has the discipline of economics become that when individuals and groups want to know the value of things, they usually turn to economics to find out. Yet, economists are only able to tell them what the exchange value of a thing is and what it will exchange for in actual or imagined markets. This is not wrong per se, but it brings strongly into focus a fundamental distinction between the disciplines of economics and sustainability.

The former is narrowly focused on understanding a single element of the molecular construct of value, its exchange value component. While this is useful analytically, it should not form the basis for policy. Sustainability, on the other hand, brings into focus value's pluralistic nature and in doing so recognizes the need to balance the realization of exchange value through markets with at least three other critical value components: the value things have in use, the value of the labor they embody, and the value of the ecosystem functions they serve.

Value in City Branding Literature

The branding literature generally, and the city branding literature especially, has a long tradition of applying political economic analyses, both conventional and critical, to understand how geographic spaces get transformed over time. For example, DuPuis and Greenberg (2019) study the effects of city branding policies through the lenses of the growth machine theory, a reference to Molotch's account of the growth machine (1976), which is one of the earliest and most-cited analyses of urban development.

Building on elite theory, Molotch (1976) argued urban development was an inter-elite struggle by property owners operating through the local government apparatus to obtain individual economic benefits from such things as changes to zoning regulations, road construction, internal and externally funded investment projects, and so forth. Yet, despite the intensity of these struggles that pitted landowners against each other, all were united in promoting urban growth as this was the glue that cemented the community together. A growing city was one that was increasing in population size benefitting residents (as a form of trickle-down economics) through higher house prices, real estate agents through increased sales, main street retailers through increased purchases, and associated service providers through more golf club, tennis club, and Rotary club members. This translated into cities being run by a special group of people as Molotch describes (1976, p. 214):

> The people who participate with their energies, and particularly their fortunes, in local affairs are the sort of persons who—at least in vast disproportion to their representation in the population—have the most to gain or lose in land-use decisions. Prominent in terms of numbers have long been the local businessmen, particularly property owners and investors in locally oriented financial institutions, who need local government in their daily money-making routines. Also prominent are lawyers, syndicators, and realtors who need to put themselves in situations where they can be most useful to those with the land and property resources. Finally, there are those who, although not directly involved in land use, have their futures tied to growth of the metropolis as a whole. At least, when the local market becomes saturated one of the few possible avenues for business expansion is sometimes the expansion of the surrounding community itself.

The political economy of city branding has also been analyzed from a more explicitly Marxist-informed political economy. For example, and building on the ensuing four decades of globalization that occurred since Molotch's analysis, Anttiroiko (2014) explores "the tension between accumulation of capital and the reproductive aspects of everyday life" (p. 5) under a capitalist mode of production, to advance the hypothesis that cities compete globally to extract their share of value from global flows, which then shapes the city's functions and labor systems. This shifts the drivers of the growth machine from a local, land-owning elite to the structures and dynamics of the global capitalist system itself. Cities, like individuals, are forced to compete for a share of the global economic surplus. Branding is one important way cities can render themselves attractive to insiders and outsiders alike: to ensure the city obtains its rightful share of the global flows of finance, foreign direct investment, creative talent, tourists, and international students.

While the city branding literature is replete with references to value and values, it is in the political economy literature on the topic that one sees distinctions being made between economic value (Mommas 2002; San Eugenio Vela et al. 2020), use value (Cleave and Arku 2020), and public value (Florek and Insch 2020; Hereźniak and Anders-Morawska 2020), with occasional references to labor value (Campbell 1996) and amenity value (Van Assche et al. 2016). To make sense of these usages, we employ the philosophical distinction outlined earlier between valuing ends (construed as being either exclusively exchange value or some combination of exchange and non-exchange values) and valuing means (construed as being either top-down elitist or bottom-up participatory) to generate the simple two-by-two matrix set out in Table 1.

This typology enables four different models of place branding to be identified. These are a top-down, exchange value exclusive, growth machine model; a top-down, balanced exchange/non-exchange, public value model; a bottom-up, exchange value co-option model; and a bottom-up, balanced exchange/non-exchange value, participatory place branding model. We discuss the major features of each of these models of city branding below teasing out with examples drawn from the literature how they are conceptualized and operationalized.

Table 1 Place branding values' typology

		Ends of place branding	
Means employed to realize place branding ends		Grow exchange value	Optimize exchange and non-exchange values
	Elite means	*Growth machine model* Top-down, market-driven, place branding processes (ignoring non-business stakeholders)	*Public value model* Top-down, market- and municipality-driven, place branding processes (including municipal representatives) to integrate exchange value with non-exchange values
	Participatory means	*Co-option model* Bottom-up, market-driven, place branding processes (co-opting all stakeholders into exchange value growth)	*Participatory place branding model* Bottom-up, participatory place branding and marketing (balancing exchange value with non-exchange values through inclusive stakeholder processes)

The Growth Machine Model

As discussed above, the critical literature on the growth machine (Cleave and Arku 2020; Logan and Molotch 1987; Molotch 1976) highlights how place marketing and place branding are used to relentlessly promote urban growth in the interests of the business community, which is viewed as the originator of exchange value and the key stakeholder to be consulted and accommodated. As Cleave and Arku (2020, p. 3) note, the logic of the growth machine is applied to city branding since: "if places can project strong and positive images into the marketplace of potential consumers, then perhaps there is an improved probability of attracting investments over competing markets."

They note the implication of this perspective is that exchange value is the *end* of place branding and that the consultative *means* employed "does not focus on the values and place identity held internally by long-term residents, but rather focusing on how a place is perceived by potential external investors" (Cleave and Arku 2020, p. 10). This relationship between place branding, the growth machine, and exchange value is made entirely explicit by DuPuis and Greenberg (2019, p. 353) who state: "even more progressive city administrations are controlled by an urban 'growth machine': one that privileges the 'exchange value' of local space over the use value needs of regular, non-wealthy citizens, hobbling the ability of progressive politicians to be accountable to the citizens who elect them."

From a values perspective, the key elements of the growth machine approach are as follows. Firstly, the end goal of place branding is narrowly conceptualized as the realization of exchange value. Regarding cities, the focus is on how to increase the volume of trade occurring within them and the prices of the products being exchanged. Exchange value is not considered a means to an end, but it becomes the key desired end from which all other good things are expected to flow. While growth machine advocates do not normally embrace the concept to describe their practices, they most likely associate more and better jobs, greener and cleaner environments, improved social services, reduced crime, and a range of other goods as flowing from a narrow and exclusive focus on growing exchange value. This is certainly the perspective preferred by economists, one where growth and development are treated as self-evidently good, only requiring regulation in special cases linked to market failures and externalities (e.g., Schumpeter 1942; Stiglitz 1993). Such a perspective downplays the negative consequences of the operation of the growth machine in the production and reproduction of social inequality, environmental damage, community dislocation, and elite governance. Indeed, to the extent that such negative outcomes are recognized to occur, they are viewed as being the trade-offs necessary for generating net benefits to the community overall.

Once exchange value is perceived to be the intrinsically desirable end goal to which city branding efforts should be directed, proponents of the growth machine logically view the means to achieve it as lying mostly in the business community and to treat other stakeholder groups, including residents, warily as potential blockers (Kavaratzis 2012). The means to achieving exchange value growth quickly becomes elitist as stakeholders are not equally situated in their capacity to contribute to value realization. Indeed, from this exchange value perspective, some stakeholders can be explicitly regarded as value detractors. There are heritage groups that lobby against big box store developments, green groups that lobby against subdivision applications, residential groups protesting against loss of amenity, and so forth. While such groups may need to be placated through consultative processes at the lower end of Arnstein's ladder (Arnstein 1969), and planning projects tweaked to take into account a limited number of concerns to maintain the fiction that consultative processes work (Kalandides et al. 2013), such accommodation strategies are tactical and designed to contribute toward the ultimate final strategy – more investment, more exports, more visitors, more foreign students, and thus more exchange value creation (Gertner 2011). This conception of the means employed by the fetishization of exchange value is well described by Cleave and Arku (2020, p. 14, our emphasis):

> Urban areas are sites of meaning and use value for multiple individuals and groups with diverse backgrounds. Exclusion of these groups in the development of place brand favor of a limited number of urban-elites will limit how representative a brand will be. The social and spatial tensions that exist make it difficult for place branding growth machines to create something representative—if this is even a goal. The exclusion of the public and the focus on a brand delineated by political boundaries suggests that *representation is not a goal, and instead is only a tool to drive consumption and investment from external sources.*

Co-option Model

An important variation on the growth machine model is the co-option model where the end remains basically unchanged as the realization of exchange value but the top-down, elite-driven, therapeutic consultative model (Arnstein 1969) gives way to one that is designed to be much more consultative and participatory. The adoption of new and more participatory means to place branding is driven by a recognition that "places are not just 'baskets' of products" (Ashworth and Kavaratzis 2010, p. 1276) but are socially constructed entities consisting of residents and stakeholder groups with strong views and allegiances to their own neighborhoods and communities with potential positive and negative effects on the brand (Ripoll González and Gale 2020). By enlisting these groups into place branding processes, it may be possible to deepen their enmeshment in branding's market logic as well as disrupt the emergence of counter-branding narratives.

The discourse of participatory place branding gets co-opted in this model to foster a pluralization of means to the monistic end of exchange value. This is because branding as a general activity has a strong market focus. For example, according to Askegaard (2006, p. 91), a brand is "a strategically produced and disseminated commercial sign (or a set of signs) that is referring to *the value universe of a commodity*," one according to Hislop (2001) where the desired outcome of the branding process "is the development and strengthening of a relationship or connection between a company's product and the *emotional perception of the consumer*" (cited in Sadler et al. 2016, p.18, our emphasis). When stakeholders are invited into participatory place branding processes to discuss their visions and hopes for the city, they are subtly steered by the underlying logic of place branding toward the contribution they can make to realizing exchange value. Green spaces, wetlands, concentration camps, nuclear meltdown sites, battlefields, urban art, antique rows, virtually everything that might be valued for other reasons gets recast through a branding lens for the purpose of growing exchange value.

Van Assche et al. (2016, p.1280) grasp the essential elements of this co-option model in the following observations on place branding while defending the participatory place branding element because they believe it makes a partial contribution to controlling its worst abuses:

Many critics of place branding have argued that it is essentially a neo-liberal activity that reduces everything, everyone, or every place to economic value. (Kavaratzis and Kalandides 2015; Van Assche and Chien Lo 2011)

From this critical perspective, place branding is seen as an advanced form of commodification, making everything circulate as 'material' things with a value attached, in a globalized economy. (Vanolo 2017)

We, however, would say that indeed the economy is globalized, and people are willing to pay for a variety of goods and services, including experiences and places, but that such does not imply that everything is reduced to a price. Value creation takes place, whether one is aware of that or not, and both planning and branding have partial sets of tools to manage the relation between value creation and places such as cities and regions. (Kavaratzis and Kalandides. 2015)

Putting a price on things is not a panacea, but under certain circumstances, it can help to create (dis-)incentives to act, to buy, to travel, and (dis-)incentives to maintain certain narratives of value. (Cleave, Arku, and Chatwin 2017)

Being blind to processes of value creation in place, and refraining from even attempting to manage it, does not stop it, and does not prevent the uglier consequences of capitalism. (Van Assche et al. 2016)

From an emphasis on growing exchange value evidenced by the growth machine and co-option models, the next two sections observe place branding practices aimed at optimizing exchange *and* non-exchange, usually, use values which we characterize as the public value model and the participatory place branding model.

Public Value Model

What distinguishes the public value model of place branding from the growth machine and co-option models is it explicitly recognizes there are other non-exchange value bases of value it seeks to validate in place branding outcomes. The core other value recognized in this approach is public value(s), although the concept requires further unpacking as it is used in two rather different senses. In the original conception of public value put forward by Bennington and Moore (2010), the determination of public value was the prerogative of governments who, composed of elected representatives of the people, were especially well placed to deliberate over its meaning and reach a consensus on what it was and how it might be delivered.

Building on this, civil servants and bureaucrats who advised the political apparatus were also considered to be repositories of knowledge regarding the meaning of public value, based on combinations of disciplinary expertise, agency knowledge, and consultative processes. While the public value literature often stresses that practices become more participatory, there is a lack of empirical research to demonstrate this is the case (Hartley et al. 2017). And even if practices have become more participatory, the quality of that participation can be questioned, as it has been above with regard to the growth machine model: the concern is that it continues to sit at the lower end of Arnstein's (1969) model of participation and leaves decision-making power in the hands of policy elites.

Nonetheless, and importantly, the public value model of city branding breaks with the growth machine model by explicitly recognizing there are other values beyond exchange value needing consideration in branding and planning decisions. However, the literature is not precise as to what the nature of these other values are. For some, they are use values, which references the use things have to individuals and communities independent of what they might exchange for in markets. This meaning is evident in a quotation from Cleave and Arku (2020, p. 14, emphasis added) that "urban areas are sites of meaning and *use value* for multiple individuals and groups with diverse backgrounds." It also is represented in Dupuis and Greenberg's (2019, p. 353, emphasis added) concern that the growth machine is one that "*privileges the 'exchange value' of local space over the use value needs of regular, non-wealthy citizens*, hobbling the ability of progressive politicians to be accountable to the citizens who elect them."

For others, however, the non-exchange values being referenced in public remain unspecified or are interpreted as public values – values the public expresses they have for their neighborhood, community, or region derived perhaps by stakeholder consultations, townhall meetings, or questionnaire surveys (Fukumoto and

Bozemann 2018). From this perspective, public value is made equivalent to what the public values, although the answers that can be obtained are not independent of the methodologies employed for assessing it. Moreover, this approach places us in the realm of assigned values (what values publics are assigning to the things around them), and as we know these can be virtually anything and, undeliberated on, may not be a good basis for public policy making.

Participatory Place Branding Model

Proponents of the participatory branding model argue "a place brand by nature belongs to the place and its people" (Aitken and Campelo 2011, p.917). The participatory brand paradigm resulted from developments in the marketing literature, in particular, a turn toward a service-dominant logic of marketing, emphasizing the role of stakeholders in value creation and an emphasis on brand experiences as "value is always uniquely and phenomenologically determined by the beneficiary" (Vargo and Lusch 2008, p.9).

Building on Aaker's (1996, p. 68) definition of brand as "multidimensional construct, consisting of functional, emotional, relational and strategic elements that collectively generate a unique set of associations in the public mind," Kavaratzis and Kalandides (2015) suggested a stakeholder-centered approach to place brands would indeed be more apt given the dynamism and complexity of place brands. A renewed focus on the associations and meaning of places led to a reconceptualization of brands as the result of a multilogue among a variety of stakeholders interacting with the culture and environment of place (Berthon et al. 2007 in Aitken and Campelo 2011, p.1). From this perspective, city brands are therefore considered to be the result of communicative interactions integrating people around a shared identity (Ballantyne and Aitken 2007, p.366). In their interactions, individual and social identities meet in a constant state of flux, often leading to contestation and counter-branding by detracting stakeholder groups (Trueman et al. 2004).

Public administration scholars, building on this idea of the need for greater representation in decision-making about cities, have argued urban branding is indeed linked to more democratic forms of governance and in particular network governance approaches (Braun et al. 2018; Eshuis and Klijn 2012; Zenker and Erfgen 2014). They argue a top-down approach similar to the one used in corporate branding is at odds with:

> the reality of place branding where political and democratic legitimization of brand values, brand policies and the necessary investment to develop and pursue those is vital. First (like in all other policy measures) because local authorities have to explain, justify and defend their place branding related actions against several types of political control imposed by democratic systems in the Western world. Secondly (and perhaps uniquely in the case of place branding), because the place brand, its values, its propositions and all measures that communicate the brand must have the agreement, support and assistance of local people in order to be effectively developed. (Braun et al. 2010, p.7)

Despite a common agreement that stakeholder involvement contributes to more positive outcomes (Braun et al. 2018), the jury continues to be out as per what the role of place branding should be in the advancement of places (Boisen et al. 2018). This may be partly due to an inherent conflict between public values and the common interest (as portrayed in public policies) and brand values aimed at increasing competitiveness and, therefore, focused on exchange value (Hospers 2020).

Discussion and Conclusion

Sustainability identifies two important aspects of the idea of economic value. Firstly, it points to the pluralistic ends toward which economic life should be directed. In producing and consuming our means of subsistence – air, water, food, clothing, shelter, energy, security, health, education, and other critical requirements of individual and social survival – we need to pay attention not only to the value things have when they are exchanged but also to the value they have in their direct uses, in the labor embodied in them and in the materials that compose them, and in whether and how accessing those materials has protected or damaged natural ecosystems.

From a sustainability perspective, economic life is not about accumulating exchange values but is about balancing that accumulation with the accumulation of use values, labor values, and environmental function values. These values often need to be traded off against each other. The production of exchange, use, and labor values often comes at a cost to function values, while the protection of function values often comes at a cost of these other humanistic values, and robust governance processes are required to ensure the trade-offs made do, in fact, achieve an acceptable balance.

This points to an important second feature of the sustainability: the need for more multi-stakeholder deliberative governance processes informed by science to assess how to make the required trade-offs to ensure each obtains its minimum requirements. In terms of means, then, sustainability endorses participatory place branding as the most appropriate model for delivering it. However, as it stands, theorists and practitioners of place branding for sustainability are less aware than they need to be of how sustainability has reoriented the ends and means of value. Consequently, as yet, the participatory place branding model is not delivering on its promise.

To challenge it to do a better job, we develop the following simple model based on the tetravaluation concept Gale (2018) introduced. As the term implies, place branding for sustainability will seek to achieve a balance in outcomes across the four core components of economic value: economic exchange, community use, worker labor, and environmental function values. This pluralistic concept of value as the *end* of place branding would be explicitly stated as the objective of place branding visioning exercises and planning processes.

Linked to this, practitioners of place branding would explicitly seek to bring together stakeholders into a tetravaluation process where those with an interest in exchange value (business elites) engage with those with a claim to defend municipal use value (city officials, heritage groups, sporting bodies), labor values (unions, social welfare agencies, artistic and cultural groups), and environmental function values (local natural resource management groups, environmental groups, parks, and recreational groups) to negotiate the meaning of the place brand in the context of input from universities and other independent information sources.

In this scenario, governments will act as facilitators, and business elites will be sitting alongside myriad other stakeholders, providing a level playing field for the negotiation of the identity and vision of a place. Such negotiations would take place through time and would be appropriately resourced to ensure all parties have access to high-quality support based on good governance models (Cadman 2011) with outcomes agreed as a package based on the trade negotiations model of a single undertaking. What this means is all concessions offered are conditional on a final deal being done: it enables provisional concessions to be reached to advance negotiations without locking people into a final overall package they might object to.

While some will doubt the feasibility of the tetravaluation approach, we argue there is really no alternative in an era of sustainability in which the future of places depends on our collective ability to respond to the great challenges of climate change as the ultimate footprint of the Anthropocene era. And those theorizing and increasingly also practicing participatory place branding have a great deal of experience in many dimensions of the tetravaluation process, arguably perhaps engaging in an element of praxis avant la lettre. What remains to be done is to more deeply connect the practice of participatory place branding with the underlying concepts of value that appear to be driving it and more directly to the idea of sustainability that could more explicitly inform it. Cities can only thrive on a planet that is itself thriving. And for that to happen, all action must be bent toward delivering a sustainable future at every level. Participatory place branding for sustainability embracing a new concept of economic value and a new means to its achievement offers a possible pathway forward.

References

Aaker, D. A. (1996). Measuring brand equity across products and markets. *California Management Review, 38*(3), 102–120.

Aitken, R., & Campelo, A. (2011). The four Rs of place branding. *Journal of Marketing Management, 27*(9-10), 913–933.

Anttiroiko, A. V. (2014). *The political economy of city branding*. Abingdon: Routledge.

Arnstein, S. R. (1969). A ladder of citizen participation. *Journal of the American Institute of Planners, 35*(4), 216–224.

Ashworth, G., & Kavaratzis, M. (Eds.). (2010). *Towards effective place brand management: Branding European cities and regions*. Cheltenham: Edward Elgar Publishing.

Askegaard, S. (2006). Brands as a global ideoscape. In J. E. Schroeder & M. Salzer-Mörling (Eds.), *Brand culture* (pp. 91–101). New York: Routledge.

Assche, K., Beunen, R., & Lo, M. (2016). Place as layered and segmentary commodity. Place branding, smart growth and the creation of product and value. *International Planning Studies, 21*(2), 164–175.
Ballantyne, D., & Aitken, R. (2007). Branding in B2B markets: Insights from the service-dominant logic of marketing. *The Journal of Business & Industrial Marketing, 22*(6), 363–371.
Benington, J., & Moore, M. H. (Eds.). (2010). *Public value: Theory and practice* (pp. 1–30). New York: Macmillan International Higher Education.
Boisen, M., Terlouw, K., Groote, P., & Couwenberg, O. (2018). Reframing place promotion, place marketing, and place branding-moving beyond conceptual confusion. *Cities, 80*, 4–11.
Braun, E., Kavaratzis, M., & Zenker, S. (2010). My city-My Brand: The Role of residents in Place Branding, 50th Congress of the European Regional Science Association: 'Sustainable Regional Growth and Development in the Creative Knowledge Economy', 19-23 August 2010, Jönköping, Sweden, European Regional Science Association (ERSA), Louvain-la-Neuve.
Braun, E., Eshuis, J., Klijn, E. H., & Zenker, S. (2018). Improving place reputation: Do an open place brand process and an identity-image match pay off? *Cities, 80*, 22–28.
Cadman, Timothy (2011) Quality and legitimacy of global governance: case lessons from forestry. International Political Economy Series. Palgrave Macmillan Ltd., Basingstoke, Hampshire, United Kingdom. ISBN 978-0-333-71708-0
Campbell, S. (1996). Green cities, growing cities, just cities?: Urban planning and the contradictions of sustainable development. *Journal of the American Planning Association, 62*(3), 296–312.
Carrillo, F. J., Yigitcanlar, T., García, B., & Lönnqvist, A. (2014). *Knowledge and the city: Concepts, applications and trends of knowledge-based urban development*. Routledge.
Chan, C. S., & Marafa, L. M. (2014). Rebranding Hong Kong "Green": The potential for connecting city branding with green resources. *World Leisure Journal, 56*(1), 62–80.
Cleave, E., & Arku, G. (2020). Place branding and growth machines: Implications for spatial planning and urban development. *Journal of Urban Affairs*, 1–18. https://doi.org/10.1080/07352166.2020.1768104.
Cleave, E., Arku, G., & Chatwin, M. (2017). Cities' economic development efforts in a changing global economy: Content analysis of economic development plans in Ontario, Canada. *Area, 49*(3), 359–368.
Cohrs, C., Petzel, T., & Funke, F. (2012). Authoritarian personality. In D. J. Christie (Ed.), *The encyclopedia of peace psychology* (Vol. 1, pp. 62–66). Malden: Wiley.
Common Cause. (2012). *The common cause handbook*. Maccinllyth: Public Interest Research Centre.
Common, M., & Stagl, S. (2005). *Ecological economics: An introduction*. Cambridge, UK: Cambridge University Press.
De Chernatony, L., & Dall'Olmo Riley, F. (1998). Defining a "brand": Beyond the literature with experts' interpretations. *Journal of Marketing Management, 14*(5), 417–443.
Dobb, M. (1973). *Theories of value and distribution since Adam Smith: Ideology and economic theory*. Cambridge, UK: Cambridge University Press.
DuPuis, E. M., & Greenberg, M. (2019). The right to the resilient city: Progressive politics and the green growth machine in New York City. *Journal of Environmental Studies and Sciences, 9*(3), 352–363.
Eshuis, J., & Edwards, A. (2013). Branding the city: The democratic legitimacy of a new mode of governance. *Urban Studies, 50*(5), 1066–1082.
Eshuis, J., & Klijn, E. (2012). *Branding in governance and public management*. London: Routledge.
Eshuis, J., Klijn, E. H., & Braun, E. (2014). Place marketing and citizen participation: Branding as strategy to address the emotional dimension of policy making? *International Review of Administrative Sciences, 80*(1), 151–171.

Feldman, S. (2003). Values, ideology, and the structure of political attitudes. In D. O. Sears, L. Huddy, & R. Jervis (Eds.), Oxford handbook of political psychology (p. 477–508). Oxford University Press.
Florek, M., & Insch, A. (2020). Learning to co-create the city brand experience. *Journal of International Studies, 13*(2), 163–177.
Fukumoto, E., & Bozeman, B. (2018). Public values theory: What is missing? *The American Review of Public Administration, 49*(6), 635–648.
Gale, F. (2018). *The political economy of sustainability*. Cheltenham: Edward Elgar.
Gertner, D. (2011). Unfolding and configuring two decades of research and publications on place marketing and place branding. *Place Branding and Public Diplomacy, 7*(2), 91–106.
Gilbert, J. (2008). Against the commodification of everything: Anti-consumerist cultural studies in the age of ecological crisis. *Cultural Studies, 22*(5), 551–566.
Giovanardi, M. (2012). Haft and sord factors in place branding: Between functionalism and representationalism. *Place Branding and Public Diplomacy, 8*(1), 30–45.
Hartley, J., Alford, J., Knies, E., & Douglas, S. (2017). Towards an empirical research agenda for public value theory. *Public Management Review, 19*(5), 670–685.
Hereźniak, M., & Anders-Morawska, J. (2020). Public value-driven place branding: The way forward? *Place Branding and Public Diplomacy*. https://doi.org/10.1057/s41254-020-00185-0.
Hirose, I., & Olson, J. (2015). Introduction to value theory. In I. Hirose & J. Olsen (Eds.), *The Oxford handbook of value theory* (pp. 1–9). Oxford: Oxford University Press.
Hospers, G. J. (2020). A short reflection on City branding and its controversies. *Tijdschrift voor Economische en Sociale Geografie, 111*(1), 18–23.
Jaffe, E. D., & Nebenzahl, I. D. (2006). *National image & competitive advantage: The theory and practice of place branding*. Copenhagen Business School Press.
Kalandides, A., Braun, E., Kavaratzis, M., & Zenker, S. (2013). My city–my brand: The different roles of residents in place branding. *Journal of Place Management and Development, 6*(1), 18–28.
Kavaratzis, M. (2012). From "necessary evil" to necessity: Stakeholders' involvement in place branding. *Journal of Place Management and Development, 5*(1), 7–19.
Kavaratzis, M., & Hatch, M. J. (2013). The dynamics of place brands: An identity-based approach to place branding theory. *Marketing Theory, 13*(1), 69–86.
Kavaratzis, M., & Kalandides, A. (2015). Rethinking the place brand: The interactive formation of place brands and the role of participatory place branding. *Environment and Planning A, 47*(6), 1368–1382.
Kavaratzis, M., Giovanardi, M., & Lichrou, M. (Eds.). (2018). *Inclusive place branding: Critical perspectives on theory and practice*. London: Routledge.
Kerr, G. (2006). From destination brand to location brand. *Journal of Brand Management, 13*(4-5), 276–283.
Logan, J. R., & Molotch, H. (1987). *Urban futures: The political economy of place*. Berkeley: University of California Press.
Lucarelli, A. (2018). Place branding as urban policy: The (im) political place branding. *Cities, 80*, 12–21.
Marshall, A. (1890). *Principles of economics*. London: Macmillan.
Merz, M. A., He, Y., & Vargo, S. L. (2009). The evolving brand logic: A service-dominant logic perspective. *Journal of the Academy of Marketing Science, 37*(3), 328–344.
Molotch, H. (1976). The city as a growth machine: Toward a political economy of place. *American Journal of Sociology, 82*(2), 309–332.
Mommaas, J. T. (2002). City branding: The necessity of socio-cultural goals. In T. Hauben, G. Ball, & E. V. Brinkman (Eds.), *City branding: Image building and building images* (pp. 32–48). Rotterdam: NAi Uitgevers.
Oliveira, E. (2015). Place branding as a strategic spatial planning instrument. *Place Branding and Public Diplomacy, 11*(1), 18–33.

Rees, W. E. (1997). Is 'sustainable city' an Oxymoron? *Local Environment, 2*(3), 303–310.
Ripoll González, L., & Gale, F. (2020). Place branding as participatory governance? An interdisciplinary case study of Tasmania, Australia. *SAGE Open*. https://doi.org/10.1177/2158244020923368.
Rokeach, M. (1973). *The nature of human values*. New York: The Free Press.
Sadler, R., Cleave, E., Arku, G., & Gilliland, J. (2016). A comparative analysis of place branding in Michigan and Ontario. *Urban Research & Practice, 9*(1), 16–36.
San Eugenio-Vela, Jordi, X. G., & Kavaratzis, M. (2020). The critical role of stakeholder engagement in a place branding strategy: A case study of the Empordà brand. *European Planning Studies, 28*(7), 1393–1412.
Schumpeter, J. (1942). *Capitalism, socialism and democracy*. New York: Harper & Brothers.
Schwartz, S. H. (2012). An overview of the Schwartz theory of basic values. *Online Readings in Psychology and Culture, 2*(1), 2307–0919.
Schwartz, S. H., & Rubel, T. (2005). Sex differences in value priorities: Cross-cultural and multimethod studies. *Journal of Personality and Social Psychology, 89*(6), 1010–1028.
Sevin, E. (2011). Thinking about place branding: Ethics of concept. *Place Branding and Public Diplomacy, 7*(3), 155–164.
Smith, A. (1937). *The wealth of nations*. [1776].
Steedman, I. (Ed.). (2003). *Socialism & marginalism in economics 1870–1930*. Routledge London.
Stiglitz, J. (1993). *Economics*. New York: W.W. Norton & Company.
Trueman, M., Klemm, M., & Giroud, A. (2004). Can a city communicate? Bradford as a corporate brand. *Corporate Communications: An International Journal, 9*(4), 317–330.
Van Assche, Kristof, Raoul Beunen, and Ming Chien Lo. (2016) Place as layered and segmentary commodity. Place branding, smart growth and the creation of product and value. International Planning Studies. 21(2), pp. 164–175.
Van Assche, Kristof, and Ming Chien Lo (2011) Planning, preservation and place branding: A tale of sharing assets and narratives. *Place Branding and Public Diplomacy* 7.2 16–126.
Vanolo, Alberto (2017) City branding: The ghostly politics of representation in globalising cities. Routledge, Abingdon.
Vargo, S. L., & Lusch, R. F. (2008). Service-dominant logic: Continuing the evolution. *Journal of the Academy of Marketing Science, 36*(1), 1–10.
Warnaby, G. (2009). Towards a service-dominant place marketing logic. *Marketing Theory, 9*(4), 403–423.
Yigitcanlar, T., & Kamruzzaman, M. (2018). Does smart city policy lead to sustainability of cities? *Land Use Policy, 73*, 49–58.
Zavattaro, S. M. (2018). What's in a symbol?: Big questions for place branding in public administration. *Public Administration Quarterly, 42*(1), 90–119.
Zavattaro, S. M., Daspit, J. J., & Adams, F. G. (2015). Assessing managerial methods for evaluating place brand equity: A qualitative investigation. *Tourism Management, 47*, 11–21.
Zenker, S., & Erfgen, C. (2014). Let them do the work: A participatory place branding approach. *Journal of Place Management and Development, 7*(3), 225–234.

The (Almost) Forgotten Co-producers: Intermediaries Co-creating Brand in Public Services

Afsan Chowdhury and Arjuman Naziz

The concepts of co-production and co-creation have found their places in the public management and marketing literatures since the 1980s (Terblanche 2014). After some time, these concepts have reemerged more recently, giving way to large volumes of literature in the field of both public management (Alford 1998; Brandsen and Honingh 2016; Brandsen et al. 2018; Nabatchi et al. 2017; Osborne et al. 2016) and marketing (Etgar 2008; Gronroos 2011; Lusch and Vargo 2006; Prahalad and Ramaswamy 2004; Solveig 1996; Vargo and Lusch 2004b). In the branding literature more specifically, interesting discussions have been taking place on co-creating brands and brand value (Boyle 2007; Gregory 2007; Payne et al. 2009; Ramaswamy 2008; Tynan et al. 2010). The term co-production is commonly used in the context of public services, while co-creating is a more commonly used term in commercial services and has found its way to public administration literature (Brandsen and Honingh 2018).

The recent interest in co-production and co-creation can be viewed as an influence from the service literature – the service-dominant logic to be more specific. Such influence is acknowledged in public administration (Alford 2016; Osborne and Strokosch 2013; Osborne et al. 2013; 2016) and branding literatures (Grönroos 2006; Merz et al. 2009; Vargo and Lusch 2004a, b). From the service perspective, co-production is intrinsic to public services, and service users play the central role in creating value of such services (Osborne et al. 2016) through their input, interactions, and expectations. Branding scholarship also has started to recognize dialogue and negotiations among the stakeholders as critical to brand value creation (Brodie 2009; Merz et al. 2009). Acknowledgement of the stakeholders as critical to the branding process and creating value indicates development of a so-called new brand

A. Chowdhury
BRAC Institute of Governance and Development, BRAC University, Dhaka, Bangladesh

A. Naziz (✉)
Department of Public Administration, Jahangirnagar University, Dhaka, Bangladesh
e-mail: arjuman@juniv.edu

S. M. Zavattaro (ed.), *Public Branding and Marketing*,
https://doi.org/10.1007/978-3-030-70505-3_4

logic (Merz et al. 2009). We employ similar reasoning to understand public branding. We add to the conceptualization by arguing there can be intermediary actors, between the service provider and service user, who contribute to the value creation by facilitating interactions between them.

To understand public branding, it is imperative to understand the service context, the actual process through which services reach to the users. In the context of developing countries, most policies and programs are formulated through a top-down process and implemented within considerable resource constraints. Service providers' reach to service users and service users' access to the services can be limited. In such context, there can be intermediary actors linking between the service and service users, making the service delivery possible.

In this chapter, we view co-creation of brand value of public services as an integral part of co-production of such services. Specifically, brand value is created through the interactions taking place at different stages of service delivery. We present two cases of social programs in Bangladesh, one delivered by the government and another by a non-profit organization. We show the layers of interactions that took place between the services and the service users and how they have possibly co-created brand value. The key focus in our framework is the intermediary actors and their interactions to link the service producer to the service users. We argue, the chain of social relationships that develops and ideally expands through the process is the embodiment of the brand.

Specifically, we examine the Expanded Program on Immunization (EPI), which the Government of Bangladesh has been delivering since 1979. In terms of achievement, EPI has been dubbed as near miracle (Huq 1991). It has been the largest single contributor to reducing under-five mortality rate (White 2005), raising the life expectancy at birth in Bangladesh higher than India, Pakistan, and Indonesia (Quasem 2008).

The Challenging the Frontiers of Poverty Reduction/Targeting the Ultra-Poor (CFPR/TUP) program is a targeted development program delivered by a non-profit organizations, BRAC, formerly known as the Bangladesh Rehabilitation Assistance Committee and later Bangladesh Rural Advancement Committee. (Currently the organization simply uses the BRAC acronym.) This program has been a globally recognized graduation model to address extreme poverty. It has been adapted to 114 sites in 45 countries, by the governments, non-government organizations (NGOs), and multilateral development institutions (BRAC, n.d.). Both cases provide interesting context to understand public services and public branding from the service user's perspective.

Despite the differences in the service providers and service types, both EPI and CFPR/TUP delivery demonstrate the critical role played by intermediary actors in linking between services and service users. Illuminating their role in co-producing the services and creating brand value is the purpose of this chapter. Before we examine the cases, we first provide a brief overview of the key premises of the service perspective, co-creation of brand value, and co-production of public services. Drawing on the cases, we then seek to explain how the social relationships built and expanded through interactions among service producer, service users, and the intermediary actor dynamics can become the embodiment of the public brand.

The Service Perspective, Co-creating Brand Value, and Co-producing of Public Services

The fundamental proposition of the service-dominant logic is the customers are the co-creators of value, and the key concern is the service (Vargo and Lusch 2004a, b, 2008). While both goods and services are to render service (Gummesson 1995), the service-dominant logic argues the latter should not be viewed as binary subsets of products (Vargo and Lusch 2004a, b). Goods and services have a nested relationship where goods are the vehicles used in the service provision (Lusch and Vargo 2006).

In contrast to the goods-dominant logic where the producer is the creator of value and the consumer is the user of it, service-dominant logic holds both as resource integrators (Merz et al. 2009). This idea reflects in Kotler's (1986) use of the term prosumer. However, little attention was paid to the concept of co-production and co-creation by marketing scholars until recently (Terblanche 2014).

Co-creating Brand Value

In some marketing research, co-creation refers to the idea that customers can take part in the production of goods and services they consume and create value for the company (Etgar 2008; Gronroos 2011; Lusch and Vargo 2006; Prahalad and Ramaswamy 2004; Solveig 1996; Vargo and Lusch 2004a, b). From the perspective of co-creation, a brand needs to be viewed as a continuous social process (Muniz and O'guinn 2001). Value creation starts with the supplier's understanding of the customer's value-creating processes and understanding how to support it (Payne et al. 2009). Both the producer and the customer participate in the process of co-creating value, which in turn results in the value-in-use (Lusch and Vargo 2006; Merz et al. 2009). Dialogue and negotiations are critical to the process. Value is created not only from the products and services but also from the experiences (Prahalad and Ramaswamy 2004) resulting from the interactions among organizations, related systems or processes, service employees, and the customers (Bitner et al. 1997; Brodie 2009). It is important to note that creating value is not so much about the conventional advertising and branding but more about building the processes to support customer experiences (Pine and Gilmore 1999).

Similar to the branding thinking, the service perspective has influenced the public services thinking as well. It has given way to interesting discussions on the service users' role in co-producing public services. However, most conceptualizations of co-production are not significantly different from the conventional view that participation can be added in to the core activity to enhance the performance of the activity and bring benefits to the engagers.

Co-producing Public Services

The need for institutionalized participation, especially at the local level, started to surface since the 1960s (Aleshire 1972; Greitens 2018; Nabatchi and Leighninger 2015). However, the way participation was defined and implemented by the governments was still problematic (Nabatchi and Leighninger 2015). Arnstein's (1969) criticism was influential in raising questions on the distribution of power to the citizen in such conceptualization of participation. A substantial body of literature followed since, giving way to different perspectives on participation (Duţu and Diaconu 2017). In development theories, alternative conceptualizations of participation started to emerge. The practice of controlled participation, especially in aid-depended projects that viewed participation within the framework established by authorities and official institutions, and bureaucracy as the main actor (Martinussen 1999), faced criticism. However, efforts to secure participation have not always been effective (Cleaver 1999).

While citizen participation is common now, it was for long considered undesirable or unimportant and sometimes still is (Brandsen et al. 2018). Administrators generally are ill-equipped to facilitate participation, and as a result, they might think it is problematic. They can be ambivalent about public involvement (King et al. 1998). The formal bureaucratic structure also cannot provide the space for participation. In the practitioner-client hierarchy (Fischer 1993), participation is often ineffective and conflictual, and it happens too late in the process, often way after most decisions related to it are made (King et al. 1998). As the formal space does not allow for incorporating participation, informal negotiations take place to determine the success of the delivery.

The recently popularized concept of co-production, as both a discourse and a practice, dates back to the 1970s (Sorrentino et al. 2018) having origin primarily in the work of Elinor Ostrom and her colleagues (e.g., Parks et al. 1981, cited in Brandsen and Honingh 2018). The interest in the topic died down in the 1980s (Brandsen et al. 2018) but has reemerged (Alford 2002; Bovaird 2007; Brandsen and Pestoff 2006). It signifies revival of the interest in citizens' engagement in producing public services (Brandsen et al. 2018; Brudney and England 1983; Jo and Nabatchi 2018; Nabatchi et al. 2017; Sorrentino et al. 2018). The revival is perhaps also associated with the fiscal constraints like those of the 1970s and 1980s (Nabatchi et al. 2017).

Co-production as defined by Elinor Ostrom (1996) is "a process through which inputs from individuals who are not 'in' the same organization are transformed into goods and services" (p.1073). It involves a critical mix of activities public service providers and citizens contribute to the provision of public services (Alford and Freijser 2018; Brudney and England 1983). It is about the interactions between the service users and the service providers, which however should not be viewed as similar to consumerism or user empowerment (Osborne et al. 2018).

Co-production, as defined by Loeffler and Bovaird (2016, p. 1006), is "public services, service users and communities making better use of each other's assets and resources to achieve better outcomes or improved efficiency." Inherent in these conceptualizations is how each party has resources critical to the production of the service in question. At the heart of co-production is how the quality and performance of a service process are shaped primarily by the expectations of the users, their active or passive role in the service delivery, and their subsequent experience of the process (Osborne et al. 2018). It is intrinsic to many services and can exist even when an organization consciously did not plan it or wanted it (Alford 2009).

From the service management perspective, co-production is intrinsic to the service experience; however, in the public management literature, co-production still explores how to add in the inputs of the service recipients into planning and delivery of the service (Osborne et al. 2018). Alford (2009) has observed client co-production is still viewed as an option that "can add to the repertoire of institutional arrangements available to public sector organizations in seeking to achieve their purposes" (p. 10). As it appears, most of the conceptualizations of co-production still do not challenge the basic premises of public management that services are rendered and controlled exclusively by the service professionals (Brandsen and Pestoff 2006). One exception is the conceptualization by Osborne et al. (2018) that aligns with the service management perspective.

Drawing on service-dominant logic, Osborne et al. (2013) proposed the public service-dominant approach that was later reframed by Osborne (2018) as the public service logic. The key essence of the public service logic framework is the central role of the service users in co-creating value of public services. The key argument is "it is public service users who create value through their interaction with public services – and it is public service organizations that co-produce this with them, not vice versa" (Osborne et al. 2018, p. 21).

Regardless of whether an organization recognizes co-production or not, in many cases, it is already engaged in it (Alford 2009), and often it is impossible to deliver the service without it. There can be several actors as co-producers in the process of co-production. In the public administration literature co-production primarily includes efforts from the "regular producers" and "consumer producers" (Parks et al. 1981, p. 1002). Regular producers are the government agencies and employees (health, education, law enforcement, etc.); the clients include individuals outside the agency (Ostrom 1996). In the service literature, through the low, moderate, and high levels of participation, consumers can take the role of productive resource such as contributor to quality, satisfaction and value, as well as competitor to the service organization (Bitner et al. 1997).

Nabatchi et al. (2017) identified two broad groups of actors as co-producers: the state actors, the regular producers such as the agents of government serving in a professional capacity, and the lay actors, voluntary members such as citizens, clients, and/or customers. Alford (2002) gives more specific categories of co-producers: the client, volunteers, and citizens. Clients mainly act individually and receive private value. Volunteers mainly bring value to others, while citizens work with others to bring collectively enjoyed benefits.

In the following discussion, we take a service users' perspective to identify the co-producers in the EPI and CFPR/TUP delivery. Our key focus is on the intermediary co-producers – field workers and rural elites – and their role in creating brand value for the services. These actors are difficult to put into any specific category as they create both private value and welfare value. Our view complies more with Andreasen's (1981) conceptualization of middlemen cooperation in social marketing. The cases highlight the interactions among the service producers, and service users, facilitated by the field workers and rural elites. Interestingly, both EPI and CFPR/TUP show similar pattern of social relationships, despite being delivered by different providers and having different services.

Expanded Program on Immunization (EPI) in Bangladesh

The Expanded Program on Immunization (EPI) has been one of the largest public health interventions worldwide. Six vaccine-preventable diseases including tuberculosis, polio, diphtheria, whooping cough, tetanus, and measles have been the leading causes of high childhood morbidity and mortality in developing countries (WHO 2000). The EPI was launched in the 1974 World Health Assembly resolution to protect the children of the world against these diseases by 2000 (Chowdhury et al. 2003).

In Bangladesh, the EPI was launched in 1979 (Chowdhury et al. 2003). It started as a pilot project in eight sub-districts or *thanas* (Sarkar et al. 2015) with six vaccines for the six vaccine-preventable diseases. The services initially were available only through the district- and sub-district-level facilities (Arifeen et al. 2013). Because vaccination centers were few and primarily located in the health-care facilities in urban areas, the EPI coverage remained less than 2% by the year of 1984 (Sarkar et al. 2015).

The process gained momentum as the government committed to the Global Universal Child Immunization Initiative (UCI) program and began a phase-wise process of EPI intensification from 1985 to 1990 (Chowdhury et al. 2003; Jamil et al. 1999; Sarkar et al. 2015). Eventually, the program became a major component of the maternal and child health interventions in the primary health-care approach in the government's third 5-year plan (1985–1990). Vaccines were provided through a new and systematically designed outreach system (Arifeen et al. 2013). By 1990, EPI was taken to 476 sub-districts or *upazilas*, 92 major municipalities, and 6 city corporations and made available to many infants and pregnant women (Sarkar et al. 2015) in the localities. Currently, EPI in Bangladesh is implemented through both clinics and outreach sites such as the EPI spots and satellite clinics (Jamil et al. 1999). The most critical element in the EPI that often did not get deserved credit was the outreach centers that allowed taking EPI directly to the people.

The success of the program has largely been attributed to the public-government-NGO participation. Initially, the program was delivered through government health workers (HWs). Gradually by the 1990s, many NGOs joined as partners in the program, playing a crucial part in mobilizing the community to attend EPI delivery sites (Arifeen et al. 2013). The role of the NGOs in mobilizing the community to attend the EPI services played a significant role in the program's success (Perry 2002; Perry and Chowdhury 2020). Altogether, the alliance between the formal and informal networks helped the program to deliver at a scale that Huq (1991) dubbed as a "near miracle." By 1989, near universal access to immunization services was reached, with an increase in coverage from less than 2% in 1985 to 65% in 1992 (Huq 1991; Perry 2002; Sarkar et al. 2015). This was laudable because Bangladesh was not expected to reach such targets given the weakness of its formal health services networks.

The EPI delivery involved stakeholders at the international, national, and local levels. The key international stakeholders included the World Health Organization (WHO), UNICEF, USAID, the Japan International Cooperation Agency (JICA), Rotary International, the Swedish International Development Agency, and the Global Alliance for Vaccine and Immunization (GAVI) (Osman 2008). The national program office remained in charge of the design and planning, management, procurement, and distribution of vaccines and other supplies, as well as of monitoring and evaluation of the program (Arifeen et al. 2013). At the local level, it was the HWs, primarily the vaccinators and the community mobilizers, who ensured the delivery of EPI to people.

The health workers, including health assistants (HA) and family welfare assistants (FWA), under the Ministry of Health and Family Welfare (MHFW), have been responsible for informing, motivating, registering, and convincing mothers to attend the immunization sessions (Nandy 1997). They were the first contacts between the program and community. Mobilization took place during household visits. The HWs built a rapport and trust with the community (Nandy 1997), and to most people, they were the main sources of information (Bhuiya et al. 1995). The main concern in this program was motivating people and achieving their trust (Wolffers 1988).

In parallel to the service component advocacy, social mobilization activities were carried out to promote EPI. National-level advocacy plans were designed to generate organizational and mass support in favor of the EPI. Television and radio advocacy, television spots with film and television celebrities, UNICEF's special ambassador for sports Imran Khan's visit to Dhaka, and Hollywood actress Audrey Hepburn's presence are examples of the publicity of the campaign.

The government, donor agencies, and NGOs conducted an extensive and effective campaign of social mobilization during the late 1980s (Kahn and Yoder 1999; Nandy 1997). A great deal of communication materials was produced based on the logo "Moni" – a term of endearment for one's child – as a marker of the program identity. In the Moni logo, the child (moni) is surrounded by a ring for protection from six vaccine-preventable diseases with six arrows around it symbolizing six vaccines (Abed et al. 1991). While Moni is now a widely known symbol, according to one author's personal experience (Afsan Chowdhury), at the primary stage, the parents did not understand logo in the manner that the logo designers did. They thought that the red circle protecting the child represented the diseases and the arrows around it were killing the diseases. It was the HWs who translated the meaning of the logo to them.

The branded message or the slogan for EPI was "immunize your child," which was produced on posters, leaflets, and other materials. The state-endorsed messages worked as an assurance to people of the government's/program's intent, suspending suspicions that it was part of family planning programs (Nandy 1997). The branded message and the logo were displayed widely, even on rickshaw carts and boats. The campaign is considered one of the most successful social communication programs in the history of Bangladesh (Abed et al. 1991).

The EPI is an example of large-scale coverage achieved through community-based outreach that evolved to meet changing conditions (Arifeen et al. 2013). Technical experts including the health planners pointed out that to achieve the immunization cycle, it was necessary to have sufficient outreach sites in all the villages of Bangladesh. In its absence, not enough children could be reached on time. The program, therefore, could not be successful on its merit or plan. As this became a major crisis, the national program office and local offices under the MHFW began to consult the health workers who were making door-to-door visits and understood the local context better. This coincided with an evaluation report on several communication materials' usage, which reported that most of the mothers had heard about the immunization from the field HWs and not mass media communication materials (UNICEF field report).

The ministry allowed some flexibility to the vertical structure of EPI. In an effort to find alternatives to reach the target of child immunization, field offices started to consult the health workers to explore if some of the village elites would allow their home space for outreach sites facilities once a month. Within a month, the most successful health mobilization initiative had been undertaken in Bangladesh. Almost every village in the plain land area had an immunization outreach site.

The negotiation and dialogue between the health workers and the rural elites were crucial to the success setting up the outreach centers. According to one author's (Afsan Chowdhury) field experience, the key arguments that HWs gave to the rural elite to convince them were generally as follows:

- If the *matbar* (rural elite) allowed the homes to be used as sites, his prestige would be high, and his reputation would be higher.
- His work would save many lives, and everyone would be grateful to him, and this would be his gain in every sense including rural politics.
- His family, other families, and the village would gain as children would not die from dreaded diseases, and divine benefit and people's love would be gained.

The HWs reported the *matbars* became interested and even spent their own money to decorate their houses on the designated immunization day. It was a radical act of social mobilization that literally saved thousands of lives. In this, the local elite played a role as a custodian by providing space and thus created legitimacy and trust for the service that included an unsought product like vaccine. This support from the rural elites has largely enabled the program to reach most children in the villages. However, they have not gained deserved recognition despite being critical actors to deliver immunization by allowing their home spaces to be used as outreach sites. Without those outreach sites, the EPI program could hardly reach the community as rapidly as it did.

Figures 1 and 2 provide rough sketches of the key interactions that took place at different points of EPI delivery in rural Bangladesh. Figure 1 shows the connections between the government as service provider and parents as service users. The EPI services delivery mechanism from central to local level is considered as the formal space of the service. It primarily consists of the following: i. at the central level, the National Program Office of EPI of the Directorate General of Health (DGH) under the Ministry of Health and Family Welfare (MHFW); ii. at the District level, the Civil Surgeon's Office; iii. at the sub-district or Upazila level, the Upazila Health and Family Planning Officer's Office (UHFPO); and iv. at the Union level, the Union Health Centers.

The service user's space or the broader community is viewed as the informal space. While the field health worker forms part of formal space and the elite of the informal, together they constitute a space that shares element from both spaces.

Figure 2 suggests how HWs convinced parents to vaccinate their children and bring them to the outreach sites for vaccination. The rural elite, mobilized by the HWs, supported EPI or the government to take the service to the community, which otherwise would not have been possible.

The relationships in the process evidently developed in layers, and the intermediary actors connected the services to the parents – the service users. These outreach sites are considered visible evidence of the extent of social mobilization achieved by the program (WHO 1994).

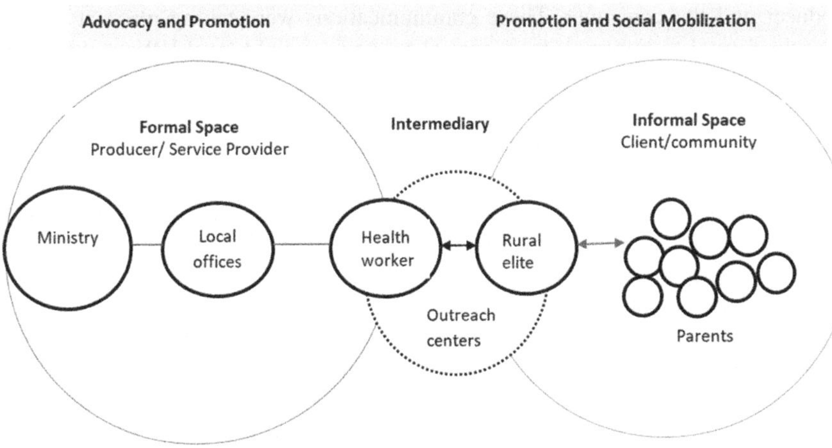

Fig. 1 The intermediary actors in EPI delivery

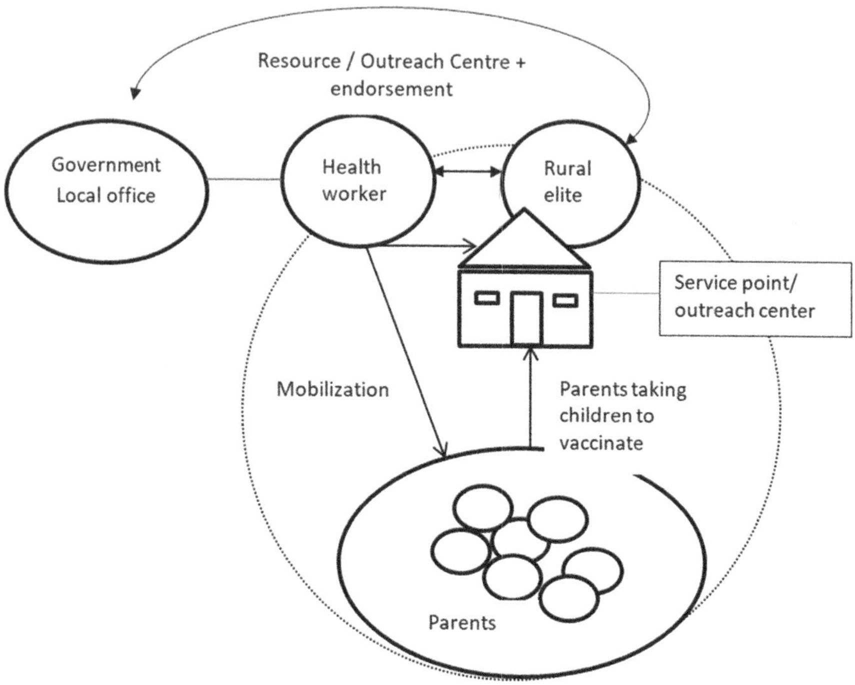

Fig. 2 Interactions among actors for service delivery

The extensive social communication efforts including the logo, branded messages, and mass media campaign through television commercials to promote vaccination were the formal branding efforts by the government – the service producer. However, these were only add-ons to the interactions taking place among critical actors, including health workers and *matbar*, the intermediary actors between the producer and the consumer. These communications were taking place at layers, from the formal to informal, from central program to field-level HWs and to *matbars* and parents as service users or consumers.

Among the different communication strategies, the most convincing one to the mothers was interpersonal contact with government health workers, neighbors, or relatives and NGO workers (Abed et al. 1991). The health workers were found to be the most trusted and relied upon communicators and rural health service providers. The most useful communication tool to the community has been the EPI site flag and signs displayed prominently and can be seen for miles away (Abed et al. 1991). Without the mobilization efforts by the community-level health workers, from both the government and NGOs, and endorsement from the local elites (*matbars*), the formal branding activities – the logo and television commercials – could achieve little.

Trust was built through the social networks started from the formal space to the informal space, often making it difficult to identify where the formal interactions stop and the informal interactions start. Health workers have frequently become the part of informal space where the *matbar* becomes the part of formal space. The local elites and the health workers have translated the branded messages, logo, and television commercials as trustworthy to the parents, the service users. In this process, the brand becomes more than the product – vaccines or the program. It becomes more of an integral part of the social relationships, a social brand that kept the program progressing and sustainable.

Here both the HWs and the rural elites worked as intermediary actors. The co-production was possible through the mobilization capacity of the HWs who ended up creating a long chain of participation from the bottom to the top.

In the following, we examine another case of social services delivered by BRAC, one of the largest non-profit organizations in the world. This program has also been internationally applauded for its impact in reducing extreme poverty in Bangladesh. We seek to understand how, despite having different provider and different types of services, similar patterns of intermediary space developed.

BRAC Targeting the Ultra-Poor (CFPR/TUP) Program

Bangladesh has a comprehensive portfolio of public safety net interventions to assist the poor. However, many of the programs fail to reach the chronically poor – the ultra-poor segment of the society (Ahmed et al. 2009). BRAC started the Challenging the Frontiers of Poverty Reduction/Targeting the Ultra-Poor (CFPR/TUP) program, currently known as the Ultra-poor Graduation Program (UPG), in 2002. It is a targeted development program for the extremely poor population group in Bangladesh living on less than $1.90 a day and consuming less than 1500 calories per day. These households are mostly female headed, and conventional anti-poverty programs are ill-equipped to support these households. The aim of the CFPR/TUP program was primarily to build a more sustainable livelihood for the targeted population by providing them a solid economic, social, and humanitarian foundation (Hossain and Matin 2007) and enable them to graduate from their extreme poverty.

The key component of the UPG or the CFPR/TUP program has been asset transfer. Here, asset transfer refers to a high-value asset package or interest-free loan provided to the targeted households to build their livelihood on it. Targeting or selection of the households eligible for asset transfer has been conducted jointly with the community through participatory rural appraisal (PRA) that includes wealth ranking to identify the poorest households.

Effective targeting has been a hallmark of BRAC's CFPR/TUP program (Sulaiman and Matin 2006). Through rigorous targeting, the program selected households that (a) were dependent upon female domestic work or begging as income source, (b) owned less than ten decimals of land, (c) had no male adult active member in the household, (d) had children of school-going age engaged in paid labor, and (e) possessed no productive assets (Ahmed et al. 2009). Throughout the process, BRAC field workers or the program organizers (PO) play a critical role in engaging villagers in the PRA, planning for asset transfer and ensuring social support in favor of the program.

The program faced challenges at its early stages. Several studies identified community suspicion that the ulterior motive of BRAC behind free asset transfer was to convert them to Christianity (Ahmed et al. 2009; Das et al. 2013; Mallick 2013). Such rumors, often encouraged by local religious leaders, resulted in a period of stigmatization of some targeted households until the community realized that the rumors were false (Das et al. 2013).

It was anticipated that the envious elite could even take away the resources; thus, vulnerability of the poor women would be enhanced due to the asset transfer (Hossain and Matin 2007). Based on the assumption that the local elite would be against the program, the field workers of BRAC began to consult them, trying to identify what the initial responses of such groups would be. As reported by field workers, BRAC personnel came to learn there was no hostility of the rural elite toward the extreme poor. They were not considered a threat as they were so impoverished, and hence no ill feelings existed regarding asset transfer. The elite were helping the poor, as it showed them in a favorable light and increased their clout in the village. They wanted to play a role in poverty alleviation in the villages. If given an opportunity, these elite would be ready to be directly involved. Later research conformed to the findings of the field workers regarding the motivation of the rural elites and their local ownership (Hossain and Matin 2007).

To ensure local buy-in for the program, rural elites were accommodated by forming village assistance committees locally known as Gram Shahayak Committees (GSCs) or the Gram Daridro Bimochon Committees (GDBC). This attempt, however, was largely a response to the perceived threats to the success of the program (Hossain and Matin 2007). The GDBCs were generally multi-class informal village bodies expected to protect the assets transferred to the target groups, provide advice, and support access to the government and other resources (Hashemi and De Montesquiou 2011). It also monitored the progress of program participants. Links to village elders also built confidence and social networks for the village poor (Hashemi and Umaira 2011). The involvement of the elites was considered extraordinary. Their main motivation in being involved was scope for heightened reputation and prestige in the community contributing to long-term socio-political assets. As viewed by Hulme and Moore (2008), "the most adventurous partnership of the program was engagement with the local elites, conceptually which is an extraordinary step" (p. 162).

Figure 3 provides a rough sketch of the key interactions in the process of CFPR/TUP delivery. The context of CFPR/TUP delivery was different because of its service type. Transfer of assets to the poorest could have affected the power relations in the community, giving rise to resistance against the program. The interactions in the process show that in parallel to taking the services to the chronically poor, CFPR/TUP delivery aimed at reducing the restraining factors and enhancing the facilitating factor for the program. The robust targeting was crucial. By engaging villagers in publicly identifying the target households, field workers ensured local buy-in for the program. After selection direct service in the form of asset, training

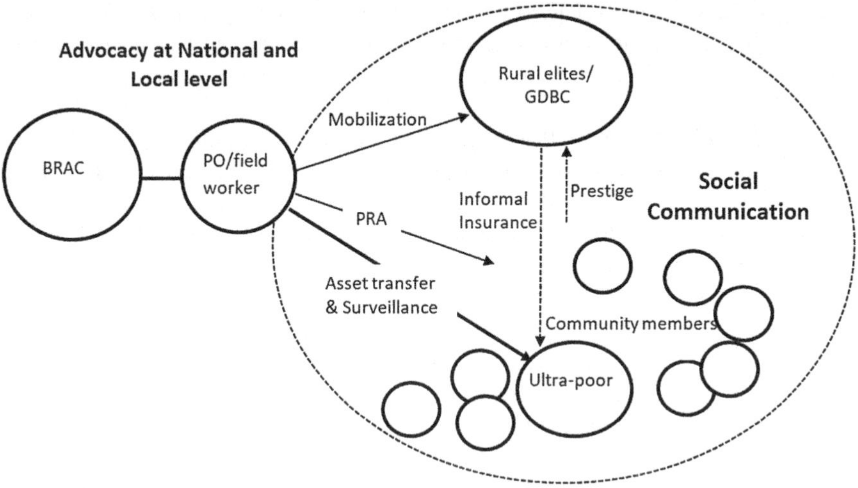

Fig. 3 Key interactions among the actors in the process of CFPR/TUP (UPG) delivery

and other supports were provided to the targeted households. The GDBC helped to monitor that the transferred asset is not stolen or sold off by the household members.

In parallel to the core program activities, BRAC set up a communication program to support the CFPR/TUP. The Advocacy and Human Rights Unit of BRAC designed and implemented the policy advocacy and social communication strategy for promoting the program at the national and local levels. Policy advocacy aimed at influencing government policies and practices in favor of extreme poor. The key focus of the social communication interventions at the community level aimed at influencing the community decision-making in terms of supporting the extreme poor.

In its approach, the social communication intervention for CFPR/TUP program contrasted with the conventional social communication approach where the key focus remains on individual-level behavior change. Instead of focusing on the communication through materials and messages, social communication for CFPR/TUP aimed at creating an enabling environment through participation, interaction, and mobilization. It was essentially based on the premise that behavior and decision-making in rural Bangladesh is a collective process and collective decision-making process needs to be aimed to change individual behavior. The field communication worker, or the *jogajog karmi*, played the role of the catalyst in the process translating communication intervention and initiating dialogue.

The *jogajog karmi* aimed to engage community members using Interactive Popular Theater (IPT), video, and audio programs. IPT is an interactive form of performing medium that has been widely used by various NGOs for raising awareness of various issues (Rafi and Rahman 2009). It provides a scope for the audience to feel part of the story. IPT was used to stimulate the audience about the condition of the chronically poor households in their community, followed by community meetings to discuss on extreme poverty. The plot of the performance generally was based on stories collected from the local context and performed in a local dialect so that the audience could relate to the performance and take part in the discussions after the IPT. Here, the communication materials were not an end in themselves. Instead, the materials worked as media for initiating dialogue among the community members.

The CFPR/TUP or the UPG program worked within a multi-actor framework including the extreme poor households, the service provider/producer (BRAC) and rural elite as the key intermediary stakeholder. The village was the collective unit that represented the collective decision-making by the community. The field workers, or the program organizers, and the *jogajog karmis*, although were part of the service providers, played the role of mediator among different stakeholders.

In both EPI and CFPR/TUP, the service provider had limited reach to the client, and the service user's access to the services were limited. The key constraints were resources and limited understanding of the social access culture. The field mobilizers and rural elites facilitated in strengthening both the service producers and the service users. These intermediary actors developed and expanded the existing social channels through the dialogue and negotiations among the stakeholders to bring the services closer to people and take service users to the service provider.

The field mobilizer's acceptance in the community helped them to achieve the service objective. The rural elite achieved heightened reputation in the community that contributed to their socio-political capital. They also aided in achieving better health and living standards for the community they are part of. Together, they supported the service user's experiences and wellbeing and added to their own experience and wellbeing simultaneously.

The co-production of services was possible through the mobilization capacity of the field workers and endorsement by elites, creating a long chain of participation from the bottom to the top. While the field workers have been part of the service provider, their relationship with the community and the rural elite goes beyond the formal service delivery space. The role of rural elite, on the other hand, could be categorized as volunteers, as they are part of the community. However, they needed to be mobilized by the field workers to be co-producers of the services. This implies they created both private value for themselves and welfare value for the service users and the broader community. In this, they are difficult to fit into the client-volunteer-citizen category suggested by Alford (2002).

The key observations from the cases can be summed up as follows: (a) people-centered programs work better when stakeholders co-produce the program; (b) the opportunity for co-production or the capacity of public cooperation tends to be higher when followed by dialogue and negotiations, instead of assuming that it would be possible without any such actions; (c) the proximity of the intermediary actor(s) to consumers; and (d) program communication elements including logo, television commercials, IPTs, and audio visual promotional activities are add-ons to the actual branding process that take place through the social interactions at different layers.

As we move toward further understanding of the branding process in these programs, it becomes more evident that in a collective decision-making context, public branding needs to account for the informal interactions taking place in parallel to the formal interactions. The key implication is the two-tier equation between service producer and service user needs adjusting, as both case studies show that the intermediary actors have been critical player in strengthening both service producer and service user. The producer cannot directly reach the service user to supply services effectively, and the service user has low access to the producer as well. In this context, the product (social services) cannot be marketed without the intermediary or middlemen. The promotional and branding efforts can only complement and reinforce the image that is being developed through the layers of dialogues, collaborations, and negotiations.

The brand, therefore, gains value not just on service value or consumer satisfaction but effective facilitation of the social forces. Hence, the value creation chain itself gains brand value. So, it can be argued that public sector branding needs to be seen beyond the product or service providing brand only. It needs to be an integral part of the entire chain of relationships among the co-producers. The social relationships that are built through interactions among service producer, service users, and the intermediary actor dynamics become the embodiment of the program brand.

Conclusion

In this chapter, we argued co-creation of brand value of public services is an integral part of co-production of such services and public sector branding needs to be seen beyond the product or service providing brand only. Brands need to be seen as part of the interactions taking place at different stages of service delivery. Taking two social service cases from Bangladesh, where one was provided by the government and another by non-profit, we show that in the context where service provider's reach to service users is limited, intermediary actors can facilitate to build the relationship.

The two cases detailed in the chapter show that between the producer and the primary consumers, there were intermediary actors who mediated in the process of program delivery. While both programs had parallel promoting and branding activities, it was the intermediary actors and their interactions that determined how those message and materials were translated. The promotional and branding efforts complemented and reinforced the image that was developing through co-production of services.

As we move toward further understanding the branding process in these programs, it becomes increasingly evident that in a collective decision-making context, public branding needs to take into account the informal interactions that take place, in parallel to the formal interactions. The key implication is that the two-tier equation between service producer and service user needs to be expanded to include intermediary actors, as they strengthen those relationships. When the service cannot be marketed without the support of the mediating actor(s), it is likely that the brand gains value not just on service or service user's experience but also by effective facilitation of them. Thus, the value creation chain itself gains brand value and not just the services. The social relationships that are built through the layers of interactions among the service producer, the service users, and the intermediary actor become the embodiment of the program brand.

References

Abed, F. H., McKee, N., Chowdhury, A., Chowdhury, A. M. R., & Rahman, R. (1991). Social mobilization for EPI in Bangladesh. In M. Huq (Ed.), *Near miracle in Bangladesh* (pp. 9–34). Dhaka: University Press Limited.

Ahmed, A. U., Rabbani, M., Sulaiman, M., & Das, N. C. (2009). The impact of asset transfer on livelihoods of the ultra poor in Bangladesh. *BRAC Res. Monogr. Ser, 39*, 339–380.

Aleshire, R. A. (1972). Power to the people: An assessment of the community action and model cities experience. *Public Administration Review, 32*(Special Issue), 428–443.

Alford, J. (1998). A public management road less travelled: Clients as co-producers of public services. *Australian Journal of Public Administration, 57*(4), 128–137.

Alford, J. (2002). Why do public-sector clients coproduce? Toward a contingency theory. *Administration & Society, 34*(1), 32–56.

Alford, J. (2009). Public value from co-production by clients. *Public Sector, 32*(4), 11–12.

Alford, J. (2016). Co-Production, interdependence and publicness: Extending public service-dominant logic. *Public Management Review, 18*(5), 673–691.

Alford, J., & Freijser, L. (2018). Public management and co-production. In T. Brandsen, T. Steen, & B. Verschuere (Eds.), *Co-production and co-creation: Engaging citizens in public services* (pp. 40–48). Routledge.

Andreasen, A. R. (1981). *"Power potential" channel strategies in social marketing* (BEBR faculty working paper no. 0743). University of Illinois.

Arifeen, S., Christou, A., Reichenbach, L., Osman, F. A., Azad, K., Islam, K. S., & Peters, D. H. (2013). Community-based approaches and partnerships: Innovations in health-service delivery in Bangladesh. *The Lancet, 382*(9909), 2012–2026.

Arnstein, S. R. (1969). A ladder of citizen participation. *Journal of the American Planning Association, 35*(4), 216–224.

Bhuiya, A., Bhuiy, I., & Chowdhury, M. (1995). Factors affecting acceptance of immunization among children in rural Bangladesh. *Health Policy and Planning, 10*(3), 304–311.

Bitner, M. J., Faranda, W. T., Hubbert, A. R., & Zeithaml, V. A. (1997). Customer contributions and roles in service delivery. *International Journal of Service Industry Management, 8*(3), 193–205.

BRAC (n.d.). Ultrapoor graduation Programme Bangladesh. BRAC. http://www.brac.net/program/ultra-poor-graduation/

Brandsen, T., & Honingh, M. (2016). Distinguishing different types of coproduction: A conceptual analysis based on the classical definitions. *Public Administration Review, 76*(3), 427–435.

Brandsen, T., & Honingh, M. (2018). Definitions of co-production and co-creation. In T. Brandsen, T. Steen, & B. Verschuere (Eds.), *Co-production and co-creation: Engaging citizens in public services* (pp. 9–17). Routledge.

Brandsen, T., & Pestoff, V. (2006). Co-production, the third sector and the delivery of public services: An introduction. *Public Management Review, 8*(4), 493–501.

Brandsen, T., Steen, T., & Verschuere, B. (2018). Co-creation and co-production in public services: Urgent issues in practice and research. In T. Brandsen, T. Steen, & B. Verschuere (Eds.), *Co-production and co-creation: Engaging citizens in public services* (pp. 3–8). Routledge.

Brodie, R. J. (2009). From goods to service branding: An integrative perspective. *Marketing Theory, 9*(1), 103–107.

Brudney, J. L., & England, R. E. (1983). Toward a definition of the coproduction concept. *Public Administration Review, 43*(1), 59–65.

Bovaird, T. (2007). Beyond engagement and participation: User and community coproduction of public services. *Public administration review, 67*(5), 846–860.

Boyle, E. (2007). A process model of brand cocreation: Brand management and research implications. *Journal of Product & Brand Management, 16*(2), 122–131.

Chowdhury, A. M. R., Bhuiya, A., Mahmud, S., Salam, A. A., & Karim, F. (2003). Immunization divide: Who do get vaccinated in Bangladesh? *Journal of Health. Population and Nutrition, 21*(3), 193–204.

Cleaver, F. (1999). Paradoxes of participation: Questioning participatory approaches to development. *Journal of International Development: The Journal of the Development Studies Association, 11*(4), 597–612.

Das, N., Yasmin, R., Ara, J., Kamruzzaman, M., Davis, P., Behrman, J., Roy, S., Quisumbing, A. R. (2013). *How Do intrahousehold dynamics change when assets are transferred to women? Evidence from brac's challenging the frontiers of poverty reduction — targeting the ultra poor program in Bangladesh* (IFPRI Discussion Paper 01317). Washington, DC: International Food Policy Research Institute.

Duţu, A., & Diaconu, M. (2017). Community participation for an open public administration: Empirical measurements and conceptual framework design. *Cogent Business & Management, 4*(1), 1287980.

Etgar, M. (2008). A descriptive model of the consumer co-production process. *Journal of the Academy of Marketing Science, 36*(1), 97–108.

Fischer, F. (1993). Citizen participation and the democratization of policy expertise: From theoretical inquiry to practical cases. *Policy Sciences, 26*(3), 165–187.

Gregory, A. (2007). Involving stakeholders in developing corporate brands: The communication dimension. *Journal of Marketing Management, 23*(1–2), 59–73.

Greitens, T. J. (2018). Citizen participation in public management. In A. Farazmand (Ed.), *Global Encyclopedia of public administration, public policy, and governance* (pp. 744–749). Cham: Springer Nature.

Grönroos, C. (2006). Adopting a service logic for marketing. *Marketing Theory, 6*(3), 317–333.

Gronroos, C. (2011). Value co-creation in service logic: A critical analysis. *Marketing Theory, 11*(3), 279–301.

Gummesson, E. (1995). Relationship marketing: Its role in the service economy. In W. J. Glynn & J. G. Barnes (Eds.), *Understanding services management* (pp. 244–268). Wiley.

Hashemi, S. M., & De Montesquiou, A. (2011). Reaching the poorest: Lessons from the graduation model. *Focus Note, 69*(1), 1–15.

Hashemi, S. M., & Umaira, W. (2011). *New pathways for the poorest: the graduation model from BRAC (Report No. 10).* Centre for Social Protection (CSP), Institute of Development Studies.

Hossain, N., & Matin, I. (2007). Engaging elite support for the poorest? BRAC's targeted ultra poor programme for rural women in Bangladesh. *Development in Practice, 17*(3), 380–392.

Hulme, D., & Moore, K. (2008). Assisting the poorest in Bangladesh: Learning from BRAC's 'targeting the ultra-poor' programme. In A. Barrientos & H. David (Eds.), *Social protection for the poor and poorest: Concepts, policies and politics* (pp. 194–210). London: Palgrave Macmillan.

Huq, M. (Ed.). (1991). *Near miracle in Bangladesh.* University Press Limited.

Jamil, K., Bhuiya, A., Streatfield, K., & Chakrabarty, N. (1999). The immunization programme in Bangladesh: Impressive gains in coverage, but gaps remain. *Health Policy and Planning, 14*(1), 49–58.

Jo, S., & Nabatchi, T. (2018). Co-production, co-creation, and citizen empowerment. In S. Osborne, T. Brandsen, T. Steen, & B. Verschuere (Eds.), *Co-production and co-creation: Engaging citizens in public services* (pp. 231–239). Routledge.

Kahn, M. M., & Yoder, R. A. (1999). Bangladesh: Actions for sustainable immunization services. *Health Reform Prior Serv*, 9–11.

King, C. S., Feltey, K. M., & Susel, B. O. N. (1998). The question of participation: Toward authentic public participation in public administration. *Public Administration Review, 58*(4), 317–326.

Kotler, P. (1986). The prosumer movement: A new challenge for marketers. *Advances in Consumer Research, 13*, 510–513.
Loeffler, E., & Bovaird, T. (2016). User and community co-production of public services: What does the evidence tell us? *International Journal of Public Administration, 39*(13), 1006–1019.
Lusch, R. F., & Vargo, S. L. (2006). Service-dominant logic: Reactions, reflections and refinements. *Marketing Theory, 6*(3), 281–288.
Mallick, D. (2013). How effective is a big push to the small? Evidence from a quasi-experiment. *World Development, 41*, 168–182.
Martinussen, J. (1999). *Society, state & market: A guide to competing theories of development.* Dhaka: University Press Limited.
Merz, M., & Vargo, S. (2009). The evolving brand logic: A service-dominant logic perspective. *Journal of the Academy of Marketing Science, 37*, 328–344.
Merz, M. A., He, Y., & Vargo, S. L. (2009). The evolving brand logic: A service-dominant logic perspective. *Journal of the Academy of Marketing Science, 37*(3), 328–344.
Muniz, A. M., & O'guinn, T. C. (2001). Brand community. *Journal of Consumer Research, 27*(4), 412–432.
Nabatchi, T., & Leighninger, M. (2015). *Public participation for 21st century democracy*. Wiley.
Nabatchi, T., Sancino, A., & Sicilia, M. (2017). Varieties of participation in public services: The who, when, and what of coproduction. *Public Administration Review, 77*(5), 766–776.
Nandy, S. (1997). *Immunization in Bangladesh: studying sustainability through social science approaches (Report No.)*. BRAC Research and Evaluation Division (RED).
Osborne, S. P. (2018). From public service-dominant logic to public service logic: Are public service organizations capable of co-production and value co-creation? *Public Management Review, 20*(2), 225–231.
Osborne, S. P., & Strokosch, K. (2013). It takes two to tango? Understanding the co-production of public services by integrating the services management and public administration perspectives. *British Journal of Management, 24*, 31–47.
Osborne, S. P., Radnor, Z., & Nasi, G. (2013). A new theory for public service management? Toward a (public) service-dominant approach. *The American Review of Public Administration, 43*(2), 135–158.
Osborne, S. P., Radnor, Z., & Strokosch, K. (2016). Co-production and the co-creation of value in public services: A suitable case for treatment? *Public Management Review, 18*(5), 639–653.
Osborne, S. P., Strokosch, K., & Radnor, Z. (2018). Co-production and the co-creation of value in public services: A perspective from service management. In T. Brandsen, T. Steen, & B. Verschuere (Eds.), *Co-production and co-creation: Engaging citizens in public services* (pp. 18–26). Routledge.
Osman, F. A. (2008). Public-private partnership in health service delivery: lessons from Bangladesh. http://crossasia-repository.ub.uni-heidelberg.de/144/1/Osman_Partnership_in_health_service_delivery.pdf
Ostrom, E. (1996). Crossing the great divide: Coproduction, synergy, and development. *World Development, 24*(6), 1073–1087.
Parks, R. B., et al. (1981). Consumers as coproducers of public services: Some economic and institutional considerations. *Policy Studies Journal, 9*(7), 1001–1011.
Payne, A., Storbacka, K., Frow, P., & Knox, S. (2009). Co-creating brands: Diagnosing and designing the relationship experience. *Journal of Business Research, 62*(3), 379–389.
Perry, H. B. (2002). Primary Health Care in Bangladesh: Challenges, approaches, and results. In R. Jon & W. John (Eds.), *Community-based health care: Lessons from Bangladesh to Boston*. Harvard School of Public Health.
Perry, H. B., & Chowdhury, A. M. R. (2020). NGO contributions to community health and primary health care: Case studies on BRAC (Bangladesh) and the Comprehensive Rural Health Project, Jamkhed (India). In *Oxford Research Encyclopedia of Global Public Health.*
Prahalad, C. K., & Ramaswamy, V. (2004). Co-creation experiences: The next practice in value creation. *Journal of Interactive Marketing, 18*(3), 5–14.

Pine, B., and Gilmore, J. (1999). *The Experience Economy*. Boston: Harvard Business School Press.
Quasem, M. A. (2008). *Development strategies and challenges ahead of Bangladesh*. Dhaka: Palok Publishers.
Rafi, M., & Rahman, H. (2009). Popular theatre and BRAC. *BRAC Res. Monogr. Ser, 40*, 1–86.
Ramaswamy, V. (2008). Co-creating value through customers' experiences: The Nike case. *Strategy & Leadership, 36*(5), 9–14.
Sarkar, P. K., Sarker, N. K., Doulah, S., & Bari, T. I. A. (2015). Expanded programme on immunization in Bangladesh: A success story. *Bangladesh Journal of Child Health, 39*(2), 93–98.
Solveig, W. (1996). The customer as co-producer. *European Journal of Marketing, 30*(4), 6–17.
Sorrentino, M., Sicilia, M., & Howlett, M. (2018). Understanding co-production as a new public governance tool. *Policy and Society, 38*(3), 277–293.
Sulaiman, M., & Matin, I. (2006). *Targeting effectiveness of CFPR/TUP in scale-up environment* (CFPR/TUP Working Paper Series No.8). BRAC Research and Evaluation Division.
Terblanche, N. S. (2014). Some theoretical perspectives of co-creation and co-production of value by customers. *Professional Accountant, 14*(2), 1–8.
Tynan, C., McKechnie, S., & Chhuon, C. (2010). Co-creating value for luxury brands. *Journal of Business Research, 63*(11), 1156–1163.
Vargo, S. L., & Lusch, R. F. (2004a). The four service marketing myths: Remnants of a goods-based, manufacturing model. *Journal of Service Research, 6*(4), 324–335.
Vargo, S. L., & Lusch, R. F. (2004b). Evolving to a new dominant logic for marketing. *Journal of Marketing, 68*(1), 1–17.
Vargo, S. L., & Lusch, R. F. (2008). Service-dominant logic: Continuing the evolution. *Journal of the Academy of Marketing Science, 36*(1), 1–10.
White, H. N. (2005). *Maintaining Momentum to 2015? An impact evaluation of interventions to improve maternal and child health, nutrition and population in Bangladesh*. The World Bank.
WHO. World Health Organization. (1994). Expanded Programme on Immunization : Programme review = Programme élargi de vaccination : Examen du programme. *Weekly Epidemiological Record = Relevé épidémiologique hebdomadaire, 69*(12), 87–90.
Wolffers, I. (1988). Limitations of the primary health care model. A case study from Bangladesh. *Tropical and Geographical Medicine, 40*(1), 45–53.
World Health Organization. (2000). *The world health report 2000 - Health systems: improving performance*.

The Brussels-Capital Region: Toward a More Integrated Strategy for Europe's Capital?

Vincent Mabillard

Introduction

Throughout the world, many places have now developed some sort of marketing strategy. Among them, a large number of cities have even established place brands, from "Be Berlin" to the iconic brands of "Iamsterdam" or "OnlyLyon" to areas with a more international focus, such as Geneva ("GenevAlive"). The case of Iamsterdam, created in 2003, is particularly interesting because it has embraced a comprehensive approach, including all numerous local and neighboring stakeholders (such as the municipality, tourism board, neighborhood, and business representatives) to develop an integrated strategy (Kavaratzis 2008). In the same vein, the French city of Lyon brought together 28 partners to promote the place through an international marketing program (Gayet 2017). At the regional level, the "Valais" brand in Switzerland is often cited as a great example of a global integrated approach, bringing most economic actors together. This Valais umbrella brand also covers aspects related to tourism and quality labels; its success can be explained by the support given by internal actors and homogeneity perceived by external actors (Vuignier 2018). In addition, other Swiss federated entities (hereafter cantons) and several French regions have also developed integrated marketing strategies, sometimes focusing primarily on economic attractiveness.

Examples such as Iamsterdam and OnlyLyon, which extend to neighboring areas, are also umbrella brands, concentrating on different sectors and target groups (residents, tourists, investors). These structures differ strongly in both institutional terms (nonprofit organization supported by regions and municipalities, public-private partnerships, city or state promotion bodies, etc.) and resources (human and financial). The importance of the brand also varies greatly from one place to another.

V. Mabillard (✉)
Solvay Brussels School of Economics & Management, Université Libre de Bruxelles (ULB), Brussels, Belgium
e-mail: vincent.mabillard@ulb.be

S. M. Zavattaro (ed.), *Public Branding and Marketing*,
https://doi.org/10.1007/978-3-030-70505-3_5

In some North American cities, such as Mississauga, Canada, local authorities have appointed a city marketing and branding manager.

Moreover, the new economic rationale has fostered deeper collaboration between distinctive institutional areas, aiming at better responding to citizens' demands on the one hand and increasing place competitiveness on the other. From an economic standpoint, this has often led to clusters of activity and cross-border collaboration, as shown in Nordic countries (Hospers 2004). Place brand strategies have then become a trend worldwide, through a rapid dissemination of such practices. Although these strategies take different forms – and, paradoxically, often use similar programs – they share a common goal: attractiveness and retention of target groups (Mabillard and Vuignier 2017).

In this sense, it is particularly surprising to note Brussels has not been much commented in the place branding literature. Indeed, a closer look at this case shows Europe's capital is a geographical and cultural crossroad, and one of the world's most cosmopolitan cities, given migrants account for more than 50% of its residents (IOM 2015); 183 nationalities are represented and 108 languages are spoken in the Brussels' 19 municipalities (visit.brussels 2019). According to the Euromonitor database (Yasmeen 2019), Brussels is the 60th most visited city in the world (20th in Europe) in 2019, with more than 4.2 million visitors – a growth of 8.5% compared to the previous year. Although tourism is just one indicator of attractiveness, it offers marketers a platform for selling their places. Regarding overnight stays and business travels, Brussels is in a particular position as it attracts numerous employees of European institutions. Additionally, many people work temporarily in the city or attend events on a regular basis. In a way, these individuals can be regarded as place ambassadors (Jansen-Verbeke et al. 2005) given their connection to the place. Business remains the main motive for trips to Brussels, but reasons for leisure have gained importance during the last few years. A survey among employees of European institutions from 2003 shows the place image is largely shaped by the city's European, international, and multicultural features, while, somehow paradoxically, the so-called European Quarter is considered as a landmark by less than 15% of the 125 respondents (Jansen-Verbeke et al. 2005).

From an economic point of view, figures reveal the Brussels-Capital Region (BCR) is appealing for businesses because it is home of 20 institutions of the European Union (EU); 42 intergovernmental organizations, including the North Atlantic Treaty Organization (NATO); about 300 regional and local representations; more than 10,000 lobbyists; and more than 5000 diplomats (highest number in the world). This international presence generated more than 120,000 jobs in 2020, and almost 40,000 people work directly for the European institutions (Office of the Brussels Commissioner for Europe and International Organizations 2020).

As such, Europe's capital is the regulatory and diplomatic center of the EU; this specific attribute may be regarded as a strategic hub for multinationals and brands. In this vein, the Brussels Commissioner for Europe and International Organizations (CEIO) calls for a specific "regional seat" policy. How does the BCR address these issues in its promotion? What steps have been taken in terms of place marketing?

Based on these aspects, did a strategy emerge, and if so, how has it been designed? Has the BCR developed a place branding approach? The current chapter aims at providing answers to these questions and proposes paths for future research through both academic- and practice-oriented lenses. For most parts, it is the result of exchanges with Bénédicte Wilders, who is the Senior Strategic CEO Advisor and acting director of the R&D Department at hub.brussels, the BCR's agency for business support.

Background: Brussels, a Multilayered City

A better understanding of the promotion of Brussels requires a short explanation of Belgium's complex institutional system. Historically, in this federal country, the distribution of tasks and responsibilities occurred along two lines. The first line referred to culture (including language) and resulted in what is known as Communities today – the Flemish Community, the French Community (Wallonia-Brussels Federation), and the German-speaking Community. The second line was inspired by economic interests and provided considerable autonomy to the three newly created Regions – the Flemish Region, the Walloon Region, and the BCR.

Other institutional layers include 10 provinces and, at the local level, 589 municipalities in 2019 (Table 1). The BCR (1.2 million inhabitants) consists of 19 municipalities, one of them being Brussels City (185,000 inhabitants). Institutionally, Brussels has five capital city functions: for the Belgian nation, for the Wallonia-Brussels Federation, for the Flemish Region/Community, for the BCR, and for the EU. As a result, it is often referred to as a multilayered capital city (Jansen-Verbeke and Govers 2009) due to these different functions and its highly cosmopolitan

Table 1 The institutional structure of Belgium (based on Albrechts 2001)

Levels	Institutions				
Federal	National institutions: House of Representatives, Senate, King				
Community	German-speaking	French	Brussels Common Community Commission		Flemish
			French Community Comm.	Flemish Community Comm.	
Region	Walloon Region 		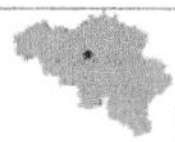 Brussels-Capital Region		Flemish Region 
Linguistic	German	French	Bilingual (French-Dutch)		Dutch
Province	5 provinces		Brussels-Capital		5 provinces
Municipality	262 municipalities		19 municipalities		300 municipalities
Inhabitants[a]	3,633,795		1,208,542		5,589,069

[a]Source: Statbel (2020) *Structure of the population* (January 1, 2019)

character. From both tourism and economic perspectives, attractiveness issues have mostly revolved around the European and international aspects of the place. This is emphasized by Elmhorn (2001), who argues Brussels is a "world city" located in the European agglomeration economy.

Reflecting on this institutional puzzle, Jansen-Verbeke and Govers (2009, p. 155–156) emphasize how "the complexity of a city like Brussels cannot be communicated easily" and that "it is hard for Brussels to create a consistent brand image." According to Hein (2008), Brussels' diverse functions further complicate attempts at creating a strong European identity attached to the city. Starting from there, one might then wonder how the city could be promoted, what brand(s) should be used, and how collaboration between multiple agencies, at different levels, can be organized. Part of the answer relates to the region's more extensive approach compared to the previous one and the elaboration of a flexible system of "baseline" (the domain extension ".brussels"). This approach was preferred to a single logo, which would not reflect the region's diversity. The domain name in English was chosen to ease tension between the linguistic communities (Baele 2012), to avoid several domain names (French, Dutch, and English), and to position Brussels as an international capital.

This preoccupation echoes the tension between acknowledging the multifaceted character of a large city and simplifying its image for marketing purposes, as underlined by Kalandides (2006). This flexible approach was designed to "make it possible to enter into dialogue with a whole variety of different target groups and [to provide] an environment that fits everyone's needs" (be.brussels 2019). The main idea was then to create a logo that is highly adaptable and immediately recognizable while remaining at the same time entertaining and multilayered. Partners can adapt the brand to the region's new image, which is also reserved for the region's institutions and used for city marketing purposes (Table 2).

Although it gives freedom to the different actors that can use the brand as they like (e.g., "be everywhere, be.brussels" for taxis), this approach ensures a global

Table 2 Differences between the institutional brand, city marketing, and the "shared" brand (based on Bruxelles-Capitale 2012)

Institutional brand	City marketing	"Shared" brand
Valorization of the actions and services of authorities through a new logo	Valorization of the whole territory and its public, private, and associative actors (extension ".brussels")	Visual consistency in all partners' communication/valorization of the partner's regional affiliation
Main target groups: residents, taxpayers, and citizens	Main target groups: residents, but the brand gives more visibility for international actors as well	Main target groups: internal and external actors (tourists, companies, and investors)
Example RÉGION DE BRUXELLES-CAPITALE BRUSSELS HOOFDSTEDELIJK GEWE BRUSSELS-CAPITAL REGION	*Examples* be a winner be.brussels be ...action! be.brussels	*Examples* hub .brussels visit.brussels

visual identity common to all BCR actors. The idea of developing this new logo (a flower with a heart in the middle) followed the global trend of city brands such as "Be Berlin" or "I love NY" to better market the city internationally, according to regional government leaders. It seems the new brand, which was to be used and "played with" by the different regional stakeholders, has been working well so far.

An analysis conducted on a large survey ($n = 592$) of .brussels websites by Waty et al. (2018) reveals that diverse sectors are currently using the new logo, such as services (hotels and education, e.g., 51.2%), culture (11.5%), transportation (10.7%), and tourism (3.8%). Apart from that, the domain names are associated with public services in 18.6% of all cases, and 28.2% of all analyzed websites are either directly or indirectly referring to the Brussels' institutions (government, parliament, etc.). To our knowledge, there has been no publication of institutional or scientific evaluations of the new branding strategy deployed by the BCR several years after its formal introduction.

However, the brand applications differ from the distinctive logos typically used in other cases, such as Amsterdam (municipality) or Vaud (canton in Switzerland). In the latter cases, an institutional brand, managed by elected officials, concerns residents primarily (City of Amsterdam, Canton de Vaud), while internal and external publics are typically targeted with a different place brand (Iamsterdam, Vaud+). The management of the brand is also different because it usually includes numerous stakeholders and favors a collective approach (Table 3). The place brand may sometimes exceed institutional borders; for instance, Iamsterdam encompasses other areas (e.g., Amsterdam beach), while Vaud+ is strictly limited to the cantonal borders (Vuignier 2018).

In the case of Brussels, the regional government recalls that "a single logo will not do to provide the BCR with a strong identity. A rich visual identity remains necessary so as to represent the region's diversity" (be.brussels 2019). However, there was a need to redefine the marketing approach for at least three reasons: (a) the willingness to follow trends initiated by other cities, increasingly viewed as competitors; (b) the necessity to establish a strong brand of the region, clearly distinctive

Table 3 Differences between institutional and place branding (based on Vuignier 2018)

Institutional branding	Place branding
Valorization of the actions and services of authorities	Valorization of the whole territory and its public, private, and associative actors
Institutional actors at the center	Collective approach (co-construction)
Mainly managed by elected officials	Multi-stakeholder governance
Institutional branding of the place (institution's name, official logo)	General branding of the place (place's borders may vary greatly, destination name)
Main target groups: internal actors (residents, taxpayers, and citizens)	Main target groups: internal and external actors
Examples City of Amsterdam canton de vaud	*Examples* I amsterdam. VAUD+

from the Brussels' city brand (where a new logo was also created in 2015 – "BXL, our city"); and (c) the willingness to solve the language tension by adopting an extension in English, as mentioned above (Baele 2012).

Compared to more integrated approaches, Brussels relies on separate organizations to promote the region. It should be reminded here that the BCR, as a distinct institution, is quite recent (it was founded in 1989). In terms of promotion, visit.brussels is the tourist agency, whose mission is to spread the image of a "world city," capital of 500 million Europeans. Created in 2015, the Office of the CEIO aims at maintaining and further developing 60 years of extensive European and international diplomatic presence in Brussels. Its mission is therefore to welcome expatriates as well as international bodies and to ensure their deployment in harmony with the needs of the residents and the growth of Brussels as a Belgian region.

Moreover, a regional institute for research and innovation was set up in 2004. As a public agency, Innoviris takes care of aspects related to stimulating (through calls for projects) and financially supporting citizens, for-profit and nonprofit companies, as well as research institutes to achieve progress in their field of activity. This mission is especially important given more than 11,000 companies are established in Brussels every year, making it the European region with the largest number of newly created businesses per inhabitant. From an economic point of view, entrepreneurship and innovation remain important, and it must be noted here that thanks to its successes in these areas, the BCR contributes up to 18% of the total Belgian growth domestic product (GDP) in 2019 (innoviris.brussels 2019).

Finally, hub.brussels is a public agency established in 2018 to help businesses to develop in Brussels and overseas. Its mission consists mainly in aiding project initiators, business founders, and firms in the process of international expansion. Other essential goals include attracting foreign investors and advising them on key aspects of their present or future economic activity in Brussels. The next section provides more details about this particular agency and focuses on its centrality for the region's economic promotion.

hub.brussels: The Case of Economic Attractiveness

A central actor of Brussels' promotion, hub.brussels is a public limited company under public law with a social purpose. It is a governmental agency because the only shareholder is the government of the BCR, which also constitutes the body's General Assembly and is represented by seven members (out of nine) sitting on the board of directors. Representatives of the local economy also can raise their concerns through a strategic commission, made up of six employers' representatives and six employees' representatives (parity is thus ensured). As of December 2020, the management board comprises the different departments of the organization: research and design (R&D), corporate services, partnerships and networks, sectoral and thematic accompaniment, international networks, and internationalization support.

The main missions of the organization relate also to the development of a strategy of attractiveness for foreign companies and to the provision of quality services to those businesses once they are established in Brussels. In addition, hub.brussels and its one-stop shop "1819" are in charge of several governmental programs, such as Women in Business (WiB), which insists on the importance of entrepreneurship for women (Législature 2019–2024). As such, WiB is a project supported by the BCR and implemented by the 1819, whose missions are to stimulate employment for women in Brussels, to promote women entrepreneurship in the region, and to monitor its evolution, typically followed by recommendations sent to the relevant authorities (1819 – wib.brussels 2020).

Among its numerous activities, hub.brussels provides customized accompaniment for new companies and project initiators. This mission translates into the motto of the agency: "our business is to help yours." Regarding inward investment, Brussels is promoted to non-EU investors as a gateway to the EU and its 500 million inhabitants. For North American and Asian companies in particular, the expansion of their activities in Europe is highlighted, as well as the existence of numerous headquarters (HQ) of European firms and EU institutions. For EU investors, the focus shifts to the different economic clusters offered by Brussels. The fact that the region hosts high-level decision centers, especially EU regulation centers, is emphasized. Clusters are a cornerstone of the strategy, welcoming companies to grow their business in the most flourishing sectors of the region. They concentrate mainly on software, the circular economy and sustainability, lifetech, audio-visual, and hospitality (hub.brussels 2019). hub.brussels also offers customized accompaniment regarding important aspects such as better integration in a multilingual and multicultural capital city and affiliation to professional European federations. Advice about the global ecosystem, the most promising sectors, the best legal form for the company, and employment conditions in Brussels are equally important. For all these aspects, the newly created website of the Business Unit Invest intends to be the main access point for actors and a showcase to the world (why.brussels 2020).

To increase support and attractiveness, the region has created "accelerators," providing investors with individual coaching, networking opportunities, and expert advice to launch their business. Among other essential actors (including universities), hub.brussels is active in these initiatives. "Incubators" give project initiators space and time to grow; spaces are rented at advantageous rates, personalized coaching is offered, and partnerships are expected to make the projects burgeon. Investors have been offered the possibility to "try Brussels for free."

The idea is also to give a welcome pack (3 months for free) to foreign companies so they can check out the region's business opportunities. The pack offers a place to work, Internet, multilingual support service, a conference room and legal information, analysis of available incentives, and location research to the involved firms (hub.brussels 2020). Moreover, the 1819 platform has been designed as a single information point for anyone interested in starting or growing a business in Brussels. Founded about 10 years ago, it was an ambitious pioneering project; through this channel, individuals were and still are receiving top-quality information about how to structure their ideas clearly, what are the opportunities and benefits, and what are the key partners to work with. Free of charge, advisors of the 1819 platform respond to more than 8500 demands annually in several languages. This number raised significantly during the COVID-19 period since the platform was playing an "helpdesk" role for companies. The situation led to the hiring of new staff and training sessions and, of course, increased the potential to attract new clients. In this regard, promotion campaigns were launched during the crisis. Finally, it should be added that the 1819 service organizes "thematic weeks" to help entrepreneurs and create networking opportunities.

The agency boasts great results since its creation 2 years ago. In 2018, more than 90% of surveyed investors found individual coaching useful. Even better, "accelerators" have been perceived as useful in 100% of the cases (hub.brussels 2019). The organizations that preexisted the creation of hub.brussels had already recorded great results in satisfaction surveys. The challenge is now to maintain these high ratings through the new agency, which is providing more services and is stimulating deeper synergies compared to the previous, more fragmented institutional configuration. Statistically, among the 61 foreign investment projects that have picked Brussels as the best location in 2018 (EY 2019), 21 have settled in the region with the full support of hub.brussels, and 8 have benefited from the "incubators" program. The agency has accompanied 2020 projects individually, including 486 through the European Programs for Internationalisation and Innovation (hub.brussels 2019). In Brussels, the National Contact Point (NCP) helps companies, universities, research centers, and other stakeholders from the region to bid for European funding. In 2018, local actors involved in research and development and innovation have obtained 7 million euros in financial support from the EU. How can these early successes be explained?

Apart from crucial collaborative projects with essential partners, which are presented below, the agency can count on a strong support of the region's government (Législature 2019–2024), a large network of economic and trade attachés around the world (37 offices by the end of 2019), and area managers in Brussels. It is remarkable that interregional agreements enable the Brussels companies to benefit from certain offices established by the three Regions overseas (at the same time, some of the 37 offices owned by the BCR are used by the Walloon and Flemish Regions). From time to time, thematic seminars and roadshows are organized abroad to promote the economic assets of the region. Sometimes, this occurs through trade missions led by the Princess and state visits (four times a year). hub.brussels took part in various events in 2019, where it was, for example, present with an exhibition booth. Around 120 actions were scheduled in the International Action Plan 2019, and 140 actions were scheduled for 2020, which were changed, recreated, digitized (e.g., webinars), and recentered on European markets due to the COVID-19 crisis. Following the Brexit vote in 2016, the agency also launched a campaign entitled "The future is close – Brussels is yours" targeting companies facing the economic consequences of the referendum. Less aggressive than Amsterdam's campaign, for instance, the message was not "leave London and relocate in Brussels" but "you are also welcome in Brussels." This campaign targeted primarily sectors related to regulation issues, in which companies need an agreement to develop their business in the EU. Finally, feedback is frequently collected in the region's municipalities and analyzed by the Research Business Unit. The main aim is to develop projects in harmony with the daily needs and expectations of Brussels' inhabitants.

Although an integrated approach has not been preferred (yet) regarding the promotion of Brussels, it is widely believed that collaboration between close partners remains crucial to achieving better results and working more efficiently. Numerous examples of such collaborations include sharing offices abroad and competencies with the Walloon Export and Foreign Investment Agency (AWEX) and Flanders Investment and Trade (FIT) when it comes to exports. Other examples include the stands shared with the Walloon and Flemish Regions during economic fairs on some occasions and to partnerships established with other cities (e.g., Tokyo Metropolitan Government). In addition, hub.brussels reaches out to business representatives, especially when companies want to create joint ventures overseas. In Brussels, the agency maintains a close relationship with the CEIO. It also coordinates the Economic Coordination Council (ECC), which is a platform for exchanging economic information between key actors of the region: Brussels Environment, the public transportation company, the Port of Brussels, etc. From a legal perspective, the Brussels Bar is obviously a privileged partner, particularly in the promotion of Brussels as an arbitration hub in Europe. Finally, hub.brussels plays a crucial role in coordinating local partnerships, supporting the creation of merchant associations, and steering construction sites or urban projects that impact citizens' life directly (hub.brussels 2019).

Increasing Coordination Between Different Agencies

These collaborations and the incentives to increase impact and efficiency have already materialized. In this sense, hub.brussels illustrates this trend because the organization results from the merger of three former entities: atrium.brussels, Brussels Invest & Export, and impulse.brussels. This was already a first step toward a rationalization of the structures in charge of the economic development. As a result, the new agency for business support of the BCR – hub.brussels – was created on January 1, 2018, to provide a one-stop shop for new businesses. Expectations focused mainly on the simplification of all aspects related to staff skills and coordination with municipalities. However, hub.brussels has no financial competence and cannot provide funding to companies (Fig. 1).

The new agency signed a management service agreement with the legislature from 2019 to 2024 and is now linked with the regional government until 2023 (the agreement was signed in 2018 in the wake of the creation of the agency). Although the establishment of a unified body indicates a desire to better integrate public organizations in charge of Brussels' promotion, this first move does not encompass other agencies and gather them under an identical banner or logo (preferred by other places, e.g., OnlyLyon). As explained above, the diversity of Brussels is, according to the region's government, better addressed through an extension (.brussels)

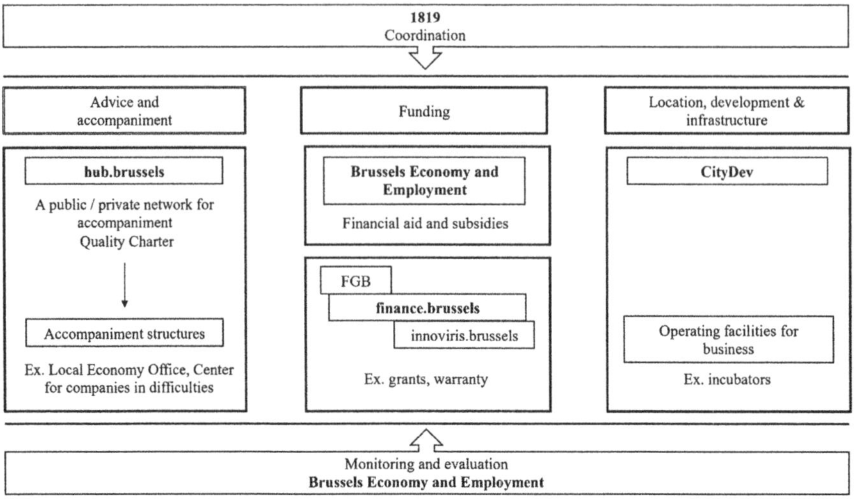

Fig. 1 Institutional structure of the economic actors in the Brussels-Capital Region (as of October 22, 2015)

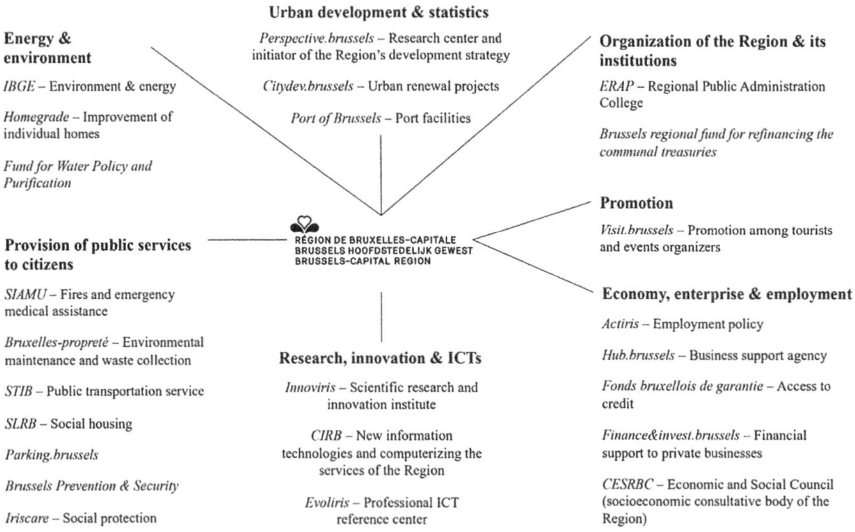

Fig. 2 Regional bodies of the Brussels-Capital Region (be.brussels 2019)

applied to different actors, which partly retain their own visual identity. The decentralized structure of public regional bodies remains in place to this day, fulfilling various tasks but potentially overlapping in certain sectors (Fig. 2). These agencies are either institutions of public benefit or nonprofit organizations (be.brussels 2019).

At the international level, several coordinated actions have also been launched between the different partners of the BCR. The first example refers to the concerted organization of seminars with visit.brussels and international.brussels, which implements the region's external policy. For instance, during events focusing on economic issues, promotional movies provide the public with information about Brussels' life-related aspects, including quality of life (green spaces, gastronomy). Conjoint missions enable the coordination strategy to materialize through deeper collaboration between different actors, including especially the BCR Secretary of State responsible for European and international relations and hub.brussels.

Second, the "Brussels Days" are a fertile ground for these coordination efforts. They aim at reinforcing the international presence and global positioning of the BCR. Through high-level networks, these events combine all aspects of place promotion: the economic dynamism of Brussels, its cultural heritage, its tourism potential, and, of course, the region's way of life. In May 2018, "Brussels Days" took place in Washington, D.C., and New York City. In the US capital, the morning session started with a seminar organized by hub.brussels, the Belgian Embassy, and an international research institute, discussing in detail the concept of high-energy performance buildings. The event offered opportunities to connect Belgian and American speakers from both the public and the private sector. Additional seminars were organized by visit.brussels to promote Brussels as a leisure destination as well as a "meeting, incentives, conferences, and exhibitions" (MICE) destination. Third, these efforts are materializing in attempts to create "Brussels Houses" abroad, which would gather tourism and economic attractiveness for foreign investors, export promotion, welcome expatriates, and serve as a "Brussels' foreign affairs" desk in a long-term effort to better promote the region. Priority will be given to cities and regions that have signed a bilateral agreement with the BCR.[1]

More generally, the government of the BCR wishes to promote a coordinated approach vis-à-vis specific projects, such as screen.brussels (development of creative industries and of the broadcasting sector), using steering committees made up of several stakeholders: visit.brussels, hub.brussels, and finance.brussels (Législature 2019–2024). At the local level, the 1819 platform and hub.brussels are mandated by the government to give practical help to merchants and artisans regarding e-commerce platforms, to ensure their visibility and their presence online. Partnerships benefiting the local companies as well as the community are therefore strongly supported by the executive authorities of the BCR. The latter also aim at implementing a more global strategy to internationalize the region's economy, bringing together the trade attachés from hub.brussels and the tourism attachés from visit.brussels – the "Brussels Houses."

In the same vein, the region's attractiveness policy shall build upon a more coordinated strategy, involving all relevant stakeholders to better promote the BCR (Législature 2019–2024). The dissemination of brochures concentrating on the diverse assets of Brussels, in specific sectors/clusters and relevant languages for the

---

[1] Around 30 cities and regions all around the world have already signed a cooperation agreement with the BCR. They include country capitals such as Beijing, Berlin, Moscow, and Seoul, urban centers (e.g., Chennai, Lille), and broader jurisdictions (e.g., Quebec Province, Region of Rabat-Salé-Kénitra).

targeted recipients, could be increased (not only disseminated during the "Brussels Days"). Moreover, the region's candidacy to become a European Capital of Culture in 2030 will be partnered with numerous public bodies, including institutional partners such as Brussels City and the 18 other municipalities. Finally, the government will collaborate more closely with the CEIO and visit.brussels to strengthen the ties with the actors of the European Quarter and to enhance the attractiveness of this particular neighborhood.

Although the region's government has published a quite clear roadmap, challenges remain numerous along the way to reach a more efficient promotion strategy. The first obstacle lies in the vagueness of the articulated projects of coordination between the different actors. A positive aspect of the Legislature 2019–2024 is that creating synergies appears as a key element for further developing city marketing and the whole promotion of the BCR. Secondly, the government emphasizes the need to better take advantage of existing networks, such as Metropolis or Eurocities, and subsequently to better defend the interests of Brussels as an urban region. Furthermore, a protocol service should be appointed to support Brussels International during the "Brussels Days" organized overseas and the trade missions led by the King or the Prince of Belgium.

The third challenge recalls an issue faced by several other cities and regions: most places have been increasingly confronted with various politico-institutional perimeters, going much beyond institutionally recognized borders (Vuignier 2018). In this sense, the government of the BCR argues that "competitiveness clusters are of regional competence (support for economic development and applied research), but in many sectors the industrial reality goes beyond the borders of the regions. As a priority and pilot project, the government will participate jointly or with one of the two regions in the creation of a competitiveness cluster around digital-related professions" (Législature 2019–2024: 60).[2]

A final difficulty corresponds to the need for increased visibility of the BCR in international publications. Certain international rankings do not present any figures about Belgium at all and, consequently, about Brussels too. Although the region and its commuting zone appears in the European Regional Competitiveness Index (EU 2019), it is, for instance, absent from appealing publications for investors such as the Global Cities Investment Monitor (KPMG 2019). However, the BCR is increasingly present in global rankings such as Hot Spots 2025 (The Economist 2013) and the report entitled fDi European Cities and Regions of the Future 2020/21 (The Financial Times 2020). These rankings, although somehow inaccurate and incomplete, are important because "they can be used as instruments for the positioning of cities as they consider the specific local characteristics of each city in a comparative way. For investors, rankings provide information to assist decisions regarding site selection" (Giffinger et al. 2010, p. 310).

In addition, the BCR projects itself externally by providing statistics about its activities and has committed to foster an open data policy and to raise transparency

[2] Translated from the original version in French to English by the author

in the next 5 years (Législature 2019–2024). These objectives are also motivated by another challenge: reaching the public at large and going beyond communication to experts' circles exclusively. Finally, visibility in mainstream and specialized media remains a high priority for the region's agencies in charge of the promotion of the place, as highlighted in their annual reports (e.g., hub.brussels 2019).

Conclusion

In a context characterized by growing competition between regions and cities, what is the approach preferred by the BCR? Are the regional authorities opting for a multi-stakeholder approach or a more integrated, long-term strategy? Taking a closer look at the General Policy Statement (GPS) of the BCR government (Législature 2019–2024), an enhanced coordination between the various agencies promoting Brussels seems to be favored during the current legislature. To do so, collaborative events will be organized more frequently, and missions abroad will be more coordinated. This ambition goes hand in hand with the idea of better using the ".brussels" extension to reinforce the BCR brand. There is a strong willingness from the government to go in this direction, as explicitly mentioned in the GPS. However, it should be remembered that the region is only 30 years old, that it has only acquired certain competencies recently, and that it possesses limited resources. Moreover, this domain extension somehow covers the need to address different audiences' interests, through market segmentation and targeted critical activities, while branding the whole organization (Kavaratzis and Ashworth 2005) – in this case the BCR.

The discussion of the Belgian institutional structure, as well as the diverse functions of Brussels as a multilayered capital presented in the first sections of this chapter, reminds us of the country's complex institutional system characterized by high levels of decentralization and autonomy. Multiple layers and numerous agencies work on the same public policies, and compromises must be found on a regular basis. Nevertheless, this way of working *à la belge* has not prevented the country from being successful in several domains, including economic clusters such as the biotechnology sector. On another note, Belgium has been a pioneering state and is now a frontrunner regarding the circular economy, with a clear aim of building a more sustainable society (European Economic and Social Committee 2019). In this regard, hub.brussels has collaborated on the publication of the *Brussels Start-up guide*. This book, which will be widely distributed to enhance Brussels' image as a "start-up" city, is part of a private initiative that provides handbooks for entrepreneurs in over 40 cities (as of 2020).[3]

The BCR is aligned in this national – and European – effort and has even launched circlemade.brussels in 2018, the first cluster for innovative actors toward a more performing circular economy in Brussels (hub.brussels 2019). Although the BCR

[3] Source: Start-up Guide Brussels (https://startupguide.com/shop/startup-guide-brussels)

has successfully managed to deal with this complex institutional system and has made substantial efforts to better promote itself as a green, promising place for investors, the government has realized that a rationalization of the structure and the processes would certainly prove beneficial in the long run. As a result, more efficiency in economic promotion is expected from the creation of hub.brussels, and future collaborations with other key stakeholders (public agencies and private sector organizations) are meant to turn Brussels into a more attractive place for residents, tourists, and businesses.

The government of the BCR has well understood that being labeled as the "Capital of Europe" falls short of an efficient city marketing strategy. Of course, Brussels is a "world city," European, multicultural, and international, and will likely remain so for quite a long time. As such, it will continue to promote itself as a regulatory hub, a "supra" capital similar to Washington, D.C., in North America or Singapore in Asia. At the same time, there is an emerging need to sell the other advantages of the region, Brussels being, for instance, a "start-up" city and, as mentioned just above, a strong supporter of economic sustainability, positioning itself as a leader of the transition to a more circular economy. Other assets of the BCR relate to the cultural activities taking place on its soil and its global quality of life: Brussels belongs to the top 30 of the Quality of Living City Ranking 2019 (Mercer 2019). A deeper collaboration with visit.brussels should certainly be envisaged over the next few years. Such a collaborative work should be reinforced by the fact that although most visitors come to Brussels for business, leisure is becoming increasingly important. This observation opens possibilities for coordination and, more ambitiously, combination of efforts targeting identical groups.

In conclusion, it should be noted that the case of Brussels, although undercommented in the literature, faces numerous challenges and opens as many paths for further research. While it remains too early for conducting impact assessments, the new branding strategy and the creation of agencies such as hub.brussels being recent, it may prove particularly interesting to study how residents respond to the approach preferred by the regional authorities. As affirmed by Kavaratzis et al. (2017), inclusive place branding and participatory processes are essential since inhabitants are affected by place branding, which reconstructs and reinforces narratives, potentially gaining support from citizens or leading, on the contrary, to disaffection. It would also be enlightening to see if city marketing and branding are treated as isolated strategies or if they are, in practice, part of a broader policymaking process, occurring at different levels, as recommended by Kalandides (2011). Because other stakeholders are also vital to the region's economy, a suggestion for researchers would be to better understand the effects of the marketing strategy (or the absence thereof) on investors and businesses. Finally, regarding the institutional actors themselves, it would be equally important to dig deeper into the strategies developed to reach more coordination between the numerous agencies involved in the promotion of Brussels. A close monitoring of the concrete projects to be implemented in the next 4 years would be an initial necessary step in that direction.

References

1819 – wib.brussels. (2020). La plateforme de l'entrepreneuriat à Bruxelles (Brussels' platform for women entrepreneurship). Retrieved https://www.womeninbusiness.brussels/.

Albrechts, L. (2001). Devolution, regional governance and planning systems in Belgium. *International Planning Studies, 6*(2), 167–182.

Baele, M. (2012). be.brussels, le nouveau logo de la Région bruxelloise. *RTBF*, July 12, 2012.

be.brussels. (2019). The Brussels-Capital Region's house style. Retrieved, https://be.brussels/about-the-region/the-brussels-capital-region2019s-house-style.

Bruxelles-Capitale. (2012). Charte graphique de la Région de Bruxelles-Capitale 2012. Retrieved, https://be.brussels/files-fr/a-propos-de-la-region/mrbc/charte-graphique/charte-graphique-de-la-region-de-bruxelles-capitale.

Elmhorn, C. (2001). *Brussels: A Reflexive World City*. Stockholm: Almqvist & Wiksell International.

EU. (2019). European Regional Competitiveness Index 2019. Retrieved https://ec.europa.eu/regional_policy/en/information/maps/regional_competitiveness/#3.,

European Economic and Social Committee. (2019). *Final Report: Circular economy strategies and roadmaps in Europe – Identifying synergies and the potential for cooperation and alliance building*. Brussels: European Union.

EY. (2019). *Belgian attractiveness survey 2019*. Brussels: EYGM Limited.

Gayet, J. (2017). *Le nouveau marketing territorial: Enjeux, pratiques, facteurs clés de succès et nouveaux modèles*. Paris: Corps & Ame Editions.

Giffinger, R., Haindlmaier, G., & Kramar, H. (2010). The role of rankings in growing city competition. *Urban Research & Practice, 3*(3), 299–312.

Hein, C. (2008). Brussels, Capital of Europe: Part of a polycentric headquarters system. In R. De Groof (Ed.), *Brussels and Europe* (pp. 73–96). Brussels: Academic and Scientific Publishers (ASP).

Hospers, G.-J. (2004). Place marketing in Europe: The branding of the Oresund Region. *Intereconomics, 39*(5), 271–279.

hub.brussels (2020) *Try Brussels for Free*. Retrieved, http://why.brussels/try-brussels-for-free.

hub.brussels. (2019). *Annual Report 2018*. Retrieved, https://hub.brussels/en/blog/activity-report-2018/.

innoviris.brussels. (2019). Mission & vision. Retrieved, https://innoviris.brussels/mission-vision.

International Organization for Migration (IOM). (2015). *World Migration Report 2015*. Geneva: IOM Publishing.

Jansen-Verbeke, M., & Govers, R. (2009). Brussels: A multi-layered capital city. In R. Maitland & B. W. Ritchie (Eds.), *City Tourism: National Capital Perspectives* (pp. 142–158). Wallingford: CABI.

Jansen-Verbeke, M., Vandenbroucke, S., & Tielen, S. (2005). Tourism in Brussels, Capital of 'New Europe'. *International Journal of Tourism Research, 7*(2), 109–122.

Kalandides, A. (2011). City marketing for Bogotá: A case study in integrated place branding. *Journal of Place Management and Development, 4*(3), 282–291.

Kalandides, A. (2006). *Fragmented branding for a fragmented city: Marketing Berlin*. Paper presented at the 6th European Urban and Regional Studies Conference. Roskilde, Denmark, September 2006.

Kavaratzis, M. (2008). *From city marketing to city branding: An interdisciplinary analysis with reference to Amsterdam, Budapest and Athens*. PhD thesis, University of Groningen, Groningen, the Netherlands.

Kavaratzis, M., Giovanardi, M., & Lichrou, M. (Eds.). (2017). *Inclusive Place Branding: Critical Perspectives on Theory and Practice*. London, UK: Routledge.

Kavaratzis, M., & Ashworth, G. J. (2005). City branding: An effective assertion of identity or a transitory marketing trick? *Tijdschrift voor Economische en Sociale Geografie, 96*(5), 506–514.

KPMG. (2019). Global Cities Investment Monitor 2019. Retrieved https://gp-investment-agency.com/wp-content/uploads/2019/06/GlobalCitiesInvestmentMonitor2019web-compressed.pdf.

Législature 2019–2024. (2019). *Déclaration de politique générale commune au Gouvernement de la Région de Bruxelles-Capitale et au Collège réuni de la Commission communautaire commune*. Retrieved http://www.parlement.brussels/wp-content/uploads/2019/07/07-20-D%C3%A9claration-gouvernementale-parlement-bruxellois-2019.pdf.,

Mabillard, V., & Vuignier, R. (2017). The less transparent, the more attractive? A critical perspective on transparency and place branding. *Place Branding and Public Diplomacy, 13*(4), 348–359.

Mercer. (2019). *Quality of Living City Ranking 2019*. Retrieved https://mobilityexchange.mercer.com/Insights/quality-of-living-rankings.,

Office of the Brussels Commissioner for Europe and International Organisations. (2020). Brussels, International Capital: The Figures 2020. Retrieved https://www.commissioner.brussels/component/fleximedia/431-413187brussels-international-capital-2020-eng?Itemid=304.,

Statbel. (2020). *Structure of the population: On 1st January 2019, Belgium had 11,431,406 inhabitants*. Retrieved, https://statbel.fgov.be/en/themes/population/structure-population.

The Economist. (2013). *Hot Spots 2025. Benchmarking the future competitiveness of cities*. London, UK: The Economist Intelligence Unit.

The Financial Times. (2020). *fDi European Cities and Regions of the Future 2020/21*. London, UK: The Financial Times Ltd..

visit.brussels. (2019). *Why Brussels?*. Retrieved https://visit.brussels/site/en/article/why-brussels.

Vuignier, R. (2018). *Attractivité des territoires et place branding: Etude exploratoire de la sensibilité des décideurs d'entreprise à la marque territoriale*. PhD thesis, University of Lausanne, Lausanne, Switzerland.

Waty, M., Van Hooland, S., Hengchen, S., Coeckelbergs, M., De Wilde, M., & Decroly, J.-M. (2018). How hot is .brussels? Analysis of the uptake of the .brussels top-level domain name extension. *The E-Journal for Academic Research on Brussels, 119*, 1–17.

why.brussels. (2020). *Why Brussels?*. Retrieved http://why.brussels/.

Yasmeen, R. (2019). *Top 100 City Destinations 2019*. Euromonitor International, December 2019.

Municipal Branding and Social Media: Comparing Poland, Norway, and Ukraine

Wawrzyniec Rudolf and Arild Wæraas

Municipalities in many countries seek to define themselves as "brands" to help position themselves in competition for taxpayers, tourists, state funding, reputation, and business ventures (Cassel 2008; Glińska and Rudolf 2019; Lockert et al. 2019b; Moldenæs 2016). They are pressured to explain to their citizens why they are a better place to live and work, why tourists should visit, and why investors should seek business opportunities there. However, empirical research has thus far paid limited attention to the many visible ways in which municipalities attempt to build an attractive brand image of themselves. In particular, the current tendency for municipalities to build their brands on social media has remained a largely unexplored area of inquiry (Glińska and Rudolf 2019).

Although Facebook has become immensely popular among individuals with more than 2 billion active users globally (Statista 2020), and it currently seems to be a major means by which all kinds of organizations connect with their constituents (Shen and Bissell 2013; Treadaway and Smith 2012), the lack of research into the relationship between social media and municipality branding leaves a gap in our understanding of how these entities brand themselves in practice.

The purpose of this chapter is to examine the contents of municipalities' Facebook posts as part of their branding strategies and to assess the relationship between branding and social media in a comparative perspective. Specifically, we combine the brand personality dimensions of Aaker (1997) later adapted to the cities studied (Glińska and Rudolf 2019) and the three overarching strategies for municipal branding proposed by Wæraas, Bjørnå, and Moldenæs (Wæraas et al. 2015) as a theoretical lens through which to analyze the posts. Brand personality

W. Rudolf (✉)
Faculty of Management, University of Lodz, Lodz, Poland
e-mail: wawrzyniec.rudolf@uni.lodz.pl

A. Wæraas
Faculty of Theology, Diaconia, and Leadership, VID Specialized University, Oslo, Norway
e-mail: arild.waeraas@vid.no

S. M. Zavattaro (ed.), *Public Branding and Marketing*,
https://doi.org/10.1007/978-3-030-70505-3_6

traits are considered essential in brand management (Keller and Richey 2006), and they can be expected to materialize in official self-presentations (Balmer and Greyser 2003), including Facebook posts (Glińska and Rudolf 2019).

Thus far, however, it is not clear which personality traits tend to be invoked with which strategy: the *Place* branding strategy, the *Organization* branding strategy, or the *Democracy* branding strategy (Wæraas et al. 2015). Accordingly, we seek to understand how these strategies entail the use of different personality traits in official Facebook posts. Our approach is comparative and longitudinal, examining differences and similarities in the contents of Facebook posts of 16 Polish, 10 Norwegian, and 18 Ukrainian municipalities to bring out cross-national contrasts and similarities. Our research questions are:

- What dimensions/traits of the city's brand personality are used in the contents communicated by cities in Poland, Norway, and Ukraine through their Facebook accounts?
- What is the leading Facebook city branding strategy in each of the countries, and which personality dimensions and traits are communicated by each of them?
- To what extent are the personality dimensions/traits associated with each of the Facebook city branding strategies similar in the three investigated countries?

The research questions are primarily motivated by a theoretical interest in the relationship between a given branding strategy and espoused brand personality traits. By connecting the two, a better understanding of how the strategies are translated into specific brand messages at the municipal level of government could be obtained. Secondarily, the research questions are motivated by the potential impact on brand messages of prevailing public management regimes, economic development, and political circumstances. The comparison of the three countries, which are situated in different cultural, political, and economic environments, could provide valuable insights into how branding strategies are shaped by the institutional context.

Theoretical Observations

Branding means to instill a unique and attractive image of an entity in the minds of external observers. In the context of the public sector, brand images could be associated with the overall identity of the organization (organizational branding) (Whelan et al. 2010), the services offered (service branding) (Knowles and Howley 2000), and the policies for which the organization is responsible for implementing (policy branding) (Eshuis and Klijn 2012; Evans and Hastings 2008). Empirical observations suggest that attempts at creating such images have flourished in the public sector and that branding could have beneficial effects because it creates awareness, leads to a better reputation, improves effectiveness, and increases revenues (Leijerholt et al. 2018).

Empirical research has also examined the significance of the visible, externally oriented ways through which public sector brands are expressed including slogans, taglines, websites, logos, core values, and films (Dholakia and Acciardo 2014; Karens et al. 2016; Lockert et al. 2019a; Moldenæs 2016; Sataøen and Wæraas 2015; Tschirhart 2008). Although previous studies have examined aspects such as brand strategies (Lockert et al. 2019a, b; Wæraas 2014) and communicated brand images (Cassel 2008; de Rosa et al. 2012; Moldenæs 2016; Nielsen and Salomonsen 2012), the use of social media as a means of building brands at the municipal level has been largely ignored – despite the fact that "every" municipality currently seems to use Facebook as a means of connecting with constituents (Cleave et al. 2017; Glińska and Rudolf 2019).

In the following theory section, we review existing research on municipal branding and point to a need for more research into the role of social media in municipal branding. We then provide an overview of municipal branding strategies and the personality traits that we expect to emerge in the municipalities' brand messages. We close our theory section by addressing the context for the municipal branding strategies in Poland, Norway, and Ukraine.

Social Media in Municipal Branding

Social media are forms of electronic communication enabling users to share information, ideas, personal messages, and various contents such as videos, audio, and pictures (Merriam-Webster Dictionaries 2020). Some of the most popular examples of social media include Facebook, Twitter, Instagram, and LinkedIn. Our focus in this chapter is Facebook, which currently has more than 2 billion users. Government agencies are popular platform users, as a study by Burson-Marsteller revealed that 87% of governments that are United Nations members are on Facebook (Techweez 2016).

A large number of empirical studies have examined local governments' communication on social media (Bonsón et al. 2012; Ellison and Hardey 2014; Mossberger et al. 2013; Omar et al. 2014; Zavattaro et al. 2015). A Swedish case study of Karlstad municipality's Facebook posts (Magnusson et al. 2012), for example, found that most of the posts concerned the marketing of different events in the municipality, followed by the promotion of the municipality as a good place to live or work. However, to our knowledge, only a few studies have seen these activities on social media from a branding perspective (Cleave et al. 2017; Glińska and Rudolf 2019; Sevin 2016). This lack of empirical research is unfortunate because social media could be even more important for municipalities in their branding efforts than for central government agencies. At least three reasons support this view.

First, municipal administration bodies interact more closely with citizens daily than central government administration. As a result, they are better positioned to create relationships and enhance civic engagement and confidence than the central government (Eshuis and Klijn 2012; Lockert et al. 2019a). Because municipalities provide public services such as schools, water, and roads, which are used every day, the general public arguably forms their impression of the municipality partly based on their own experiences with these services. Top municipal officials, therefore, could have a great deal to gain from influencing citizens' perceptions through strategic communication on social media and from providing proactive information that could boost support and prevent negative feelings toward the municipality.

Second, in contrast to central government agencies, municipalities are places where citizens live, travel, and spend their leisure time. Municipal administration bodies are not in control of every relevant feature of the municipality, but they could instill and reinforce pride in the municipality by reaching target groups where they live, highlighting important aspects of the historical background of the area, reminding citizens about important events and attractions, and generally promoting any unique and appealing feature of the municipality that could create emotional attachment "from within."

Third, social media is a potentially powerful way of generating positive perceptions about local democratic arrangements. Whereas the central political-administrative system depends on citizens' participation at national elections at regular intervals only (e.g., every second or fourth year), local governments are better positioned to interact more continuously with the local population through, for example, meetings, seminars, and consultation. By informing on social media about events at which citizens are encouraged to provide feedback, the local government could spread the perception among citizens that their municipality is a successful local democracy particularly attentive and responsive to the concerns and needs of their constituents.

Social media posts could in principle be used to invoke any kind of identity, any kind of personality trait, and any kind of values, and they could reach tens of thousands of readers in a very short time. As such, social media is a very convenient and efficient way of communicating brand messages. In the following section, we elaborate on different branding strategies that potentially could be reflected by the posts.

Municipal Branding Strategies

We distinguish between three overarching strategies for local government branding: the *Place*, *Organization*, and *Democracy* branding strategy (Wæraas et al. 2015).[1] The strategies could involve concrete change projects in urban transformation and redevelopment as a way to improve environmental, social, and economic conditions (Ma et al. 2019). More likely and important in terms of instilling positive brand images, however, are external communication initiatives directed at specific target groups. Given our interest in social media and local government branding, our focus

[1] The typology of branding strategies takes as a starting point a specific view of what municipalities *are,* not what they *do*. As a result, services branding and the branding of policies are not part of the typology.

is not on actual change but on how these strategies are reflected in official local government communication.

With the first strategy, place branding builds on the idea that cities, regions, and municipalities are geographical places whose set of attractive features can and should be marketed and "sold" to potential visitors, residents, and investors. From this perspective, the primary task of local government branding is to convince these target groups that the municipality is a good place for tourists to visit, a good place for current and future residents to live, and a good place for investors to run a business. Communicated brand images could be associated with the history and cultural heritage of the area, local attractions, and activities of interest to the target groups and the economic potential and benefits of the municipality.

The *Organizational* branding strategy corresponds to the notion of corporate branding in private sector settings (Fetscherin and Usunier 2012). In this strategy, the local government is an organization that offers and "sells" services to residents, builds and reinforces relationships with its existing employees, and makes itself attractive to future employees. The goal of the strategy is for the municipal administration to be seen as a good employer, a good service provider, and a good place to work. Conveyed brand images could reflect the excellent quality of services, the fair treatment of municipal workers and good benefits, or the friendly atmosphere and supportive work environment in the municipal organization.

Finally, the *Democracy* branding strategy aims to build perceptions in the minds of citizens and political parties of a well-functioning local democracy. Local democracies cannot function well unless they have the trust and support of the citizens, which can be achieved by repeatedly informing them about issues of importance, communicating about ways for citizens to participate by providing feedback to city officials and politicians, reminding them that their votes and voice matter, and generally promoting the municipality as attentive to constituents' needs and concerns.

The branding strategies have thus far not been significantly explored in empirical research. To provide a better understanding of their specific properties in practice, we propose to see them in conjunction with the brand personality traits outlined by Aaker and further adapted for local government branding by Glińska & Rudolf (2019).

Personality Traits in Brand Messages

The brand personality concept is related to brand identity and image. Brand personality can be defined as "a set of human characteristics associated with a brand" (Aaker 1997, p. 347). It is based on intangible (soft) features and covers the emotional side of brand image (Ţugulea 2017). Brand personality is treated as a key component contributing to the image created in the consumers', users', or inhabitants' minds (Vinyals-Mirabent and Mohammadi 2018).

Originally, the concept of brand personality, derived from psychology, was successfully applied to companies and their products (Plummer 2000), but later it was also used in place marketing processes aimed at building a destination brand.

Previous research has focused mainly on the personalities of touristic brands for cities and regions (Lee and Suh 2011; Murphy et al. 2007; Usakli and Baloglu 2011) and even more often for entire countries (Kim and Lehto 2013; Kumar and Nayak 2018a, b; Opoku 2009; Rojas-Méndez et al. 2013). The results of these studies have been mainly published in the top travel journals, such as the *Journal of Travel Research, Journal of Hospitality & Tourism Research,* and *International Journal of Contemporary Hospitality Management.* Within the destination marketing field, the concept of destination personality is developing dynamically (Baştuğ et al. 2020; Masiello et al. 2020; Sharifsamet et al. 2020), which suggests it is an emerging field.

The most recognized concept of identifying brand personality is Jennifer Aaker's (1997) brand personality scale. It is a research tool that originally consisted of 42 personality characteristics (level 2) grouped into 15 personality traits (level 1) and then into 5 following dimensions: *Sincerity, Excitement, Competence, Sophistication, and Ruggedness*. The Aaker model has been a common starting point for a number of studies concerning the measurement of personalities of places (Glińska and Kilon 2014; Kovačić et al. 2019; Usakli and Baloglu 2011), but it is also argued that the model does not adequately represent the personality of the place, as it reflects only positive characteristics of personality and also the city is a specific multidimensional product that should be described by extended scales (Kaplan et al. 2010).

As indicated, most studies about brand personality referring to places are concerned with their tourist attractiveness. This is one of the reasons why the Aaker scale, based only on positive characteristics, applies readily to places perceived as tourist destinations. There are, however, cities for which the tourism function is not the leading one, and the local authorities want to build a brand not only in relation to this target but also aimed at other target groups – residents, users of services, investors, entrepreneurs, or voters.

The study on brand personality of cities acting as regional capitals conducted in two European countries (Glińska and Rudolf 2019) provides an interesting example. The study resulted in an extended brand personality scale. The *Prosocial Attitude* dimension has been added with two new traits (*cooperative, bureaucratic*). Also, the *Excitement* dimension has been extended with the *patriotic* trait. The structure of the dimensions and following two levels of personality traits resulting from that study are presented in Table 1. All the new elements – new to Aaker's brand personality scale – have been marked with capital letters. The structure presented in the table has been applied for our empirical study described in the next sections.

It is worth noting that brand personality studies are conducted from sender and recipient perspectives. In the literature, the reception of messages is more frequently examined (Murphy et al. 2007; Sharifsamet et al. 2020). This is a legitimate approach in that it makes it possible to estimate the effectiveness of communication activities taken toward specific audiences. The alternative for this study is the perceived place brand personality approach (Govers et al. 2007; Masiello et al. 2020). It consists in identifying communication strategies carried out by municipalities. It seems to be an especially interesting approach for a cross-country analysis, but at the same time to get valuable comparative results, it requires a systematic research approach for each of the investigated countries.

Table 1 Dimensions of city brand personality and its personality traits (dimensions and traits new to Aaker's scale marked with capital letters)

Dimension	Brand personality traits – level 1 (18) + level 2 (54)
Sincerity	*Down-to-earth*: down-to-earth, family oriented, small town
	Honest: honest, sincere, real
	Wholesome: wholesome, original
	Cheerful: cheerful, sentimental, friendly
Excitement	*Daring*: daring, trendy, exciting
	Spirited: spirited, cool, young
	Imaginative: imaginative, unique
	Up-to-date: up-to-date, independent, contemporary
	PATRIOTIC: patriotic, country-loving, proud, religious, injured/wounded, historical
Competence	*Reliable*: reliable, hard-working, secure
	Intelligent: intelligent, technical, corporate
	Successful: successful, leader, confident
Sophistication	*Upper-class*: upper-class, glamorous, good looking
	Charming: charming, feminine, smooth
Ruggedness	*Outdoorsy*: outdoorsy, masculine, Western
	Tough: tough, rugged
Prosocial Attitude	COOPERATIVE: cooperative, governance, open-minded, prosocial
	BUREAUCRATIC: Bureaucratic, formal

Source: Own elaboration on the basis of Aaker (1997) and Glińska & Rudolf (2019)

A Comparative Perspective on Local Government Branding

The secondary objective of this study is, as stated earlier, to examine similarities and differences across the three countries with respect to the contents of their municipalities' Facebook posts. Our comparative perspective does not directly address specific public sector or municipal brands such as logos, names, designs, or symbols alone that could identify "one seller's good or service as distinct from those of other sellers"(American Marketing Association 2017). We only do so if these measures are used in a Facebook post and communicated using a specific branding strategy. The differences and similarities across the selected municipalities with respect to the contents of these posts are our focus to bring out comparative insights. Specifically, the geographical distance between the countries, different public management regimes and administrative cultures, economic development, and different socio-political situations in which the three countries find themselves can be expected to generate different municipal assessments of what type of message is relevant and important to share with the local population. Various municipal responsibilities within the political system and different governance structures could also generate varying priorities.

Poland

Poland is the fifth-largest European Union (EU) country with a population of more than 38 million after Germany, France, Italy, and Spain. It borders Germany on the west and Ukraine and Belarus to the east, where the latter is the external border of the entire European Union. Poland is a unitary parliamentary republic. An important complement to the state government is the territorial self-government, which operates at three levels, regional, intermediate (Polish: *powiaty*), and municipal, with elections held every 4 years.

In the years 1945–1989, Poland was within the bloc of socialist countries and at that time remained under the strong influence of Russia. Although more than 30 years have passed since the change of the political regime, Poland still needs to reduce its distance from Western economies in both living conditions and social capital. In these processes, the EU has played a huge role supporting Poland's development through European policies, especially those dedicated to regions lagging behind (e.g., cohesion policy). The support has been aimed at infrastructure development, but it is also done by promoting system and management solutions successful in other countries. This also applies to the functioning of public administration. As part of support from the European Social Fund, Poland continues to improve institutional systems, supports the construction of national and international labor networks, and offers opportunities to improve qualifications of public sector employees, including those at the lowest levels (Glinka and Hensel 2017).

Poland has been a member of NATO since 1999 and the European Union since 2004. In terms of economic development, it has been considered a leader among the countries of Central and Eastern Europe. Successive national governments have been in favor of liberal politics. In 2015, the conservative party seized power in the country, which has slowed market reforms, and began to develop the welfare state policy without first developing the economy to a sufficiently high level (Bunikowski 2018). It should be emphasized that the mayors of most large cities in Poland identify themselves with parties in opposition to the central government and are thus in favor of the liberal approach to the functioning of the economy. In these cities, executives and constitutive authorities are often in conflict with each other, which makes it much more difficult for the mayor to govern their city. Today, politics is beginning to dominate public discourse, and its subject is an ideological dispute, which pushes the daily affairs of residents to the background.

Returning to the public administration and its transition in Poland, it must be acknowledged that in the last 30 years, and especially in the period after the country's accession to the EU, much has been done toward the market approach in the public sector (Czaputowicz and Sakowicz 2017). The evidence-based policy promoted across the EU has also been promoted in Poland from the beginning of the 2014–2020 programming period. On the other hand, actions aimed at promoting governance processes complementing democratic governance in Poland despite pressure from the EU have proved less effective at this stage. It should be noted the conservative central government in Poland is highly efficient and effective in achieving its intentions and election promises, but it does so by hindering democratic governance processes (Bugarič 2019). In addition, the structural political conflict at the national level between the ruling Conservatives and the Liberals in opposition is shifting to the lower levels of power and to society as a whole, which is not conducive to the introduction of a governance model in managing public issues.

Norway

Norway is the smallest country in our study, with a population of 5.3 million. It is a parliamentary democracy and constitutional monarchy but with only symbolic responsibilities given to the monarch. Norway's post-war period can be characterized as peaceful and successful in terms of economic growth and social equality. Oriented toward the West, Norway has been a NATO member since 1949 and enjoys tight integration with the EU through the European Economic Area membership. This membership gives access to the EU's single market and allows for the free movement of goods, services, capital, and people between Norway and other EU member states. Membership in the EU is still a debated issue in Norway, although two referendums (1972 and 1994) ended with a rejection of the proposal to join.

Since the 1980s, a number of large-scale public administration reforms often referred to as New Public Management have been implemented in the country (Christensen et al. 2008). Partly driven by escalating welfare state costs and partly by neoliberal ideologies calling for more efficiency, quality, and more value for every krone spent, these reforms have transformed the way central and local government entities are organized and managed. Reforms include outsourcing and privatization of previously centralized services, the introduction of competition between public and private service providers (especially in municipal administration), the creation of single-purpose organizations, and the introduction of managerial techniques from business settings. Interestingly, the latter has opened up the Norwegian public sector to branding and reputation management practices (Wæraas and Sataøen 2015). This has made public sector organizations more user and service oriented, hence leading to (at least seemingly) more transparent, digitalized, and open municipal governments that increasingly communicate with their constituents on the Internet through social media such as Facebook.

The administrative structure of Norway consists of the central government, 11 counties, and 356 municipalities as of January 1, 2020. Municipalities and cities are usually the same, although some municipalities may consist of several cities or towns. This has increasingly become the case because of large-scale municipal amalgamations. Counties are not superordinate to municipalities as they have their own separate functions. County and municipal elections are held every fourth year.

Norway has a consensus-oriented political culture with long traditions for negotiation and compromise. Transitions from one cabinet to another have been peaceful since parliamentarism was introduced in 1884 (except for the Nazi party takeover in

1940). This is especially true in local government affairs. The aldermanic system ensures that political positions to the municipal chairmanship are allocated among the main political parties according to their strength in the municipal council. As a result, all parties are held accountable for political decisions, and they all have a joint interest in reaching good decisions relating to municipal services.

Ukraine

Ukraine has 42 million inhabitants (as of 2019), although it should be added that during the past 5 years, unfavorable demographic processes and economic migration have led to a 10% decrease in the country's permanent population. Ukraine's political system is a unitary semi-presidential republic. Like neighboring Poland, the country liberated itself from the Russian influence, albeit a little later (1991). It should be noted, however, that Ukraine after World War II became one of the republics of the USSR, which was associated with a much stronger influence of the socialist system on all areas of life of the country. For almost 30 years, with less or more involvement, the Ukrainian authorities have been trying to integrate into European and global structures. Dialogue is still underway toward Ukraine's integration into NATO structures. Ukraine, on the other hand, has been associated with the EU since 2017. However, the lack of effective structural reforms and Ukraine's incriminating military conflict on the border with Russia have effectively slowed down positive developments and led the country's economy to recession (Leitch 2019). It results in unprecedented economic migration to other European countries, which has been made possible by easing EU visa restrictions.

The dynamics of Ukraine's economic reality go hand in hand with political instability. Many political parties often fail to reach consensus, which hinders governance processes and consistent pursuit of stability in all dimensions. The previous socialist system and the current difficult economic situation negatively affect the level of public trust/confidence. It is not easy to get the attention of voters in this situation, which is used by populist parties.

Ukraine has taken up the challenge of public administration reforms in the context of its aspirations for a full EU membership. However, the reforms undertaken have proved ineffective, corruption has not been reduced, and most of the planned changes are in the initial stage of implementation (Danko et al. 2020). This makes Ukraine today a country with a hierarchy-based administration system, dependent on the political system, and with a low level of public confidence.

The self-government in Ukraine is established at three levels: regions (Ukrainian: *oblasti*), intermediate units (Ukrainian: *rajony*), and local (Ukrainian: *rada*). The authorities at all levels are elected every 5 years. It is worth indicating that at both regional and intermediate levels, only constitutive organs (councils) are elected in the democratic election procedure. The executive organs are appointed in both cases by the President of Ukraine and subordinate to this office. They are also controlled by councils, but only within the scope of selected tasks, and the possibilities for the dismissal of the governors by the Council are severely limited.

Local self-governments in Ukraine are represented by councils at rural, town, and city levels and their executive bodies. Despite the constitutional and statutory guarantees, Ukraine is facing the problem of inefficient functioning of executive bodies of local authorities. The role of local government in the management of municipalities in Ukraine is still small. Local governments are characterized by a lack of financial self-sufficiency, and a great deal depends on the executive authorities at the *oblast* and *rajon* level appointed by the president of the country. As part of the broader decentralization process implemented in Ukraine since 2014, the lowest-level self-government reform is being carried out, aimed at merging the smallest units into larger ones, which are later expected to have more potential to take on new tasks. There is also a plan to strengthen self-government at the regional level, which is being prepared in the context of Ukraine's entry into the European Union.

To sum up, the real local government in Ukraine currently operates at the level of cities, where both bodies – the council and the mayor of the city – are chosen by the local community. However, it should be noted cities rely heavily on state funding, which makes it difficult to run the local policy in a fully independent manner (Table 2).

To show the situation of investigated countries with respect to their level of governance, the Worldwide Governance Indicators (2020) developed by the World Bank were used. The WGI have reported aggregate and individual governance indicators for most of the countries in the world since 1996. According to the WGI, *governance* is defined as a set of traditions and institutions by which authority in a country is exercised. Within the scope of this term, there are three elements (Worldwide Governance Indicators 2020): (1) the process by which governments are selected, monitored, and replaced, (2) the capacity of the government to effectively formulate and implement sound policies, and (3) the respect of citizens and the state for the institutions that govern economic and social interactions among them. The WGI use perceptions-based data – opinions coming from both firms and households. It is a valuable tool for capturing the realities of governance outcomes "on the ground." It relies on the experiences and views of informed respondents.

The WGI are based on six dimensions of governance, voice of accountability, political stability and absence of violence, government effectiveness, regulatory quality, rule of law, and control of corruption, which all have been presented for the countries included in the study (Table 3). Results (2018) indicate how significant are the differences within selected European countries regarding governance practices. Norway is positioned here as leading, Poland as less advanced, and Ukraine as far behind in all categories. The data are presented together with figures for the USA to better recognize the situation in a comparative manner.

The authors are aware the above indicators apply to entire countries. However, at the same time, they are aware that the facts observed also describe governance processes at the level of local authorities, which are part of the public administration of the countries examined.

Table 2 Selected characteristics of Poland, Norway, and Ukraine

	Poland	Norway	Ukraine
Population	38,411,148	5,367,580	41,980,000
Density per km^2	124	15	77
GDP per capita (PPS)	$31,939	$74,356	$9283
Political system	Unitary parliamentary republic	Parliamentary democracy and constitutional monarchy	Unitary semi-presidential republic
Political-administrative culture	The system is officially based on collaboration but in practice mainly the formal approach is employed	Consensus-oriented, involvement of affected stakeholders and unions	The hierarchical approach. The authorities do not share their power with any stakeholders
Municipal governance	Direct elections every fourth year for the mayor and city council separately	Aldermanic with elections every fourth year (Bergen and Oslo have a parliamentary system)	Direct elections every 5 years for the mayor (cities) and the municipal council separately
Relation to EU	European Union member since 2004	European Economic Area membership since 1992	Associated with the European Union since 2017
Average municipal population	15,507	15,077	1413
Number of municipalities	2477	356	29,718

Sources: National census/statistics agencies for Poland, Norway, and Ukraine, International Monetary Fund 2019, World Bank (2019/2020)

Table 3 Worldwide Governance Indicators for 2018 for the selected countries

Indicator	USA	Poland	Norway	Ukraine
Control of corruption	1.32	0.64	2.09	−0.87
Government effectiveness	1.58	0.66	1.89	−0.42
Political stability and absence of violence	0.48	0.55	1.15	−1.83
Regulatory quality	1.76	0.88	1.76	−0.22
Rule of law	1.45	0.43	1.97	−0.72
Voice and accountability	1.04	0.72	1.73	−0.01

Source: Worldwide Governance Indicators 2020

Methods

Directed content analysis (Hsieh and Shannon 2005) of Facebook posts was used as the research method. Within this approach, the researchers make use of codes formulated based on existing theory. In our study, we used the scale of city personality dimensions/traits (Aaker 1997; Glińska and Rudolf 2019) presented in Table 1 and the typology of the municipal branding strategies (Wæraas et al. 2015). The three strategies were associated with the three target groups each: the *Place* branding strategy (local residents, tourists, and investors), the *Organization* branding strategy (employees, users of services, and job seekers), and the *Democracy* branding strategy (voters, future voters, and political parties). The study falls into the projected place brand personality approach. It was based on the researchers' assessment of posts, but the content analysis leads to quantification of these contents in terms of predetermined categories in a systematic and replicable manner.

The authors wanted to select the cities from each investigating country that play the specific role of administrative regional capitals, being important local economic, educational, and cultural centers. As a result, they are similar in terms of the scope of the municipal services they deliver (i.e., primary education, local roads, water and sewage infrastructure, primary healthcare, welfare/social services, cultural services, etc.), as well as the administrative function they possess. It was also important to select relatively large cities in each country in which the communication on social media is not carried out by one single person but as a result of the work and cooperation of various individuals and entities. Therefore, 16 cities in Poland, 10 in Norway, and 18 in Ukraine[2] were selected to conduct the qualitative research.

The official Facebook accounts of 44 cities from Poland, Norway, and Ukraine were included (Appendix 1). The posts published during a similar period, starting from January 15, 2020, were coded. The observation was carried out until the official Facebook account of a given city reached 30 posts. The authors assumed that such a number of posts would be sufficient to know the typical personality traits of the city presented in their Facebook messages. Consequently, 480 posts from Polish cities, 300 from Norwegian ones, and 540 from Ukrainian ones were evaluated. The 30-post limit was reached in an average Polish city in as few as 9 days and in a Ukrainian average city in 17 days but lasted much longer (39 days) for an average Norwegian city.

The full content analysis (photos, text, video, and audio) of Facebook posts was implemented using the inter-coder reliability approach. The coders were native speakers (Polish, Norwegian, Ukrainian) cooperating (in English) on a day-to-day basis. The following steps were taken within the content analysis process:

1. Learning about the content of a post with all forms of content presentation (photos, text, video, and audio).
2. Coding the post by 0, 1, 2, or 3 personality traits out of the 18 traits contained within the modified Aaker scale (level 1). They are presented as traits (in bold) in Table 1. To make an accurate assessment of the contents of the posts, the

[2] It is worth noting that Ukraine is divided into 24 regions. However, for the six cities that are the capitals of the eastern regions due to the armed conflict on the border with Russia, it was not possible to collect data. Three cities (Donetsk, Lugansk, Simferopol) at the time of the research were occupied by enemy forces, and the other three located in the neighboring regions to the territories where the military processes were ongoing were not active on Facebook.

coders also used the full list of 54 traits (level 2), as well as the general description of all the 6 personality dimensions presented in Appendix 2.
3. Coding the posts by the target groups associated with branding strategies (more than one selection possible).
4. In case of any doubts how to code a particular post, a discussion among the coders was initiated to choose the most acceptable interpretation of the content.

The following procedure has been used to identify statistically significant differences at the level of personality dimensions of municipal brand between Poland, Norway, and Ukraine. For every selected municipality, the number of observations for each of the 18 personality traits of city brand was added up. The value of each variable for a given municipality informs us how many specified content items indicating a given personality trait have been found within the posts published in the municipalities starting from January 15, 2020. Additionally, three branding strategies (*Place, Organization, Democracy*) were assessed (nine variables). Next, unitarization of all 27 partial variables, using formula (1) accounting for the maximum and minimum of all 44 cities being considered, was performed (Aksoy and Haralick 2001):

The range of variation for each of the 27 partial variables was [0;1].

$$Z_{ij} = \frac{X_{ij} - \min_i \{X_{ij}\}}{\max_i \{X_{ij}\} - \min_i \{X_{ij}\}}, \; i = 1,2,\ldots,n; j = 1,2,\ldots,m \tag{1}$$

X_{ij}– variable value Z_{ij}– variable value after unitarization

In the next step, the average value of variables for each dimension (based on data resulting from the unitarization) was identified. In this manner, six variables measuring the level/intensity of city brand personality were established: *Sincerity, Excitement, Competence, Sophistication, Ruggedness,* and *Prosocial Attitude.* Additionally, using the same methods, three variables related to the strategies were conducted. The range of variation of every variable fit within [0;1].

Descriptive statistics were used to evaluate the level of each dimension within the three countries. Taking into consideration the quantitative-level measurement of the nine synthetic variables, one-way ANOVA was used for a comparative analysis of the three countries. Using the Shapiro-Wilk normality test, we were able to confirm non-normality for each variable in at least one country. That is why the Kruskal-Wallis test was used to compare the distribution of the individual dimensions within the three countries. If differences were statistically significant, the post hoc test (with the Bonferroni correction) was used for comparison of pairs of countries. It was assumed that differences between countries are statistically significant if $p < 0.05$. Calculations were performed in Microsoft Excel and IBM SPSS Statistics (PS IMAGO 4.0).

Findings

Dimensions and Traits of City Brand Personality Across the Countries

Table 4 presents totaled values for personality traits describing posts of municipalities from Poland, Norway, and Ukraine published on their Facebook official accounts. These values have been classified into six dimensions of city brand personality. It demonstrates dimensions showing significant differences between cities from Poland, Norway, and Ukraine. In comparing brand personalities of

Table 4 Descriptive statistics, normality test, and Kruskal-Wallis test results for the six city brand personality dimensions in the three countries being investigated

		Min	Max	M	Me	SD	S	K	p SW	p KW	Remarks
Sincerity	PL	0.31	0.74	0.44	0.41	0.11	1.36	2.55	0.039[a]	0.025[a]	UA < PL ($p = 0.028^a$)
	NO	0.28	0.59	0.37	0.33	0.10	1.33	1.04	0.035[a]		
	UA	0.09	0.55	0.33	0.33	0.11	−0.28	1.06	0.858		
Excitement	PL	0.13	0.70	0.28	0.26	0.13	2.04	5.77	0.004[a]	0.047[a]	NO < PL ($p = 0.041^a$)
	NO	0.02	0.27	0.18	0.20	0.07	−1.62	3.68	0.074		
	UA	0.05	0.32	0.22	0.23	0.07	−0.88	0.31	0.174		
Competence	PL	0.11	1.00	0.27	0.21	0.21	3.34	12.31	<0.0001[a]	0.124	n.a.
	NO	0.00	0.26	0.16	0.17	0.09	−0.50	−0.78	0.544		
	UA	0.07	0.43	0.23	0.23	0.08	0.57	2.15	0.374		
Sophistication	PL	0.00	0.81	0.35	0.32	0.22	0.57	−0.09	0.663	<0.0001[a]	UA < PL ($p < 0.0001^a$)
	NO	0.00	0.40	0.17	0.15	0.14	0.40	−0.81	0.542		
	UA	0.00	0.17	0.06	0.06	0.06	0.50	−1.19	0.004[a]		
Ruggedness	PL	0.00	0.58	0.24	0.17	0.21	0.60	−1.24	0.026[a]	0.031[a]	NO < UA ($p = 0.032^a$)
	NO	0.00	0.26	0.09	0.07	0.09	0.98	0.00	0.084		
	UA	0.00	0.56	0.23	0.20	0.15	0.81	0.44	0.136		
Prosocial Attitude	PL	0.01	0.60	0.17	0.14	0.16	1.80	3.04	0.002[a]	<0.0001[a]	NO < UA ($p < 0.0001^a$)
	NO	0.00	0.13	0.06	0.05	0.04	0.16	−0.57	0.630		PL < UA ($p = 0.017^a$)
	UA	0.07	0.57	0.36	0.37	0.15	−0.53	−0.46	0.391		

44 municipalities, 1320 posts, 3229 observations; *PL* Poland, *NO* Norway, *UA* Ukraine
M mean, *Me* median, *SD* standard deviation, *S* skewness, *K* kurtosis, *p* probability in: *SW* Shapiro-Wilk test, *KW* Kruskal-Wallis test
[a]results statistically significant ($\alpha = 0.05$)
In the Remarks column, results of the post hoc test (with the Bonferroni correction) are presented. Only statistically differences are described. Source: on the basis of the authors' research

municipalities from each country, a synthetic assessment of each one was completed for individual cities ($n_1 = 16$ for Poland, $n_2 = 10$ for Norway, and $n_3 = 18$ for the Ukraine). The higher the level of a variable (closer to 1), the higher the level (intensity) of a given city brand personality trait. Comparing the average values for the municipalities (the average from the normalized number of posts within a given dimension of the particular city from a given country), we can observe dimensions of brand personalities of cities from Poland, Norway, and Ukraine. The same procedure was conducted in relation to the three branding strategies (Table 5).

Data presented in Table 4 confirm clear differences between the level of the six city brand personality dimensions in the three countries being investigated. Only the *Competence* dimension is similar in each three countries (in the Kruskal-Wallis test, $p = 0.124$).

Comparing individual countries, for cities located in Ukraine, the *Prosocial Attitude* level was significantly higher than in the other two countries, as well as *Ruggedness. Sincerity* and *Sophistication* are at a significantly higher level than in Poland. *Excitement,* on the other hand, is at a significantly higher level in Poland than in Norway (here, in turn, Ukraine does not stand out from the other countries).

Comparing individual dimensions, their value varies from country to country (Fig. 1). Firstly, in the case of Poland, the highest level was recorded for *Sincerity* (Me = 0.41) and then *Sophistication* (0.32) and *Excitement* (0.26). Also, the

Table 5 Descriptive statistics, normality test, and Kruskal-Wallis test results for the three city branding strategies in the three countries being investigated

		Min	Max	M	Me	SD	S	K	P		
									SW	KW	Remarks
Place	PL	0.14	0.86	0.35	0.30	0.17	1.87	5.37	0.007[a]	0.012[a]	NO < PL ($p = 0.009$[a])
	NO	0.00	0.40	0.16	0.17	0.12	0.76	0.51	0.636		
	UA	0.08	0.47	0.26	0.24	0.12	0.30	−0.81	0.331		
Organization	PL	0.04	0.36	0.15	0.13	0.09	0.99	0.76	0.126	<0.0001[a]	PL < UA ($p = 0.001$[a])
	NO	0.02	0.57	0.16	0.12	0.15	2.33	6.44	0.002[a]		NO < UA ($p = 0.006$[a])
	UA	0.08	0.56	0.35	0.35	0.13	−0.19	−0.54	0.862		
Democracy	PL	0.00	0.13	0.02	0.00	0.04	2.04	3.35	<0.0001[a]	<0.0001[a]	PL < UA ($p = 0.048$[a])
	NO	0.00	0.36	0.19	0.22	0.13	−0.55	−1.02	0.161		NO < UA ($p < 0.0001$[a])
	UA	0.05	0.67	0.39	0.38	0.20	−0.22	−1.10	0.386		

Source: on the basis of the authors' research
44 municipalities, 1320 posts, 2187 observations
In the Remarks column, results of the post hoc test (with the Bonferroni correction) are presented. Only statistically differences are described
PL Poland, *NO* Norway, *UA* Ukraine
M mean, *Me* median, *SD* standard deviation, *S* skewness, *K* kurtosis, *p* probability in: *SW* Shapiro-Wilk test, *KW* Kruskal-Wallis test
[a]results statistically significant ($\alpha = 0.05$)

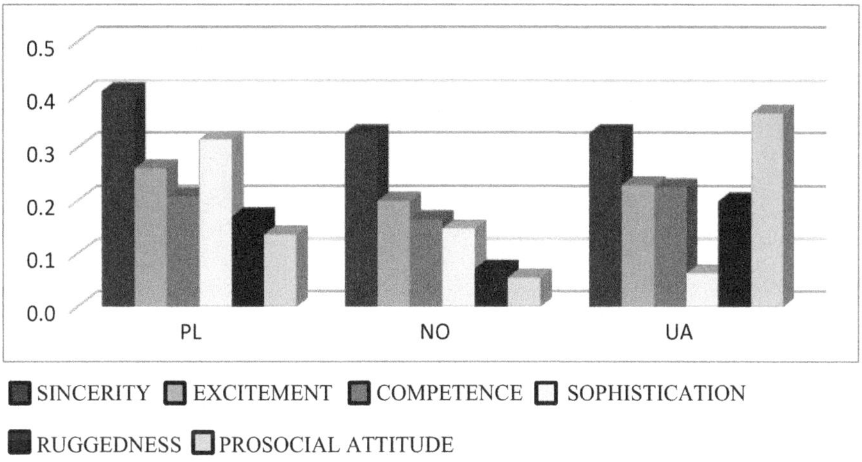

Fig. 1 The median level of each six city brand personality dimensions in the three investigated countries (according to countries). Polish municipalities (PL) $n = 1409$; Norwegian municipalities (NO) $n = 509$; Ukrainian municipalities (UA) $n = 1311$. (Source: own elaboration on the basis of the authors' research)

dimension of *Sincerity* (0.33) was a dominant one in the case of Norway, followed by *Excitement* (0.20) and *Competence* (0.17). The weight of these dimensions decomposes in the case of Ukrainian municipalities where the *Prosocial Attitude* dimension (0.37) is the leading dimension followed by *Sincerity* (0.33). The examples of the posts are given in Appendix 3.

The authors have taken a closer look at the structure of individual personality dimensions, paying special attention to the differences between cities representing the three countries: Poland, Norway, and Ukraine (Fig. 2). Although the components of the *Sincerity* dimension have a similar total value, their structure is different. In Poland, within this dimension, the posts were usually marked with the *cheerful* trait (10.3%), while in the case of Norway by the *down-to-earth* (16%) and *honest* (15%) traits. In the case of Ukraine, the dominant personality trait within *Sincerity* was *down-to-earth* (13.3%), which was mainly related to the difficult social and economic situation causing problems for the local community.

The *Excitement* dimension, although in total has a similar average score in each country, represents a completely different distribution of the traits. In Poland, within the *Excitement* dimension, the posts were marked by the *imaginative* trait (12.4%), in the case of Norway by *spirited* (16%), and in Ukraine by *patriotic* (9%). This means that cities in different countries try to influence their audience through emotions, but these emotions are linked to different sources. In Poland, these were positive emotions related to innovative urban solutions, ideas taken by the local community, and creative photographs of the city's assets. Also, the posts of Norwegian cities with the *Excitement* dimension generate positive emotions. Most often they were marked with the *spirited* trait, as their message was more thrilling and dynamic. In the case of Ukraine, among the *Excitement* dimension, the

Norway

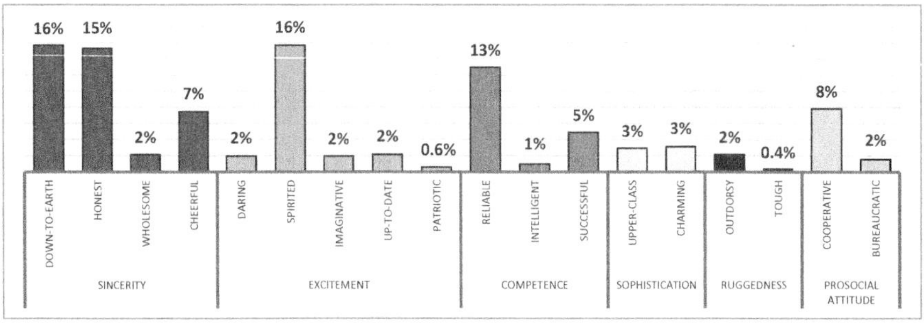

Ukraine

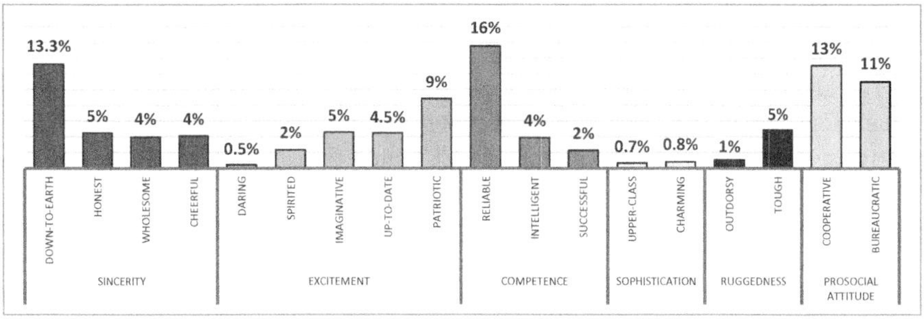

Fig. 2 Summary of personality traits identified in posts published on Facebook accounts of Polish (n = 1409), Norwegian (n = 509), and Ukrainian (n = 1311) municipalities (%). (Source: own elaboration on the basis of the authors' research)

characteristics referring to national patriotism dominate (*patriotic*). However, they present mainly negative emotions toward the national enemy in the form of Russia, which has been mainly associated with the open armed military conflict ongoing since 2014. Some examples of posts presenting the *Excitement* dimension have been given in Appendix 4.

Within the *Competence* dimension, the most frequent trait in each country was *reliable*. This trait dominates also in the case of Poland and Ukraine. In the case of *Sophistication*, both traits within that dimension are equally represented in posts from each of the country being investigated. The *Ruggedness* dimension is represented by the two personality traits: *outdoorsy* and *tough*. In the case of Norway, a measurably greater representation of *tough* can be noted. The structure of the *Prosocial Attitude* dimension is more diversified.

In the case of Poland, posts marked with the *cooperative* trait (14.6%) strongly dominate, while the alternative *bureaucratic trait* was found in a small number of posts of Polish cities (1.6%). This may be because Polish posts, as well as Norwegian ones, did not contain the characteristics of the formal message as opposed to Ukrainian posts (*bureaucratic* – 11%). The latter were characterized by a formal style based on regular transfer of information about official meetings or events without adding emotional content to it. The content of Ukrainian posts consisted of documents, agreements, and contracts, which does not fit into the informal character of social media. The bureaucratic manner of these posts is also reflected in their lengthiness – they were much longer than posts of cities in Poland and Norway.

Branding Strategies Across the Countries

Comparing the countries in terms of the Facebook branding strategies used by their municipalities, cities located in Ukraine provide significantly higher levels of the *Organization* and *Democracy* strategy than in the other two countries. The *Place* strategy, on the other hand, is at a significantly higher level in Poland than in Norway. In this case, Ukraine does not stand out from the other countries.

When comparing individual branding strategies, their popularity varies from country to country (median at Fig. 3 and Me values at Table 5). Firstly, in the case

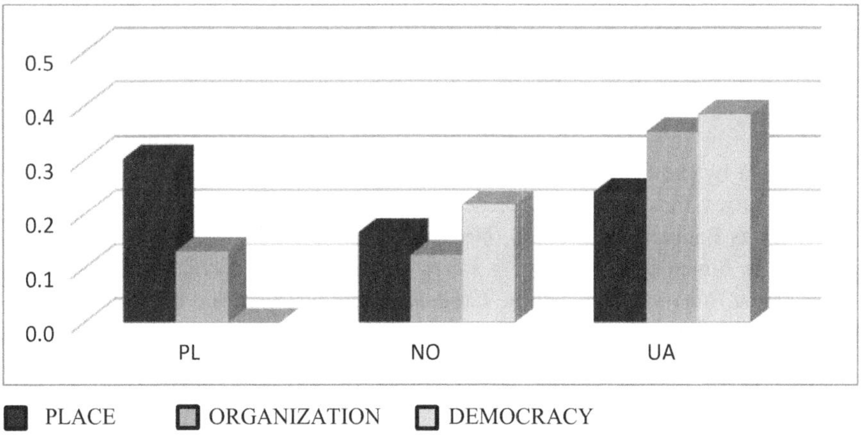

Fig. 3 The average level (median) of the usage of the three branding strategies (*Place*, *Organization*, *Democracy*) in the Facebook posts published by the municipalities in Poland $n = 794$, Norway $n = 359$, and Ukraine $n = 1034$ (according to countries). (Source: on the basis of the authors' research)

of Poland, the highest level was recorded for the *Place* strategy (Me = 0.3). This strategy appeared relatively more often in the posts of Polish municipalities. The *Organization* strategy represents a much lower level (0.13). At the same time, there were almost no posts representing the *Democracy* strategy during the period considered.

The weight of these strategies is distributed in a quite opposite way in the case of Ukrainian municipalities. The *Democracy* strategy (Me = 0.38) comes to the fore here followed by the *Organization* strategy (0.35) and then the *Place strategy* (0.24). In the case of Norway, the most important was also *Democracy* (0.22), followed by the *Place* (0.17) and *Organization* (0.12) strategies.

The authors have taken a closer look at the structure of recipients to whom messages are addressed via Facebook. In the case of Polish cities, the *Place* strategy dominates. Posts are aimed primarily at locals but also at tourists. On the other hand, far fewer communications were addressed to entrepreneurs and investors. Some differences among the cities studied can be observed within the country. Some Polish cities carry out intensive Facebook activities with a few posts a day, and many of them present images of beautiful places located within the city. There are also those that carry out fewer intensive activities and represent more diversified themes.

The structure of posts according to their recipients is less diverse for cities in Norway and Ukraine. In both countries, the number of posts targeted at voters/future voters was relatively higher than in Poland. Among the municipalities in all the three countries, however, the relatively highest number of posts was directed to users of services in the case of Ukrainian cities (22%). The examples of the posts presenting different branding strategies have been given in Appendix 5 (Fig. 4).

Correlations Between the Branding Strategies and the Personality Dimensions of Cities

To identify the relations between branding strategies and personality dimensions represented by the Facebook posts issued by the researched municipalities in the three countries, the correlations, presented in Table 6, have been verified.

As far as Poland is concerned, *Democracy* is the only strategy not linked with any of the dimensions. In turn, a fairly strong positive correlation takes place between the *Place* strategy and the *Excitement* dimension (rho = 0.660, $p = 0.005$), as well as between the *Organization* strategy and both *Competence* (rho = 0.534, $p = 0.033$) and *Prosocial Attitude* (rho = 0.563, $p = 0.023$). So, the more posts representing these strategies, the average more posts representing these dimensions. And vice versa, the more posts regarding the dimensions, the average more posts representing the listed strategies.

In the case of Norway, the relationship relates to other combinations of branding strategies and personality dimensions. A fairly strong correlation was observed

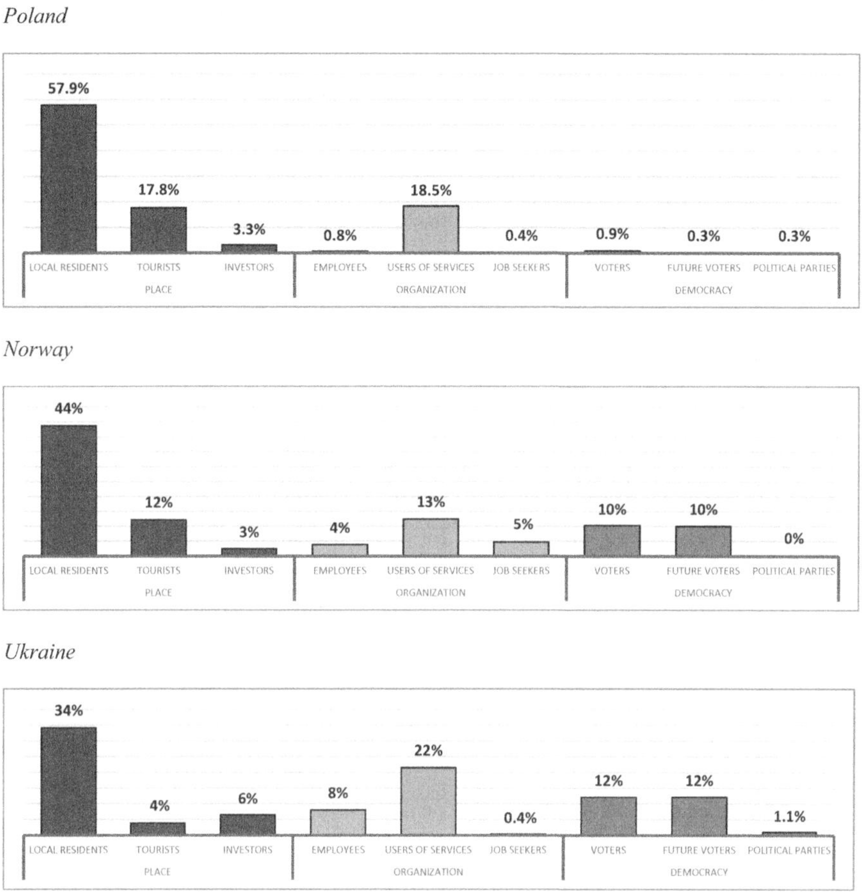

Fig. 4 Summary of target markets (grouped into the *Place*, *Organization*, and *Democracy* branding strategies) identified in the posts published on Facebook accounts of Polish (n = 794), Norwegian (n = 358), and Ukrainian (n = 1034) municipalities (%). (Source: on the basis of the authors' research)

between the *Place* strategy and the *Sophistication* dimension (rho = 0.640, p = 0.046) as well as between the *Democracy* strategy and the *Ruggedness* dimension (rho = −0.677, p = 0.031). In the first case, this correlation is positive (a stronger emphasis on the *Place* strategy involves more attention to *Sophistication*, and vice versa – the greater the role of *Sophistication*, the average greater the importance of the *Place* strategy). In the second case, the correlation is negative (cities' posts classified as published within the *Democracy* strategy focus less on *Ruggedness*, and vice versa). However, no relationship between the *Organization* strategy and individual dimensions was observed.

As far as Ukraine is concerned, there is an analogous relationship between the *Place* strategy and the *Sophistication* dimension (rho = 0.530, p = 0.024) and the

Table 6 Correlations between the branding strategies and the personality dimensions (of posts published on official Facebook profiles of the researched municipalities) by country

Specification			*Sincerity*	*Excitement*	*Competence*	*Sophistication*	*Ruggedness*	*Prosocial Attitude*
Poland	*Place*	rho	-0.072	0.660	-0.041	0.339	0.053	-0.349
		p	0.791	0.005*	0.879	0.199	0.845	0.186
	Organization	rho	0.189	-0.395	0.534	-0.492	0.054	0.563
		p	0.484	0.130	0.033*	0.053	0.843	0.023*
	Democracy	rho	-0.011	-0.351	0.159	-0.226	0.237	0.144
		p	0.969	0.183	0.556	0.401	0.377	0.596
Norway	*Place*	rho	0.091	0.467	0.018	0.640	-0.056	0.073
		p	0.803	0.174	0.960	0.046*	0.878	0.840
	Organization	rho	-0.139	-0.139	-0.049	-0.061	-0.267	-0.055
		p	0.701	0.701	0.894	0.867	0.456	0.880
	Democracy	rho	-0.422	0.128	0.092	0.145	-0.677	0.191
		p	0.224	0.724	0.800	0.690	0.031*	0.596
Ukraine	*Place*	rho	0.251	0.384	0.262	0.530	0.060	-0.166
		p	0.315	0.115	0.293	0.024*	0.812	0.510
	Organization	rho	-0.537	-0.313	-0.193	0.123	-0.162	0.508
		p	0.022*	0.206	0.442	0.628	0.520	0.031*
	Democracy	rho	-0.107	0.451	0.097	0.383	0.105	0.177
		p	0.674	0.060	0.703	0.117	0.678	0.483

Source: on the basis of the authors' research
rho rho Spearman rank coefficient, *p* probability in t test
[a]correlation statistically significant ($\alpha = 0.05$)

same relationship between the *Organization* strategy and the *Prosocial Attitude* dimension (rho = 0.508, $p = 0.031$) as in Poland. Additionally, there is a statistically significant negative correlation between the *Organization* strategy and the *Sincerity* dimension (rho = −0.537, $p = 0.022$), that is, an increase in the importance of *Sincerity* does not go hand in hand with growth, but with a decrease in the importance of the *Organization* strategy. As in Poland, *Democracy* is the only strategy that is not related to any of the brand personality dimensions.

Discussion and Conclusions

How municipalities are perceived and how they present themselves to different target groups could affect their long-term prosperity and legitimacy. Presenting favorable brand images and characteristics on social media is a step in this direction for many municipalities, including the ones studied here. Unfortunately, little is still known about how municipal branding occurs, i.e., which characteristics are promoted in specific social media posts and which branding strategies are reflected through these characteristics. Moreover, to our knowledge, no studies have approached municipal branding in social media comparatively.

Our study has sought to advance insights into these issues in the following ways. First, the study is one of the first to connect municipal branding with social media. Whereas previous studies have examined municipal branding with respect to strategies (Lockert et al. 2019a, b; Wæraas 2014) and the use of strategic communication in branding and reputation management (Cassel 2008; de Rosa et al. 2012; Moldenæs 2016; Nielsen and Salomonsen 2012), our study combines a focus on strategies *and* specific messages in the context of social media. This approach, we believe, is sorely needed, as the use of social media has proliferated in municipal administration and offered an opportunity for municipalities to reach out to their constituents not only with public administrative information but also with brand messages promoting specific personality traits that aim to create emotional attachment to the municipality.

Indeed, our study shows that branding on social media consistently involves the projection of personality traits that primarily portray municipalities as sincere, exciting, sophisticated, and prosocial rather than displaying the more technical and hard traits such as *Competence* and *Ruggedness*. The communicated brand personality traits thus highlight how municipalities seek to promote themselves as people-oriented and relational in their orientation toward their citizens. In turn, emotional attachments created by these projected personalities could yield a long-term beneficial impact on citizens' willingness to participate in local elections, run for office, and support other local democratic arrangements. The comparative aspect of our study offers support to this interpretation given the different institutional contexts of the three countries.

Although, as indicated above, at the level of dimensions, the posts of cities of different countries do not show significant differences, interesting differences can be seen if we look at the structure of different dimensions and the traits represented in them. This can be seen well in the case of *Excitement*, which is the dimension with the second (Norway) or third (Poland, Ukraine) average score among the six city brand personality dimensions. It is worth recalling that it is represented by five personality traits. Each country is dominated by a different trait, in Polish posts, *imaginative*; in Norwegian posts, *spirited*; and in Ukrainian posts, *patriotic*. This shows that seemingly similar personality brands of cities are based on different personality traits, which may be the result of the specific macro-institutional context of the countries under investigation.

Second, our study finds the *Democracy* strategy – a previously much disregarded municipal branding strategy – is relatively strong on social media compared to the *Place* and *Organization* strategies (Poland being an exception). This finding stands in contrast to many previous studies predominantly highlighting the *Place* branding (Cleave et al. 2017; Eshuis and Edwards 2013) and *Organization* branding strategies (Wæraas et al. 2015) as the most relevant for municipalities. Whereas the study by Wæraas et al. (2015) found *Democracy* branding to be marginally represented in the strategies expressed by municipal officials, our study shows that it is more prevalent in practice. For example, in the case of Norwegian municipalities, Facebook was used to portray the municipalities as attentive to citizens' concerns and needs and to inform them of opportunities in which they could express their concerns and participate in information meetings. In the case of Ukraine, the *Democracy* strategy was also recorded at a high level. In the content of many posts of Ukrainian cities, local politicians, mayors, and city councilors appear. The posts report facts from the city council, committees, meetings, and other activities in the city where these politicians exercise their functions. They seemingly present a clear message of responsibility of local government, about working together for the benefit of the inhabitants.

Thus, social media seem to offer an important opportunity for municipalities to strengthen citizens' appreciation of and participation in local government. Overall, the findings suggest social media might be more connected to *Democracy* branding in practice than what previous studies of municipal branding strategies have found. In turn, this suggests municipal branding could be a way of strengthening local governance processes (Eshuis and Edwards 2013). However, this is under the condition that the communication is followed by real engagement of the local government in citizen involvement.

Our analysis of the contents of Ukrainian city posts gives rise to assumptions about the open attitude of politicians to engage the inhabitants in the co-creation of the city, and joint resolution of local problems is an instrument to symbolically demonstrate prosocial attitudes aimed at maintaining their position and power. According to the research material, the mayors and councilors of Ukrainian cities try through posts to build their personal image as a caring hosts sensitive to the daily problems of the inhabitants and the committed national patriots. Thus, branding cities based on the *Democracy* strategy can also be a tool used to achieve not democratic but political governing goals if the country lacks a political-administrative culture based on consensus orientation and stakeholders' involvement. The lack of these values in Ukraine today seems to result in frequent political change at all levels of government.

Third, the study has not only examined the communicated personality traits and branding strategies independently, but it also extends previous research on municipal branding by examining whether a particular set of traits is correlated with the use of a particular strategy. In this respect, the findings show that the *Place* branding strategy tended to correlate with *Excitement* (Poland) and *Sophistication* (Norway and Ukraine) and that the *Organization* branding strategy tended to correlate with *Prosocial Attitude* (Poland and Ukraine) and *Competence* (Poland). The findings thus provide some support for the claim that different branding strategies entail the projection of different personality traits: If the *Place* branding strategy is chosen, social media posts are more likely to reflect traits such as *Excitement* and *Sophistication*. In this case, the municipalities want to target groups such as tourists and investors, as well as future and current residents, trying to make them perceive the municipality as an exciting and sophisticated place to live in, run a business, and visit. Conversely, if the *Organization* strategy is chosen, the findings suggest that social media posts are more likely to reflect *Prosocial Attitude* and *Competence*. In this case, the municipalities are more interested in having target groups such as users of services or current municipal employees perceive the municipality as being competent and sensitive to social problems.

Finally, the comparative aspect of our study offers an opportunity to study the significance of the macro-institutional context for municipal branding. Whereas one previous study has compared very similar countries such as Denmark and Norway (Lockert et al. 2019b), our study is the first to integrate countries from Western, Central, and Eastern Europe into a similar analysis. As such, our study has generated some insights of general relevance for municipal branding research. Most importantly, despite differences in the political-administrative culture, economic situation, and international relations across our sample, these differences seem to not affect municipal branding practices in a significant way.

The macro-institutional context has not produced notable unique patterns in the different countries concerning the municipalities' projected personality traits and branding strategies. Based on the findings, we suggest that municipal branding might be a universal phenomenon that manifests itself in brand messages that can be classified according to quite similar dimensions and similar strategies. Branding on social media is not only a marketing technique or concept that has spread to municipalities across Europe and thereby made municipal practices more similar, but the contents of the brand messages also have notable commonalities that make municipalities even more similar. These findings should be generalizable to municipalities in other European countries. Increasing the number of cities studied and the number of posts analyzed will allow for similar research, taking into account an advanced analysis on personality traits rather than just the city brand personality dimensions. Whether this is the case, however, can only be resolved in future research.

Appendix 1

List of investigating cities from Poland (16), Norway (10), and Ukraine (18)

No	Poland	Norway	Ukraine
1	Białystok	Bergen	Cherkasy
2	Cracow	Bodø	Chernivtsi
3	Gdansk	Hamar	Dnipro
4	Katowice	Kristiansand	Ivano-Frankivsk

(continued)

No	Poland	Norway	Ukraine
5	Kielce	Molde	Khmelnytskyi
6	Lodz	Oslo	Kiev
7	Lublin	Skien	Kropyvnytskyi
8	Olsztyn	Stavanger	Lutsk
9	Opole	Steinkjer	Lviv
10	Poznan	Tromsø	Mykolaiv
11	Rzeszow		Odessa
12	Szczecin		Poltava
13	Toruń		Rivne
14	Warsaw		Sumy
15	Wrocław		Ternopil
16	Zielona Góra		Uzhhorod
17			Vinnytsia
18			Zhytomyr

Appendix 2

Descriptions of the city personality dimensions assumed in the presented study.

Sincerity – describes the sensitivity of local authorities as well as the local community to local problems connected with the broadly understood concept of life within the city. The family and its care are an important aspect of this dimension. Contained within this dimension is also the city's sentimentality or its ability to recall positive facts or memories from its past. It also concerns the honesty of local communities, their ability to share with others, and local solidarity. Additionally, this dimension relates to inhabitants' characteristics such as contentment and spiritual serenity.
Excitement – demonstrates the city's courage, the determination of its citizenry, and its businesses, social organizations, and local authorities. It manifests itself through stirring events organized within the city as well as emotionally charged cultural or sporting events. The dimension additionally contains both visionary infrastructural projects occurring within the city and innovational social initiatives. All listed characteristics are connected in their desire to demonstrate strong emotions which can be communicated through youthful activity, uniqueness, or innovativeness. It can also demonstrate patriotism (national, regional, or local). It is a very emotionally charged dimension.
Competence – concerns the abilities, qualifications, and experience of the participants of local life as well as of local government. It involves a city of professional and technologically and industrially able citizenry as well as efficient public and social institutions. It relates to quality-oriented business entities that know how to successfully operate on the market. This dimension also involves the appreciation of entities' effort, ability, reliability, or creativity through the awarding of prizes or distinctions.

(continued)

Sophistication – concerns being outstanding in various dimensions – economic, educational, cultural, sports, etc. It involves being distinctive through beauty, charm, or uniqueness. An exemplification of this dimension is the ability to attract exclusive events or elites of various environments, while, at the same time, a city which is identified through this dimension is charming, delightful, fragrant, and desirable.
Ruggedness – relates to roughness, toughness, and manliness. It is opposite to femininity, sensitivity, or complexity. It concerns simple emotions, masculine lifestyle, and appreciation for nature. This dimension also involves appreciation for simplicity and power. Typical manifestations include masculine sports and appreciation for the surrounding unspoiled natural environment and animals.
Prosocial Attitude – corresponds with the dimension of *Sincerity* which shows a city's sensitivity to social problems – problems of individual people and their neighborhood, while the dimension of *Prosocial Attitude* expresses how cities operationalize activities within this scope. This style corresponds directly to a given country's accepted regime of public management (*PA* versus *NPM/NPG*). Within this dimension, the *bureaucratic* trait corresponds to the formal approach, typical of the *PA* regime, while the *cooperative* trait corresponds to the NPM/NPG regimes based on efficiency as well as an open-minded and cooperative attitude.

Source: own elaboration based on Glińska & Rudolf (2019)

Appendix 3

Selected posts representing different personality dimensions of investigated municipalities from Poland, Norway, and Ukraine

City of Lodz, Poland	Post of January 16, 2020
Coded as: *Cheerful (Sincerity)* *Imaginative (Excitement)* *Charming (Sophistication)* Local residents, tourists (*Place* branding strategy) **The post content in English:** Weekend with big events is coming 😍 ❋ Garage party 19.01 #5 🌿 We are opening a plant store on Gdańska Street! ❋ Charity concert for Madzia ❋ A brief history of tea – admission. Moderated tea tasting ❋ Bl … See more	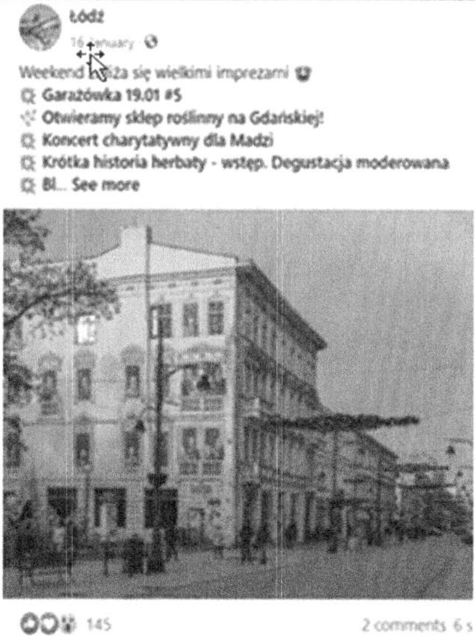

(continued)

<table>
<tr><td>City of Lodz, Poland</td><td>Post of February 17, 2020</td></tr>
<tr><td>Coded as:
Cheerful (Sincerity)
Imaginative (Excitement)
Local residents (Place branding strategy)
The post content in English:
After colorful dreams,
it's time for colorful mornings</td><td>

</td></tr>
<tr><td>City of Warsaw, Poland</td><td>Post of January 17, 2020</td></tr>
<tr><td>Coded as:
Up-to-date (Excitement)
Intelligent (Competence)
Local residents (Place branding strategy)
Users of services
(Organization branding strategy)
The post content in English:
The Capital City of Warsaw
We do not know about you but we are already digging further into the subject of the subway in Bródno.
All because of 615 ton TBM discs, fragments of which set off for the C21 station where we're building a launch shaft.
At the end of March and the beginning of April, Anna and Maria will be in full swing.
Warsaw Metro #Subway Changes Warsaw – Warsaw Public Transport</td><td>
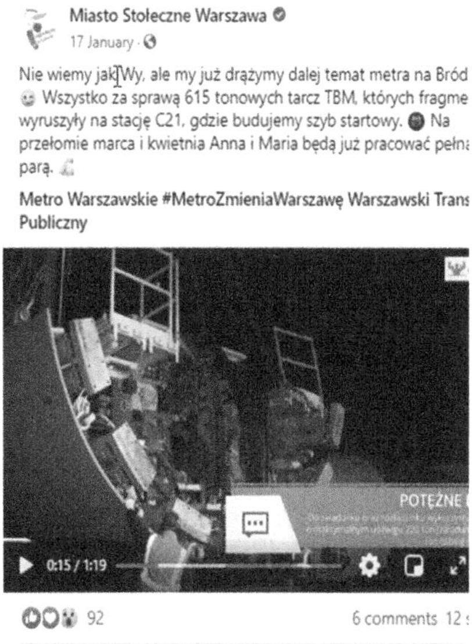

</td></tr>
</table>

(continued)

<table>
<tr><td>City of Bodø, Norway</td><td>Post of February 4, 2020</td></tr>
<tr><td>Coded as:
Daring (Excitement)
Successful (Competence)
Upper class (Sophistication)
Local residents, tourists, investors
(Place branding strategy)
The post content in English:
Bodø to become European cultural capital in 2024: Culture makes us attractive</td><td>
</td></tr>
<tr><td>City of Hamar, Norway</td><td>Post of February 7, 2020</td></tr>
<tr><td>Coded as:
Cheerful (Sincerity)
Spirited (Excitement)
Successful (Competence)
Users of services
(Organization branding strategy)
The post content in English:
The best historical result! This is going in one direction only. Forward, says the Mayor Einar Busterud to Hamar Newspaper, after Hamar scored well in a recent reputation survey.</td><td>
</td></tr>
</table>

(continued)

City of Chernivtsi, Ukraine | February 18, 2020

Coded as:
Up-to-date (Excitement)
Intelligence (Competence)
Cooperative (Prosocial Attitude)
Users of services
(*Organization* branding strategy)
The post content in English:
A free vocational course in the IT field for people with disabilities has started.

City of Kropyvnytskyi, Ukraine | January 15, 2020

Coded as:
Bureaucratic (Prosocial Attitude)
Reliable (Competence)
Users of services
(*Organizational* branding strategy)
The post content in English:
On Tuesday, January 14, in the premises of the city council of center, there was a contest for the definition of carrier on urban bus routes of general use of the city of center.
To participate in the competition submitted three enterprises-carriers: LLC "Kirovohrad-Euro-Trans" and PP "Autobaz-Kirovohrad," which declared route no. 130 "Central Market - Katranivka," and FOP Barkar C. M., which discovered the desire to work on route no. 274 "Square of Bogdan Khmelnitsky - sal. New." Vehicles on both routes will work in "minibus taxi" mode.
The decision of the PP "Autobaz-Kirovograd" was not allowed to participate in the competition, due to violations of the law of Ukraine "on car transport": the enterprise does not have enough vehicles to serve the stated route.. The other two businesses - carriers were allowed to participate in the competition and recognized as winners on the respective routes.

Source: own elaboration on the basis of official Facebook posts of the following cities: Lodz and Warsaw (Poland), Hamar and Bodø (Norway), and Chernivtsi and Kropyvnytskyi (Ukraine)

Appendix 4

Selected posts representing different personality traits of the *Excitement* dimension of municipalities from Poland, Norway, and Ukraine

City of Lublin, Poland	Post of January 20, 2020
Coded as: *Cheerful* (*Sincerity*) *Spirited* (*Excitement*) *Successful* (*Competence*) Local residents, tourists (*Place* branding strategy) **The post content in English:** *Huge congratulations to the Basket Women Arka Gdynia players who won the Polish Cup in the final tournament played at the MOSiR Arena on Zygmuntowska Avenue.* *In the finals, Gdynia basketball players defeated CCC Polkowice.* *And on behalf of the authorities of our City, the winning team was congratulated by Deputy President Beata Stepaniuk-Kuśmierzak.*	
City of Hamar, Norway	Post of February 12, 2020
Coded as: *Up-to-date (Excitement)* *Intelligent, successful* (*Competence*) Local residents (*Place* branding strategy) Users of services (*Organization* branding strategy) **The post content in English:** *Gålåsholmen has opened! The waste management facility is the first of its kind in Norway. It is ultra modern. It also offers the possibility for repairs and renovations of commodities that can be recycled.*	

(continued)

<table>
<tr><td>City of Chernivtsi, Ukraine</td><td>Post of January 22, 2020</td></tr>
<tr><td>Coded as:
Honest (Sincerity)
Patriotic (Excitement)
Local residents (Place branding strategy)
The post content in English:
The solemn laying of flowers on the occasion of the reunion day of Ukraine took place in Chernivtsi.
At the central square, representatives of the...</td><td>
</td></tr>
</table>

Source: own elaboration on the basis of official Facebook posts of the following cities: Lublin (Poland), Hamar (Norway), and Chernivtsi (Ukraine)

Appendix 5

Selected posts representing the *Place*, *Organization*, and *Democracy* branding strategies of investigated municipalities from Poland, Norway, and Ukraine

<table>
<tr><td>City of Gdansk, Poland</td><td>Post of January 18, 2020</td></tr>
<tr><td>Coded as:
Cheerful (Sincerity)
Charming (Sophistication)
Local residents, tourists
(Place branding strategy)
The post content in English:
The city is slowly coming to life. But if you have hard time waking up the Saturday will make up for everything.</td><td>
</td></tr>
<tr><td>City of Poznań, Poland</td><td>Post of January 19, 2020</td></tr>
</table>

(continued)

City of Gdansk, Poland	Post of January 18, 2020
Coded as: *Cheerful* (*Sincerity*) *Intelligent (Competence)* *Cooperative* (*Prosocial Attitude*) Job seekers (*Organization* branding strategy) **The post content in English:** Are you looking for something interesting to do to set off your professional career? Come for an internship at the City Office! Recruitment details at: http://bit.ly/stazUMP	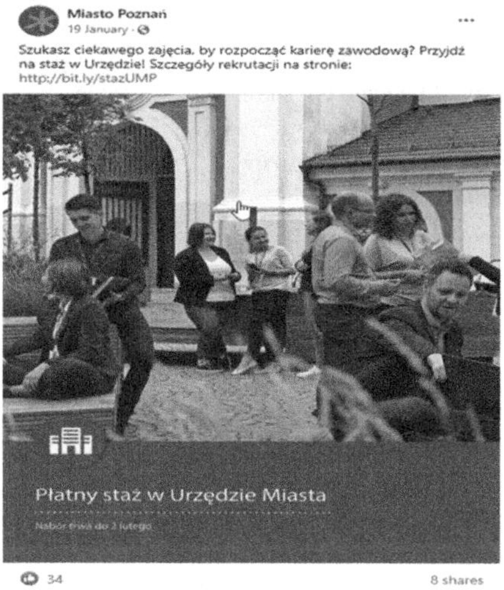
City of Kristiansand, Norway	Post of February 7, 2020
Coded as: *Cooperative* (*Prosocial Attitude*) Voters/future voters (*Democracy* branding strategy) **The post content in English:** *Kristiansand Municipality prioritizes dialogue meetings with citizens, civil society, and organizational life. 16 dialogue meetings will be held in 2020. The meetings are a three-year trial period.*	
City of Kropyvnytskyi, Ukraine	Post of January 16, 2020
Coded as: *Honest (Sincerity)* *Patriotic (Excitement)* *Tough (Ruggedness)* Local residents (*Place* branding strategy) Voters (*Democracy* branding strategy) **The post content in English:** January 16 - Memorial Day of heroes (Cyborgs) from Donetsk	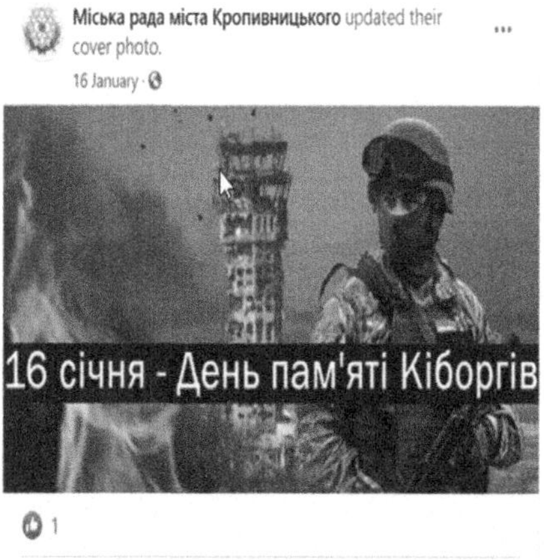

City of Gdansk, Poland	Post of January 18, 2020
City of Chernivtsi, Ukraine	16 January, 2020

Coded as:
Down-to-earth (Sincerity)
Reliable (Competence)
Tough (Ruggedness)
Local residents
(*Place* branding strategy)
Users of services
(*Organization* branding strategy)
The post content in English:
EMERGENCY ROAD PITS REPAIRS IN CHERKASY CONTINUE EVEN IN WINTER
In Cherkassy, despite it being mid-January, emergency road pits repairs continue.
According to the director of KP "CELUASH" Oleksii Melnyk, the roads in Cherkasy are now being repaired with the help of Recycler.
"In the cold season, asphalt plants do not work. However, we can eliminate the pits using old asphalt and the Oskol. We are working in every kind of weather to reduce the inconvenience for drivers from pits on the road. This method of emergency road pits repair is very economical. The temporary filling will remain on the road 1–2 month. The Recycler allows us to repair 30–40 m2 of road surface per day. We will work in this way as long as the weather permits," said Oleksiy Melnyk.
According to Komalalnyka, the pits at the following locations were repaired: on Koroleva St., Vul. Zalizniak, near the new bus station on Smila Street, at the crossroads of Narbutivska and Pashayev St. (at the 14th School), on Dakhnivska St. (from the Regional Hospital to the bridge), on Chornovola Street (from Prochemists to CHSHK), at the crossroads of Chygyrynska and Patsayev St., etc. *Work continues.*
As recently reported by the city Mayor Anatoly Bondarenko in an interview with the Antenna TV Channel, contractors are planning to launch large-scale repairs of the roads in Cherkassy in March. This year, the main emphasis of repair will be shifted to the secondary roads and intra-quarter roads.

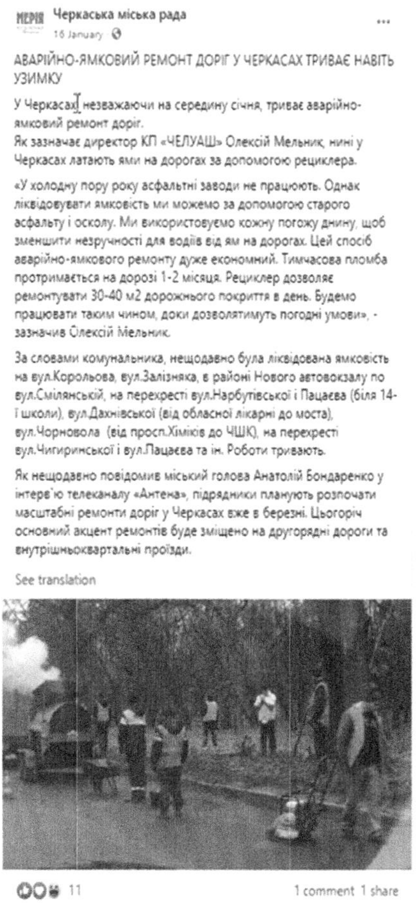

Source: own elaboration on the basis of official Facebook posts of the following cities: Gdansk and Poznań (Poland), Kristiansand (Norway), and Kropyvnytskyi Chernivtsi (Ukraine)

References

Aaker, J. (1997). Dimensions of brand personality. *Journal of Marketing Research, 34*(August), 347–356.

Aksoy, S., & Haralick, R. M. (2001). Feature normalization and likelihood-based similarity measures for image retrieval. *Pattern Recognition Letters, 22*(5), 563–582.

American Marketing Association. (2017). Definitions of Marketing. Retrieved from https://www.ama.org/the-definition-of-marketing-what-is-marketing/

Balmer, J. M. T., & Greyser, S. A. (Eds.). (2003). *Revealing the corporation. Perspectives on identity, image, reputation, corporate branding, and corporate level marketing*. London: Routledge.

Baştuğ, S., Şakar, G. D., & Gülmez, S. (2020). An application of brand personality dimensions to container ports: A place branding perspective. *Journal of Transport Geography, 82*, 102–152.

Bonsón, E., Torres, L., Royo, S., & Flores, F. (2012). Local e-government 2.0: Social media and corporate transparency in municipalities. *Government Information Quarterly, 29*(2), 123–132.

Bugarič, B. (2019). Central Europe's descent into autocracy: A constitutional analysis of authoritarian populism. *International Journal of Constitutional Law, 17*(2), 597–616.

Bunikowski, D. (2018). The constitutional crisis in Poland, Schmittian questions and Kaczyński's political and legal philosophy. *Journal of Contemporary European Studies, 26*(3), 285–307.

Cassel, S. H. (2008). Trying to be attractive: Image building and identity formation in small industrial municipalities in Sweden. *Place Branding and Public Diplomacy, 4*(2), 102–114.

Christensen, T., Lie, A., & Lægreid, P. (2008). Beyond new public management: agencification and regulatory reform in Norway. *Financial Accountability and Management, 24*(1), 15–30.

Cleave, E., Arku, G., Sadler, R., & Kyeremeh, E. (2017). Place marketing, place branding, and social media: perspectives of municipal practitioners. *Growth and Change, 48*(4), 1012–1033.

Czaputowicz, J., & Sakowicz, M. (2017). Public administration in poland public administration in post-communist countries (pp. 293–317): Routledge.

Danko Y. I., Medriv V. Y., Koblianska I. I., Kornietskyy O. V., & Raznik N. P. (2020). Territorial government reform in Ukraine: Problem aspects of strategicmanagement. *International Journal of Scientific and Technology Research, 9*(1), 1376–1382.

de Rosa, A. S., Bocci, E., & Picone, M. (2012). E-branding and institutional web sites: the "visiting card" of the Municipalities of Rome and Paris Branding and sustainable competitive advantage: building virtual presence (pp. 207–247): IGI Global.

Dholakia, R. R., & Acciardo, L. A. (2014). Branding a state university: Doing it right. *Journal of Marketing for Higher Education, 24*(1), 144–163.

Ellison, N., & Hardey, M. (2014). Social media and local government: Citizenship, consumption and democracy. *Local Government Studies, 40*(1), 21–40.

Eshuis, J., & Klijn, E.-H. (2012). *Branding in governance and public management*. New York: Routledge.

Eshuis, J., & Edwards, A. (2013). Branding the city: The democratic legitimacy of a new mode of governance. *Urban Studies, 50*(5), 1066–1082.

Evans, W. D., & Hastings, G. (Eds.). (2008). *Public health branding. Applying marketing for social change*. New York, NY: Oxford University Press.

Fetscherin, M., & Usunier, J. C. (2012). Corporate branding: an interdisciplinary literature review. *European Journal of Marketing, 46*(5), 733–753. https://doi.org/10.1108/03090561211212494.

Glinka B., & Hensel P. G. (2017). Reforms and identities: How relentless pursuit of improvements produces a sence of helplessness among bureaucrats. *Journal of Organizational Change Management, 30*(2), 142–160.

Glińska, E., & Kilon, J. (2014). Desirable traits of the city brand personality in the opinion of managers for the promotion of the city government in Poland. *Procedia-Social and Behavioral Sciences, 156*, 418–423.

Glińska, E., & Rudolf, W. (2019). City brand personality projected by municipalities from Central and Eastern Europe countries—A comparison of Facebook usage. *Sustainability, 11*(19), 5440.
Govers, R., Go, F. M., & Kumar, K. (2007). Promoting tourism destination image. *Journal of Travel Research, 46*(1), 15–23.
Hsieh, H. F., & Shannon, S. E. (2005). Three approaches to qualitative content analysis. *Qualitative Health Research, 15*(9), 1277–1128.
Kaplan, M. D., Yurt, O., Guneri, B., & Kurtulus, K. (2010). Branding places: applying brand personality concept to cities. *European Journal of Marketing*.
Karens, R., Eshuis, J., Klijn, E. H., & Voets, J. (2016). The impact of public branding: An experimental study on the effects of branding policy on citizen trust. *Public Administration Review, 76*(3), 486–494.
Keller, K. L., & Richey, K. (2006). The importance of corporate brand personality traits to a successful 21st century business. *Journal of Brand Management, 14*(1–2), 74–81.
Kim, S., & Lehto, X. Y. (2013). Projected and perceived destination brand personalities: The case of South Korea. *Journal of Travel Research, 52*(1), 117–130.
Knowles, T., & Howley, M. J. (2000). Branding in the UK public house sector: recent developments. *International Journal of Contemporary Hospitality Management*.
Kovačić, S., Milenković, N., Slivar, I., & Rancic, M. (2019). Shaping city brand strategies based on the tourists' brand perception: report on Banja Luka main target groups. *International Journal of Tourism Cities*.
Kumar, J., & Nayak, J. K. (2018a). Brand community relationships transitioning into brand relationships: Mediating and moderating mechanisms. *Journal of Retailing and Consumer Services, 45*, 64–73.
Kumar, V., & Nayak, J. K. (2018b). Destination personality: Scale development and validation. *Journal of Hospitality & Tourism Research, 42*(1), 3–25.
Lee, H.-J., & Suh, Y.-G. (2011). Understanding and measuring City Brand Personality (CPS) -In the context of South Korean market. *International Journal of Tourism Sciences, 11*(3), 1–20.
Leijerholt, U., Biedenbach, G., & Hultén, P. (2018). Branding in the public sector: a systematic literature review and directions for future research. *Journal of Brand Management*, 1–15.
Leitch D. (2019). Leaving like an Englishman: Assisting institutional reform in postcommunist Ukraine. *Development Policy Review, 37*(1), 111–128.
Lockert, Å. S., Bjørnå, H., & Krane, M. S. (2019a). Policy transfer of branding and reputation management: Motivations, challenges, and opportunities for a small rural municipality. *Public Performance & Management Review*, 1–22.
Lockert, Å. S., Bjørnå, H., Haugen, K. H., & Salomonsen, H. H. (2019b). Reputation reform strategies in local government: investigating Denmark and Norway. *Local Government Studies*, 1–22.
Ma, W., Schraven, D., de Bruijne, M., De Jong, M., & Lu, H. (2019). Tracing the origins of place branding research: A bibliometric study of Concepts in use (1980–2018). *Sustainability, 11*(11), 2999.
Magnusson, M., Bellström, P., & Thoren, C. (2012). Facebook usage in government–a case study of information content.
Masiello, B., Bonetti, E., & Izzo, F. (2020). Multiple identities of a festival: intended, communicated and perceived brand personality in the social media environment. *International Journal of Contemporary Hospitality Management, 32*(2), 749–768.
Merriam-Webster Dictionaries. (2020). Social Media. Retrieved from https://www.merriam-webster.com/dictionary/social%20media
Moldenæs, T. (2016). Story found or story lost? Storytelling in audiovisual municipality branding. *Scandinavian Journal of Public Administration, 20*(2), 59–76.
Mossberger, K., Wu, Y., & Crawford, J. (2013). Connecting citizens and local governments? Social media and interactivity in major US cities. *Government Information Quarterly, 30*(4), 351–358.
Murphy, L., Benckendorff, P., & Moscardo, G. (2007). Linking travel motivation, tourist self-image and destination brand personality. *Journal of Travel & Tourism Marketing, 22*(2), 45–59.

Nielsen, J. A., & Salomonsen, H. H. (2012). Why all this communication? Explaining strategic communication in Danish local governments from an institutional perspective. *Scandinavian Journal of Public Administration, 16*(1), 69–89.

Omar, K., Stockdale, R., & Scheepers, H. (2014). Social media use in local government: An Australian perspective. *International Journal of Public Administration, 37*(10), 666–675.

Opoku, R. A. (2009). Mapping destination personality in cyberspace: An evaluation of country web sites using correspondence analysis. *Journal of Internet Commerce, 8*(1–2), 70–87.

Plummer, J. T. (2000). How personality makes a difference. *Journal of Advertising Research, 40*(6), 79–83.

Rojas-Méndez, J. I., Murphy, S. A., & Papadopoulos, N. (2013). The US brand personality: A Sino perspective. *Journal of Business Research, 66*(8), 1028–1034.

Sataøen, H. L., & Wæraas, A. (2015). Branding without unique brands: Managing difference and similarity in a public sector context. *Public Management Review, 17*(3), 443–461.

Sevin, E. (2016). Branding cities in the age of social media: A comparative assessment of local government performance. Social Media and Local Governments. (pp. 301–320).

Sharifsamet, S., Jin, H. S., & Martin, B. (2020). Marketing destinations: the impact of destination personality on consumer attitude. *Journal of Strategic Marketing, 28*(1), 60–69.

Shen, B., & Bissell, K. (2013). Social media, social me: A content analysis of beauty companies' use of Facebook in marketing and branding. *Journal of Promotion Management, 19*(5), 629–651.

Statista. (2020). Number of monthly active Facebook users worldwide as of 1st quarter 2020. Retrieved from https://www.statista.com/statistics/264810/number-of-monthly-active-facebook-users-worldwide/

Techweez. (2016). Study reveals 87% of governments use Facebook. Retrieved from https://techweez.com/2016/01/18/governments-that-are-active-on-facebook/

Treadaway, C., & Smith, M. (2012). Facebook marketing: An hour a day: John Wiley & Sons.

Tschirhart, M. (2008). Evaluation of brand use on city government websites: a search for brand type, repetition, and consistency. *Journal of Nonprofit & Public Sector Marketing, 19*(1), 35–53.

Ţugulea, O. (2017). City brand personality—Relations with dimensions and dimensions interrelations. *Sustainability, 9*(12), 2243.

Usakli, A., & Baloglu, S. (2011). Brand personality of tourist destinations: An application of self-congruity theory. *Tourism Management, 32*(1), 114–127.

Vinyals-Mirabent, S., & Mohammadi, L. (2018). City brand projected personality: a new measure to assess the consistency of projected personality across messages. *Communications Society, 31*(4), 91–108.

Wæraas, A. (2014). Making a difference: Strategic positioning in municipal reputation-building. *Local Government Studies, 41*(2), 280–300.

Wæraas, A., & Sataøen, H. L. (2015). Being all things to all customers: Building reputation in an institutionalized field. *British Journal of Management, 26*(2), 310–326.

Wæraas, A., Bjørnå, H., & Moldenæs, T. (2015). Place, organization, democracy: Three strategies for municipal branding. *Public Management Review, 17*(9), 1282–1304.

Whelan, S., Davies, G., Walsh, M., & Bourke, R. (2010). Public sector corporate branding and customer orientation. *Journal of Business Research, 63*(11), 1164–1171.

Worldwide Governance Indicators. (2020). The World Bank Group, info.worldbank.org/governance/wgi (access: April 22, 2020).

Zavattaro, S. M., French, P. E., & Mohanty, S. D. (2015). A sentiment analysis of US local government tweets: The connection between tone and citizen involvement. *Government Information Quarterly, 32*(3), 333–341.

Part II
Internal Branding Strategies

Chapters in this part of the book give both theoretical and practical looks at the uses of internal branding strategies. Seemingly, there is a lot of focus in the scholarly literature on external reasons for and evaluations of place branding, but internal branding is equally important given its ties to organization culture, employee morale, and overall communication strategy.

Branding at its basic level is about creating meaning through connection, so internal branding strategies are no different and often find the best success when they are bottom up (employee driven) rather than top down (Asha and Jyothi 2013). One of the primary ways internal branding is studied relates to organizational citizenship behavior, simply understood as the actions of employees at work commonly thought to benefit the organization more than the individual (Ozcelik and Findikli 2014). Employees are more likely to exhibit this kind of pro-social behavior if there is a strong emotional connection to the organization (van Knippenberg et al. 2007), and a social identity approach to internal branding has been shown to strengthen these citizenship behaviors via creating those crucial brand connections (Löhndorf and Diamantopoulos 2014).

Based on a content analysis of employee interviews, King and Grace (2008) developed an employee brand commitment pyramid, with the foundation being the communication of technical information needed for people to do their jobs. The next level consists of commitment to the job, strengthened by the appropriate technical information from the first step. Next comes communicating brand-related information such as mission, vision, and values, which then ideally leads to brand commitment at the top. It is an interesting way to echo the sentiment from Asha and Jyothi (2013) that successful internal brands start at the bottom from employees rather than dictated from top leaders alone.

As Mau elucidates in his chapter below, a key reason for internal branding is developing an employer of choice model. Seen this way, internal branding becomes a core human resources function and part of an organization's overall strategy rather than only a communications tool bent on slinging logos and slogans at people (Martin et al. 2004). As Martin et al. note (2004), becoming an employer of choice is mostly a psychological process built on creating positive relationships with existing employees and attracting new ones via positive associations with the employer. Similar to branding, employer of choice strategies fail when they are "little more than sophisticated and sensitive recruitment practices" (ibid, p. 78). Brands fail when they lack an emotional and cognitive component, so there is an interesting parallel to other managerial activities, such as employer of choice, that are enacted without thought as to why and how.

Another reason to undertake employee branding strategies is because employees are a key stakeholder group vital for the organization's success. Especially corporations often focus on the bottom line via external branding and resources, but this could be an oversight if internal employee branding is ignored (Miles and Mangold 2004). In this way, internal branding can be viewed as relationship management rather than only a communications strategy. There can be ties to an organization's service quality and performance via creating a strong bond with existing employees.

Rafiq and Ahmed (2000) trace the evolution of internal branding and marketing, starting with its genesis in employee motivation and satisfaction. Internal marketing emerged as a way to reduce service inconsistencies, with the thought being more streamlined communication could reduce poor customer service interactions by getting employees "on the same page." As such, employees might become more satisfied and committed via increased communication of similar standards. This led to the next development in internal marketing, which Rafiq and Ahmed (2000) term a customer orientation. In this view, employees must be motivated and sales oriented via additional information and the ability to influence consumers. Finally, internal marketing broadened to include strategy and change management (Rafiq and Ahmed, 2000). In this view, marketing and personnel management are more tightly coupled, meaning internal marketing goes beyond only motivation but directly to organizational strategy development.

It is via this shift that more light is shed on the role internal branding plays for developing strategy. Though Rafiq and Ahmed (2000) focus on for-profit corporations, the parallels are clear with public organizations steeped in new public management (NPM) mindsets importing business-like practices into the public sector. NPM changed the way governments operate internally by focusing more on customer satisfaction rather than citizen engagement (Sataøen and Wæraas 2015). As such, the focus becomes on competition for both internal and external stakeholders, leading to the emergence of branding as a key governance strategy (Eshuis and Klijn 2012).

Internal branding, though, is not without its challenges and critiques. Employees might feel the term brand is offensive and consider it organizational brainwashing (Löhndorf and Diamantopoulos, 2014; Yaniv and Farkas 2005). Zavattaro (2011) found that employees in one city, for example, began to ignore the physical aspects of an internal brand strategy. In time, they came to ignore signs and slogans on every office wall and on items of clothing, showing how too much brand communication can be harmful to pro-social organizational behaviors and employee motivation. The disconnect can sometimes happen when employees see brands as purely slogans and logos rather than a connection to an organization's mission and vision (King and Grace 2008).

Müller (2017) examines this proposition of brands as means for normative control, finding that brand messages – good or bad – become internalized and blur the lines between a person's work and home life. For example, if someone is a brand ambassador, that means they are always representing the organization brand even when "off work" but there is never an "off duty" from embodying the brand values. Moreover, she finds employee dress is another way to physically turn employees into brand embodiments via logos and slogans on clothing. This means even if an employee is on a break but wearing organizational clothing, they are immediately on duty to serve a passing customer. Employees, then, might distance themselves from brand control by engaging in negative behavior or choosing to leave the organizational all together (Müller 2017).

Chapters in this part draw attention to the growing reliance upon internal branding strategies as it relates to place branding. As the authors show, more research is needed to better understand the success – and how that is defined – of internal branding strategies.

References

Asha, C. S., & Jyothi, P. (2013). Internal branding: A determining element of organizational citizenship behavior. *The Journal of Contemporary Management Research, 7*(1), 37–57.

Eshuis, J., & Klijn. E. (2012). *Branding in governance and public management.* New York: Routledge.

King, C., & Grace, D. (2008). Internal branding: Exploring the employee's perspective. *Brand Management, 15*(5), 358–372.

Löhndorf, B., & Diamantopoulos, A. (2014). Internal branding: Social identity and social exchange perspectives on turning employees into brand champions. *Journal of Service Research, 17*(3), 310–325.

Martin, G., et al (2004). Branding: A new performance discourse for HR? *European Management Journal, 23*(1), 76–88.

Miles, S. J., & Mangold, G. (2004). A conceptualization of the employee branding process. *Journal of Relationship Marketing, 3*(2–3), 65–87.

Müller, M. (2017). 'Brand-centered control': A study of internal branding and normative control. *Organization Studies, 38*(7), 895–915.

Ozcelik, G., & Findliki, M. A. (2014). The relationship between internal branding and organizational citizenship behavior: The mediating role of person-organization fit. *Procedia – Social and Behavioral Sciences, 150*, 1120–1128.

Rafiq, M., & Ahmed, P. K. (2000). Advances in the internal marketing concept: Definition, synthesis and extension. *Journal of Services Marketing, 14*, 449–462.

Sataøen, H. L., & Wæraas, A. (2015). Branding without unique brands: Managing similarity and difference in a public sector context. *Public Management Review, 17*(3), 443–461.

Van Knippenberg, D., Van Dick, R., & Tavares, S. (2007). Social identity and social exchange: Identification, support, and withdrawal from the job. *Journal of Applied Social Psychology, 37* (3), 457–477.

Yaniv, E., & Farkas, F. (2005). The impact of person organization fit on the corporate brand perception of employees and of customers. *Journal of Change Management, 5*(4), 447–461.

Zavattaro, S. M. (2011). An autoethnographic tale of image-generation tactics on employees. *Employee Responsibilities and Rights Journal, 23*(1), 1–16.

Branding from Within: Internal Brand Management in the Public Sector

Ulrika Leijerholt

Whenever we interact with a company or an organization, our perception of it is affected by the staff with whom we interact. This is true regardless of whether it is a private company or a public sector organization. Employee behaviors and the brand values they communicate become part of how the organizational brand is perceived (de Chernatony et al. 2006). Our personal experiences are merged with past brand-related communications such as advertising and word of mouth, along with previous experiences with the organization. Together, they help form an image of the brand in our minds.

Employees, therefore, have a vital role to play in branding. They are the organizational brand bearers who reinforce the desired brand values in stakeholder interactions. By doing so, they provide stakeholders with an understanding of the functional and emotional values people can expect to experience from the brand (de Chernatony 2002).

This is not to suggest that marketing communications and other external brand-strengthening activities do not have a role to play. Rather, it is important to realize a strong brand also depends on its employees and how well the employees convey brand values through their behaviors. Certain researchers actually suggest employees, in the form of personal encounters with external stakeholders (which in private companies would be called customers), communicate the meaning of the brand in a more powerful way than traditional mass-marketing activities (Sirianni et al. 2013). Branding, therefore, requires a more comprehensive approach in which the brand and its values are communicated through various marketing communications and employee behaviors. Together, these must send a consistent message and jointly reinforce the brand values that the organization aims to represent (Vallaster and de Chernatony 2005).

U. Leijerholt, (✉)
Umeå School of Business, Economics and Statistics, Umeå University, Umeå, Sweden
e-mail: ulrika.leijerholt@umu.se

S. M. Zavattaro (ed.), *Public Branding and Marketing*,
https://doi.org/10.1007/978-3-030-70505-3_7

When a company or an organization makes brand promises in marketing communications – whether in advertising, on its website, or on social media – it must also ensure these promises are upheld in customer interactions with the brand. Indeed, employees have a critical role to play in delivering on brand promises (Punjaisri and Wilson 2007). Failing to do so would undermine the brand, as it may create confusion among external stakeholders (Hatch and Schultz 2001). This is because there is ambiguity regarding what the organization and its brand truly stand for (Hatch and Schultz 2001).

A field of research focusing on the role of employees in branding is internal brand management, also called internal branding. Internal brand management takes an inside-out perspective (Mosley 2007) in which the underlying idea is to support greater alignment between brand values and the values and behaviors of employees (Harris and de Chernatony 2001). Thus, internal brand management aims to "communicate, express and embed brand values among employees" (Sujchaphong et al. 2015, p. 214). This supports employee values and behaviors that can help strengthen the brand. All employees need to be considered in the internal brand management efforts, no matter their function or level within the organization (Saleem and Iglesias 2016). Failing to internally support the brand can not only undermine the brand in the eyes of external stakeholders, but it can also cause internal challenges if employees perceive a misalignment between management rhetoric and employee reality (Hatch and Schultz 2001). Indeed, when internal brand management is mismanaged, it can create internal tensions ultimately posing a risk to the overall brand management process (Dean et al. 2016).

Aim and Structure of This Chapter

With a focus on the internal perspective of branding, this chapter offers some key insights into the management and implementation of branding in the public sector. A central message of the chapter is the need for public sector organizations to rethink some traditional brand approaches. This involves organizations to ensure a brand management that is in line with sector-related considerations to facilitate internal support for the brand.

The chapter is divided into the following parts: after the introduction, the chapter presents an overview of branding as it relates to the context of the public sector and to internal brand management considerations. The next part of the chapter focuses on the employee perspective of branding and presents some key insights on internal brand management in public sector organizations. The purpose is to provide an overarching understanding of the interrelated and multi-level nature of the topic. As such, this part of the chapter presents selected key factors related to the sectoral, organizational, and individual levels that influence the internal brand management of public sector organizations. The chapter ends with a brief conclusion that highlights some key messages of the chapter and some suggestions for future research.

Branding in a Public Context

Branding has traditionally been a tool for companies to promote and sell their products and services. However, in the last decade or so, branding efforts have also been noticeable within the public sector. Many public institutions have developed an organizational mission, vision, and core values, along with branding guidelines. These brand values are then publicized through various social media accounts and other marketing communications. Efforts to manage and communicate a consistent organizational brand can be seen in various public sector organizations, both national (e.g., police authorities and municipalities) and international (e.g., the United Nations and the European Union). Thus, brand logic has permeated the public sector (Leijerholt 2019).

However, the differing natures of the public and private sectors suggest that, at its core, there is a distinction between branding in these two sectors. In the private sector, the survival and success of a company depend on its ability to secure a market for its products and services (Hoggett 2006). This allows companies to make a profit and thus establish a position on the market regardless of whether their offerings provide value to society (e.g., cigarettes) (Hoggett 2006). Thus, companies can ensure their survival if they can find and sustain a market for their products and services – a market often highly promoted through marketing activities.

On the other hand, the long-term survival and legitimacy of public sector organizations depend on the societal values they provide (Hoggett 2006). In other words, public sector organizations are responsible for serving the public interest. Their governance must therefore balance a variety of interests and values as well as secure key public sector values such as equality, democracy, and security (Kallinikos 2006). For example, the police authority must maintain a service-oriented outlook by being a dependable and trustworthy organization for local residents, including crime victims. Yet, it also has the authoritative role of enforcing the law. This role must be conducted with a solid foundation in democratic values to serve the public interest. Failing this, there may be far-reaching consequences; legitimacy and citizen trust are key factors not only for individual organizations but for the entire publicly funded system (Walsh 1994; Wæraas and Byrkjeflot 2012).

This also has key implications for branding. To ensure public sector organizations first and foremost live up to the democratic values they are meant to protect, the brand cannot be the organization's main focus (Leijerholt 2019). For example, whistleblower laws or transparency laws such as the right of information, which exist in various countries, cannot be ignored in an attempt to protect the organization's brand image, nor can public sector organizations prioritize their brand over the needs and well-being of the residents they are mandated to serve.

This does not mean branding should be considered inappropriate and ineffectual within the public sector. Rather, branding can be valuable in the public sector. It can be a means of developing important organizational assets such as trust, credibility, and legitimacy, strengthening relationships with employees, and establishing relationships with local residents and other external stakeholders (Leijerholt et al.

2019a). Thus, branding can have a valuable supporting role in enhancing positive relational outcomes with stakeholders. However, it should be pursued with care.

A key consideration is the role of employees in branding. As such, their perspective is necessary for understanding the nature and suitable implementation of branding in the public sector. If we understand how employees perceive and relate to the organization, then we can begin to understand how branding, particularly internal brand management, can be successfully pursued in public sector organizations.

Internal Brand Management in the Public Sector

Internal brand management is a way for organizations to support employees to identify with, be committed to, and behave according to brand values (Burmann and Zeplin 2005; King and Grace 2012; Piehler et al. 2016). This means employee-brand alignment depends on cognitive, affective, and behavioral outcomes. Employees must not only understand what the brand represents, but they must also develop an emotional relationship to it; this relationship then translates into brand-supporting behaviors.

This cognitive outcome describes "employees' comprehension of brand-related information" (Piehler et al. 2016, p. 1580). Generally referred to as brand understanding (e.g., Piehler 2018) or brand knowledge (e.g., Löhndorf and Diamantopoulos 2014), it is a necessary component of internal brand management; employees must know not only the brand values (Piehler et al. 2016) but also what they mean for them in their daily work (King 2010).

The affective outcome of internal brand management is generally represented by brand commitment (Burmann et al. 2009b; Piehler et al. 2016). Brand commitment can be defined as "the extent of psychological attachment of employees to the brand, which influences their willingness to exert extra effort towards reaching the brand goals" (Burmann and Zeplin 2005, p. 284). However, two other affective outcomes must also be considered: brand identification (Piehler et al. 2016) and brand pride (Helm et al. 2016).

Brand identification refers to "employees' feeling of belonging to the brand" (Piehler et al. 2016, p. 1579) and is often used as a key concept within internal brand management (Burmann and Zeplin 2005; Piehler et al. 2016). From an organizational identity perspective, brand identification has both cognitive and affective components (Johnson et al. 2012). However, the affective dimension seems to be a better predictor of employee commitment (Johnson et al. 2012). Indeed, from a branding perspective, the affective component of brand identification has been linked to employee brand commitment (Piehler et al. 2016). Furthermore, brand pride, or "the pleasure taken in being associated with the (corporate) brand" (Helm et al. 2016, p. 62), has emerged as a concept within internal brand management in recent years.

Thus, research suggests there are three affective components of internal brand management outcomes helping support employees to represent the brand sincerely and successfully through their behaviors. This behavioral outcome is often expressed

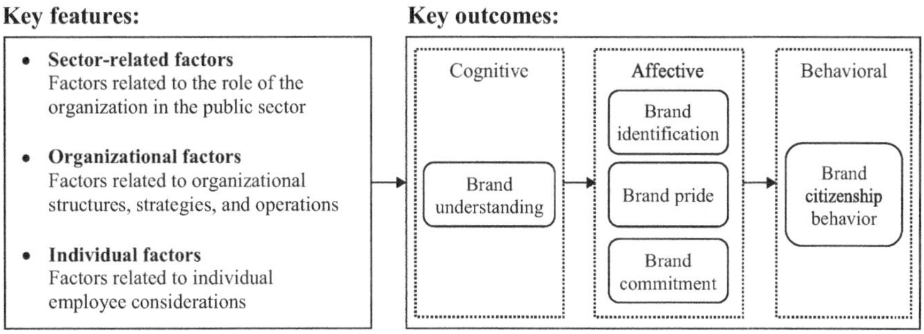

Fig. 1 Key features and outcomes of internal brand management in the public sector

as brand citizenship behavior (e.g., Piehler et al. 2016), which refers to "employee behaviors that are consistent with the brand identity and brand promise such that together they strengthen the brand" (Piehler 2018, p. 220). Brand ambassadors (e.g., Schmidt and Baumgarth 2018; Xiong et al. 2013) is another term used to describe this behavioral outcome.

To support these cognitive, affective, and behavioral outcomes, public sector organizations must take a multi-level approach that considers influences on sectoral, organizational, and individual levels, as illustrated by Fig. 1. It is important to recognize internal brand management should be an integrated and comprehensive organizational effort. Management must ensure the brand comes alive for employees (Mitchell 2002) and that there is an internal culture-brand alignment (de Chernatony and Cottam 2006). Thus, the brand must not solely be communicated through various internal marketing activities but rather lived within the organization.

Sector-Related Factors

Within the public sector, branding is not solely an organizational issue. Rather, public sector organizations should consider their distinct public nature and purpose in their branding. In this chapter, two overarching sector-related factors are presented: (1) identity and values and (2) political leadership. These influence both the brand's implementation within the organization and its core nature (Leijerholt 2019).

Identity and Values

At the core of the brand is what Saleem and Iglesias (2016) refer to as "brand ideologies," or the organizational mission, vision, goals, norms, and shared values. The culture, or the "institutional rules for behavior" (Vallaster and de Chernatony 2006, p. 768), has also been defined as essential; branding and internal organizational culture are interlinked (Hatch and Schultz 2001). Traditional branding literature generally highlights the need for consistency and coherence about identity structure and values (Harris and de Chernatony 2001; Vallaster and de Chernatony 2006). Internal brand management requires a common understanding around which the organization can align. Supporting congruency between culture and brand values is a way to ensure that the brand is presented consistently through various channels (de Chernatony and Cottam 2006).

However, public sector organizations are often rather large and fragmented. They tend to comprise a multitude of – and even conflicting – identities and values (Wæraas 2008; Wæraas and Byrkjeflot 2012; Wæraas and Solbakk 2009). Accordingly, within a public sector organization such as a tax authority, social services, or a municipality, there are multiple departments that may have their own distinct identity, culture, and values.

Furthermore, some may have a more authoritative role and objective, while others may have a more service-oriented role (Wæraas 2008). From a traditional branding perspective, this is often considered very problematic. In an organization lacking consistency and coherence regarding identity structures and values, support of brand-aligned values and behaviors among employees can be a challenge.

However, this does not necessarily seem to be the case within the public sector. In fact, attempting to achieve coherence and consistency about identity may be counterproductive for many public sector organizations. The complexity of the latter, in which many different tasks, service offerings, roles, and internal cultures run in parallel, may require such organizations to accept and embrace their heterogeneous identities (Leijerholt 2019). To ensure the brand is truly representative, public sector organizations may be better off using this diversity in their branding (Wæraas 2008) and trying to find common ground rather than emphasizing the different identities (Dholakia and Acciardo 2014). One suggestion is to keep the brand quite general, which would make it more relatable and better suited to most of the organization (Leijerholt et al. 2019b).

This also means the organizational brand may become rather generic (Leijerholt 2019). Deliberately developing a generic brand runs counter to traditional brand knowledge. Indeed, branding is conventionally considered a means of differentiation and customer preference (Aaker 1996; Knox and Bickerton 2003). Therefore, it would make sense to develop a unique and differentiated brand in the private sector. This enables customers to build relationships with the brand (Fournier 1998) and to choose one brand over another.

However, because the brand is a symbol that represents the functional (what the customer receives) and emotional values (how the customer receives it) that people can expect to experience from the organization (de Chernatony 2002), adopting the same logic for the public sector may be problematic. Should public sector organizations focus on ensuring the uniqueness of their brand, or should they instead provide a certain uniformity in the stakeholder experience, which represents underlying values that align with common values and standards within the overall sector?

At their core, public sector organizations are intended to ensure societal benefit (Wæraas 2008). As such, they must balance a variety of interests and values (Wæraas

2008) and secure the sector's ethos (Kallinikos 2006). In other words, their purpose is not to achieve a unique position on the market compared to other public sector organizations. Rather, attempts to identify and promote a unique, differentiated brand could, arguably, be perceived as democratically problematic given the sector's underlying purpose.

This is not to say that competition does not exist in the public sector. Organizations must, for example, attract businesses and residents to a region or city, secure funding, and recruit employees. Rather, the point is that competition may not be the brand's main role in the public sector. Indeed, research suggests many public sector organizations use branding more as a means of achieving and maintaining a general, positive perception of themselves and their brands (Leijerholt et al. 2019b; Sataøen and Wæraas 2015). This seems especially relevant for public sector organizations in which competition is not a main concern (Leijerholt et al. 2019b; Sataøen and Wæraas 2015). Research also suggests that achieving true brand differentiation may be easier said than done even in public sector environments subject to extensive competition, such as higher education (Chapleo 2007).

Political Leadership

Branding literature often emphasizes the considerable influence of managers on internal brand management (McCoy and Venter 2016; Punjaisri et al. 2009; Vallaster and Lindgreen 2013). This includes management at both the top and middle levels (McCoy and Venter 2016; Punjaisri et al. 2009; Vallaster and Lindgreen 2013). Leadership influences how employees perceive brand meaning (Dean et al. 2016). Leaders can also enhance employee brand identification (Burmann and Zeplin 2005) and shared values (de Chernatony and Segal-Horn 2003). However, management behavior should stem from a "genuine conviction" in brand values rather than a superficial representation (de Chernatony and Segal-Horn 2003) to do so. Thus, management must "walk the walk" to show that they truly believe in the brand values.

From a public sector perspective, the leadership component of internal brand management may be somewhat challenging. Public sector organizations are subject to an additional leadership level (i.e., the political leadership) that influences how the organization is managed (Leijerholt 2019). While traditional brand knowledge highlights the role of managers, the nature of political governance provides a dual role for leaders; political leadership is a representative both for the residents and of the top management of the organization. This dual role complicates an already complex internal brand management (Leijerholt 2019).

Political leadership can provide certain brand-related misalignment between the organization and political management (Leijerholt 2019). For example, this may arise from the sometimes-critical tone taken by politicians toward public sector organizations. Furthermore, the political leadership can be perceived to have vastly different values from the organization's values. Such issues can create a disconnect between the organization and political leadership (Leijerholt 2019). This can be detrimental to the brand and employee-brand relationships. Indeed, research suggests that while employees tend to consider the political leadership somewhat abstract and remote, the leadership also influences employee perceptions of the organization, particularly its values and management (Leijerholt 2019).

Organizational Factors

Public sector organizations must incorporate branding into organizational life. This means ensuring coherent brand messages are communicated to employees (Vallaster and de Chernatony 2005) and considering all touchpoints (Mosley 2007). Key organizational factors communicating the brand to employees include brand strategies, internal communication and marketing, human resources, internal structures and operations, and brand architecture.

Brand Strategies

According to branding literature, organizations must ensure a certain alignment between stakeholder perceptions of the brand: management, employees, and external stakeholders (Hatch and Schultz 2009). For this to occur, careful coordination of the branding effort is needed, in terms of both the messaging and delivery of the brand (King 1991). Consistency in brand communication is generally considered essential (de Chernatony 1999; de Chernatony et al. 2006).

However, public sector organizations may want to approach brand strategies less stringently than traditional advice would suggest (Leijerholt 2019). This may entail a branding strategy based on overarching brand identity and values as well as branding elements such as a common logo and style guides (Leijerholt et al. 2019b). Furthermore, organizations can allow for some internal differences and adjustments about brand values (Leijerholt et al. 2019b). Departments may be allowed to adjust the latter to align them more closely with their specific role and function within the organization (Leijerholt et al. 2019b).

While such an approach could complicate internal brand management in private companies, it may strengthen both the brand and employee-brand relationships in public sector organizations (Leijerholt et al. 2019b). This can be explained through certain sector-specific features that provide an aligning force within the organization: the purpose and values of the public sector and pro-social motivations among public sector employees (Leijerholt 2019). The latter is also known as public service motivation (PSM) and refers to "motives and action in the public domain that are intended to do good for others and shape the well-being of society" (Perry and Hondeghem 2008, p. 3). Thus, PSM is often linked to pro-social and altruistic motivations among employees but particularly relates to "specific dispositions and values arising from public institutions and missions" (Perry et al. 2010, p. 682).

This does not mean that public sector organizations should adopt an "anything goes" attitude with branding but rather they should reconsider some traditional approaches regarding their branding strategy. Indeed, a less stringent approach may be especially well-suited to public sector organizations and prevent branding efforts from infringing on citizen and employee rights or undermining organizational legitimacy (Leijerholt 2019).

Internal Communication and Marketing

To support their desired internal brand management outcomes, organizations must provide employees with information and knowledge (King 2010; Punjaisri et al. 2009). Appropriate brand-oriented communication can foster employee brand commitment (Burmann and König 2011) and strong organizational allegiance (King and Grace 2008). Newsletters, storytelling, brand books, blogs, and brand ambassadors are some examples of useful communication channels (du Preez et al. 2017). However, it is also important to acknowledge employees do not read and understand much of the information provided by the organization (Vallaster and Lindgreen 2013). Organizations should therefore ensure internal communication is approached through dialogue (Whisman 2009) and conducted in a two-way, vertical flow (Vallaster and Lindgreen 2013). This facilitates greater employee responsibility and receptiveness regarding internal brand-related information (Vallaster and Lindgreen 2013). Furthermore, the effectiveness of the communication also depends on the values being communicated. Employees must perceive the brand message as authentic and aligned with their own values (Baker et al. 2014).

Internal communication and marketing are equally important in the public and private sectors. In the public sector, useful internal and external channels for communicating the brand to employees include meetings, the intranet, PowerPoint presentations, and the organizational website (Leijerholt et al. 2019b). It is not surprising meetings are highlighted, as they align with suggestions that organizations need to ensure two-way communication (Vallaster and Lindgreen 2013) based on dialogue (Whisman 2009). Furthermore, it is imperative the values communicated by the organization are not only congruent with employee values but also perceived as valuable among employees (Leijerholt et al. 2020). As such, PSM among employees should be integrated in internal communication and marketing as a useful way to support successful internal brand management in the public sector.

Human Resources

Responsibilities of human resources (HR), which include training, support, and recruitment, are key brand-related activities. This department should therefore be considered a means of influencing desired internal brand management outcomes among employees, about both brand commitment and brand-related behaviors (Chiang et al. 2018). While there is a tendency for organizations to keep the functions of HR and communications separate, internal brand management benefits from their coordination (Aurand et al. 2005; Punjaisri et al. 2009). Coordination has been shown to affect employee attitudes and behaviors in delivering the brand promise (Punjaisri et al. 2009).

While training and support can provide a strong brand message to employees both continuously and as a means of introduction to the organization (King and So 2015; Punjaisri et al. 2009), recruitment can ensure the organization hires the right staff. This means recruiting employees whose values align with the organization's (Burmann and Zeplin 2005; de Chernatony et al. 2003) and who bring the right attitudes to the workplace (King and Grace 2012). Merely hiring people for their ability to perform the job is insufficient for building a strong brand. Instead, brand considerations should be integrated into the recruitment process by considering employee values and motivations in relation to the organizational brand.

The role of HR also includes ensuring a good internal working environment that enhances the implementation of internal brand management. The work environment is commonly considered to be closely aligned with the effectiveness of internal brand management (Punjaisri and Wilson 2007, 2011). Indeed, employee experience with the brand is strongly linked to employee brand commitment (Kimpakorn and Tocquer 2009). Work environment factors such as employee trust in supervisors, perceptions of organizational values, job satisfaction, job characteristics, extrinsic rewards, promotion satisfaction, two-way communication, and leadership have all been shown to affect employee commitment (Kimpakorn and Tocquer 2009), as have softer, relational aspects such as trust, respect, consideration, and open communication (King and Grace 2012).

From a public sector perspective, it has long been established that coordination between HR and internal communications and marketing is advantageous (Ewing and Caruana 1999). There is currently no reason to believe the role of HR in branding would significantly differ between the public and private sectors. For example, the need for a supportive organizational culture has been deemed necessary in branding within the public sector (Chapleo 2015). Furthermore, the values of job applicants, as manifested through PSM, should be considered in recruitment situations (Leijerholt et al. 2020).

However, there may also be distinct HR-related issues regarding the internal brand management of public sector organizations. For example, in a sector often characterized by complexity in the form of fragmented organizations with multiple identities and conflicting values (Wæraas 2008; Wæraas and Byrkjeflot 2012; Wæraas and Solbakk 2009), organizations must support greater integration and connections among departments (Leijerholt et al. 2019b). Such efforts can include common policies and goal management processes (Leijerholt et al. 2019b).

Internal Structures and Operations

The rules and resources available to employees in their work, or the corporate structures (Giddens 1979), allow workers to gain brand resources and knowledge. These structures, which include reward systems, infrastructure, budgeting, and control (Burmann et al. 2009b), can affect the brand message received by employees and therefore their behaviors within the organization. Organizations must recognize branding requires careful coordination throughout the organization, including tactical and operational considerations (McCoy and Venter 2016); the brand must be integrated in the everyday life of the organization (Burmann et al. 2009a; Vallaster and de Chernatony 2006).

From a public sector perspective, the past few decades have brought significant changes to corporate structures. Many of these can be attributed to the application of New Public Management (NPM) principles in the sector. The term relates to "a set of assumptions and value statements about how public sector organizations should be designed, organized, managed and how, in a quasi-business manner, they should function" (Diefenbach 2009, p. 893). From a branding perspective, NPM has arguably had a significant impact on the everyday organizational life and therefore the brand message received by employees. Studies have highlighted certain consequences of NPM on internal environments in a public context (Mawdsley and Lewis 2017) and PSM among employees (Bellé and Ongaro 2014). For example, typical NPM features such as "lean working practices, oppressive productivity monitoring systems, the public disclosure of performance indicators, and unachievable targets" (Mawdsley and Lewis 2017, p. 321) seem to complicate the internal environment. These corporate structures negatively affect weaker employee groups (Mawdsley and Lewis 2017). The link between NPM and these consequences has been noted: "NPM makes for a hostile working environment for those with physical or psychological disabilities and long-term illness" (Mawdsley and Lewis 2017, 322).

It has also been suggested that NPM may have negative consequences on PSM in public sector organizations (Bellé and Ongaro 2014). While the claims are tentative due to a lack of comparative studies prior to the implementation of NPM, it has been suggested that public sector organizations with greater NPM reforms show significantly lower levels of PSM than those that have experienced more limited NPM implementation (Bellé and Ongaro 2014). Thus, NPM may have a depleting effect on PSM. In turn, this may harm the performance of public sector organizations (Bellé and Ongaro 2014).

In summary, the influence of NPM on public sector organizations may have considerable consequences from an internal brand management perspective. Internal structures and operations play a part in the communication of the brand message to employees; employees' experiences affect their interpretation of the brand meaning (Dean et al. 2016). In a context in which PSM influences employee pro-social motivations (Perry and Hondeghem 2008; Perry et al. 2010), it is conceivable that employee experiences that undermine such motivations may also transform employees' relationships with the brand. To support favorable internal brand management outcomes among employees, public sector organizations must carefully consider how corporate structures facilitate the desired brand meaning. If these structures do not enhance brand values, then they in fact weaken the brand.

Brand Architecture

Brand architecture refers to "an organizing structure of the brand portfolio that specifies brand roles and the nature of relationships between brands" (Aaker and Joachimsthaler 2000, p. 8). It is a spectrum ranging from a branded house strategy to a house of brands strategy (Aaker and Joachimsthaler 2000). A branded house strategy features various products and services marketed under a single parent brand (e.g., Apple, which has several products incorporated under the Apple brand), whereas a house of brands strategy entails marketing products under individual, stand-alone brands (e.g., Procter & Gamble, which owns a number of distinct brands such as Pampers, Ariel, and Gillette). Public sector organizations tend to pursue a branded house approach, as can be seen in fields such as higher education (Dholakia and Acciardo 2014) and health care (Leijerholt et al. 2019b).

Traditionally, brand architecture is not a concept that has been related to internal brand management. In the public sector, however, this must be reconsidered due to the tendency for branding efforts to be pursued in multiple ways within an organization at both the organizational and departmental levels (Leijerholt et al. 2019b). However, this dynamic does not seem to create competing branding efforts; in fact, it may harmonize them (Leijerholt et al. 2019b). In effect, departmental branding efforts may reinforce the organizational brand (Leijerholt et al. 2019b). This requires that a certain alignment of support exists within the organization. In particular, four factors appear to enhance departmental branding efforts to complement and strengthen the organizational brand (Leijerholt et al. 2019a):

- The organizational core values statement, which helps unify the organization if it is anchored in the overarching purpose and values of the public sector
- Internal brand structures in the form of a common logotype, website, and communications department that help tie together individual branding efforts
- Connections within the organization that help integrate an otherwise fragmented entity through components such as common policies and goal management processes
- Public sector values, which should serve as a foundation for the organizational brand

Tapping into the purpose and values of the public sector in the definition and management of the brand seems particularly important in organizations where PSM is prevalent among employees.

Thus, the nature of the public sector appears to enable organizations to pursue a branded house approach while also allowing for individual branding efforts by different departments. Such a multiple brand effort can help organizations foster a stronger brand and greater employee brand commitment (Leijerholt 2019).

Individual Factors

Internal brand management is not solely a sectoral and organizational issue. Rather, it should also incorporate factors on an individual employee level. Two factors have especially been shown to play a central role: public service motivation and positive organizational practices (i.e., colleagueship).

Public Service Motivation

Within the public sector, PSM is an important concept – more than in the private sector (Perry and Hondeghem 2008). That does not mean it is exclusive to the public sector but rather it does not exist to the same extent nor in the same form (Perry and Hondeghem 2008). Thus, while it is not the only relevant factor because issues such as compensation and secure employment also play a part (O'Riordan 2013), the role of PSM and the power of the public purpose and ethos to attract certain people should not be underestimated:

> The notion that people are motivated to work in the public service as a result of altruism, a desire to serve, or a wish to have an impact on society is a long- standing one. It is closely associated with the idea of public service ethos, which is rooted in an understanding that the public service is different from the private sector, both because of the tasks it performs and the behaviors it expects of its employees. (O'Riordan 2013, p. 6)

Indeed, public sector purpose and values can be a strong motivator for employees. From a branding perspective, this has certain implications for both the development of the organizational brand and internal brand management.

Research shows PSM is factor in the emotional employee-brand connection of employee brand commitment (Leijerholt et al., 2020; Leijerholt et al. 2019b). When public sector organizations can tap into employee PSM and integrate it into brand values, they experience positive brand-related results. Indeed, these brand values seem to achieve a greater emotional connection and serve to unite the organization (Leijerholt et al. 2019b). Public sector organizations should therefore develop brand values with employee PSM in mind; there is an opportunity to connect brand values to the purpose and values of the public sector (Leijerholt 2019).

However, this also means the brand and its values should not be developed solely based on management aspirations for the organization. Instead, branding should incorporate the galvanizing force PSM represents among employees. To this end, organizations must integrate the purpose and ethos of the public sector into brand values (Leijerholt 2019). Therefore, branding in the public sector is not only an organizational issue (i.e., what we as an organization want the brand to stand for), but it also concerns the organization's role from a sectoral perspective. Thus, organizations can support greater employee brand commitment and arguably build a brand aligned with their public purpose.

Positive Organizational Practices

It is important to recognize employee perceptions of brand meaning are highly influenced by peers, whether managers, external stakeholders, or other employees (Dean et al. 2016). Employees learn about the brand through experiences with it, which helps them determine what the brand truly represents; these experiences are more powerful than other types of brand information (Dean et al. 2016). This is particularly true where colleagues are concerned. How employees are treated and how employees treat others with whom they work affect their brand experiences (Leijerholt 2019; Leijerholt et al. 2020). The relational aspect of internal brand management therefore deserves careful consideration.

An interesting facet of this relational aspect is positive organizational practices. The latter, which include treating people with respect, providing emotional support, inspiring each other at work, and caring about each other, have been found to support desired brand outcomes within the organization (Leijerholt et al., 2020). When employees show "colleagueship" and treat their co-workers well, these behaviors transfer to employees' relationships with the brand (Leijerholt et al., 2020). Positive organizational practices directly affect internal brand management outcomes in the public sector (Leijerholt et al. 2020).

Positive organizational practices are especially interesting from a public sector perspective due to their connection with PSM. The latter has been found to directly influence positive organizational practices and, in turn, indirectly influence affective outcomes of internal brand management (Leijerholt et al. 2020). This relationship provides an interesting opportunity for public sector organizations to strengthen their brands. Such organizations should consider the role and value of employee PSM, as it not only affects behaviors within the organization but also provides a foundation for affective employee-brand relationships.

Conclusion

Branding can be valuable for public sector organizations. It can enable increased organizational legitimacy and trust as well as improve their relationships with external and internal stakeholders (Leijerholt et al. 2019a). That said, branding also runs the risk of undermining the core purpose and values of the public sector if mismanaged. Therefore, branding in the public sector must be pursued with care.

Public sector organizations should recognize the need to adopt a multi-level approach to internal brand management that consider factors on the sectoral, organizational, and individual levels. This chapter presents some key factors, including identity and values, political leadership, brand strategies, internal communications and marketing, human resources, internal structures and operations, brand architecture, PSM, and positive organizational practices. These factors can help to ensure that public sector organizations support a brand implementation that are in line with its public nature. Further, it will help to support affective internal brand management outcomes among employees, a necessity for a strong organizational brand.

However, to do so, it is important to acknowledge branding in the public sector requires a somewhat different approach. In fact, due to the distinct nature of the sector, it may be more suitable to discuss specialized "public sector branding" rather than "branding in the public sector" (Leijerholt 2019). Indeed, the implication, and a central message of this chapter, is applying the same type of branding used in the private sector to the public sector in a cut-and-paste manner risks not only harming the brands of public sector organizations and their internal brand management but also possibly undermining the ethos and values of the sector as a whole. Instead, public sector organizations must approach branding in a manner that is appropriate for their role.

This would entail public sector organizations to consider the organizational brand as an extension of the overarching purpose, values, and role of the public sector. By careful reflection on the development and implementation of the public sector brand, the branding efforts can represent the organizational role in society rather than being solely an organizational concern. This would foster branding efforts that are more aligned with the core purposes of the sector. As such, branding can provide a way to strengthen relationships with organizational stakeholders without compromising the societal and democratic values expected from public institutions.

Suggestions for Future Research

To ensure public sector branding efforts provide value to the organization and all its stakeholders, given its distinct nature, further insights are needed into its nature and management. Such insights would provide policy makers and managers with a greater foundation to develop guidelines, policies, strategies, and operations that support a brand management that is appropriate for the sector. However, it should be recognized that such studies may potentially want to include some critical perspectives. This is to ensure that the research provides a full understanding of public sector branding, both its advantages for and its potential harmful effects on society, organizations, and stakeholders.

From an internal brand management perspective, there are several interesting avenues for future research. First, while this chapter has identified some notable features affecting the development and implementation of internal brand management in the public sector, there may also be other sectoral, organizational, and individual factors that may affect the internal brand management in the public sector. Future research may, therefore, identify other key factors that influences the internal brand management efforts in public sector organizations. This would enable public sector organizations to pursue an internal brand management that is increasingly tailored to the sector to ensure the desired outcomes.

Second, we need to understand how failure to consider these factors affect organizations and its stakeholders. For example, PSM has been identified to have a noticeable role for internal brand management in the public sector. Indeed, it is suggested that PSM is potentially a great asset to the internal brand management efforts of public sector organizations (Leijerholt et al. 2019b). However, the public sector reform of NPM seems to be depleting employee PSM (Bellé and Ongaro 2014). As such, the internal brand implications of such a shift are of key interest for future studies; if the employee motivation to do good for others and society can serve as a great asset for internal brand management efforts, how are organizations, and its relationship with external stakeholders, affected if such motivations are undermined among employees.

References

Aaker, D. (1996). Measuring brand equity across products and markets. *California Management Review, 38*(3), 102–120.

Aaker, D. A., & Joachimsthaler, E. (2000). The brand relationship spectrum: The key to the brand architecture challenge. *California Management Review, 42*(4), 8–22.

Aurand, T. W., Gorchels, L., & Bishop, T. R. (2005). Human resource management's role in internal branding: An opportunity for cross-functional brand message synergy. *Journal of Product & Brand Management, 14*(3), 163–169.

Baker, T. L., Rapp, A., Meyer, T., & Mullins, R. (2014). The role of brand communications on front line service employee beliefs, behaviors, and performance. *Journal of the Academy of Marketing Science, 42*, 642–657.

Bellé, N., & Ongaro, E. (2014). NPM, administrative reforms and public service motivation: Improving the dialogue between research agendas. *International Review of Administrative Science, 80*(2), 382–400.

Burmann, C., Jost-Benz, M., & Riley, N. (2009a). Towards an identity-based brand equity model. *Journal of Business Research, 62*, 390–397.

Burmann, C., & König, V. (2011). Does internal brand management really drive brand commitment in shared-service call centers? *Journal of Brand Management, 18*, 374–393.

Burmann, C., & Zeplin, S. (2005). Building brand commitment: A behavioral approach to internal brand management. *Journal of Brand Management, 12*, 279–300.

Burmann, C., Zeplin, S., & Riley, N. (2009b). Key determinants of internal brand management success: An exploratory empirical analysis. *Journal of Brand Management, 16*, 264–284.

Chapleo, C. (2007). Barriers to brand building in UK universities. *Journal of Nonprofit and Voluntary Sector Marketing, 12*, 23–32.

Chapleo, C. (2015). Brand 'infrastructure' in nonprofit organizations: Challenges to successful brand building? *Journal of Marketing Communications, 21*(3), 199–209.

Chiang, H.-H., Han, T.-S., & McConville, D. (2018). The attitudinal and behavioral impact of brand-centered human resource management: Employee and customer effects. *International Journal of Contemporary Hospitality Management, 30*(2), 939–960.

de Chernatony, L. (1999). Brand management through narrowing the gap between brand identity and brand reputation. *Journal of Marketing Management, 15*, 157–179.
de Chernatony, L. (2002). Would a brand smell any sweeter by a corporate name? *Corporate Reputation Review, 5*(2–3), 114–132.
de Chernatony, L., & Cottam, S. (2006). Internal brand factors driving successful financial services brands. *European Journal of Marketing, 40*(5/6), 611–633.
de Chernatony, L., Cottam, S., & Segal-Horn, S. (2006). Communicating services brands' values internally and externally. *The Service Industries Journal, 26*(8), 819–836.
de Chernatony, L., Drury, S., & Segal-Horn, S. (2003). Building a services brand: Stages, people and orientations. *The Service Industries Journal, 23*(3), 1–21.
de Chernatony, L., & Segal-Horn, S. (2003). The criteria for successful services brands. *European Journal of Marketing, 37*(7/8), 1095–1118.
Dean, D., Arroyo-Gamez, R. E., Punjaisri, K., & Pich, C. (2016). Internal brand co-creation: The experiential brand meaning cycle in higher education. *Journal of Business Research, 69*(8), 3041–3048.
Dholakia, R. R., & Acciardo, L. A. (2014). Branding a state university: Doing it right. *Journal of Marketing for Higher Education, 24*(1), 144–163.
Diefenbach, T. (2009). New public management in public sector organizations: The dark sides of 'managerialistic enlightenment. *Public Administration, 87*(4), 892–909.
du Preez, R., Benedixen, M., & Abratt, R. (2017). The behavioral consequences of internal brand management among frontline employees. *Journal of Product & Brand Management, 26*(3), 251–261.
Ewing, M. T., & Caruana, A. (1999). An internal marketing approach to public sector management: The marketing and human resources interface. *International Journal of Public Sector Management, 12*(1), 17–29.
Fournier, S. (1998). Consumers and their brands: Developing relationship theory in consumer research. *Journal of Consumer Research, 24*(4), 343–353.
Giddens, A. (1979). Central problems in social theory. Macmillan.
Harris, F., & de Chernatony, L. (2001). Corporate branding and corporate brand performance. *European Journal of Marketing, 35*(3/4), 441–456.
Hatch, M. J., & Schultz, M. (2001). Are the strategic stars aligned for your corporate brand? *Harvard Business Review*, February, 128–134.
Hatch, M. J., & Schultz, M. (2009). Of bricks and brands: From corporate to enterprise branding. *Organizational Dynamics, 38*(2), 117–130.
Helm, S. V., Renk, U., & Mishra, A. (2016). Exploring the impact of employees' self- concept, brand identification and brand pride on brand citizenship behaviors. *European Journal of Marketing, 50*(1/2), 58–77.
Hoggett, P. (2006). Conflict, ambivalence, and the contested purpose of public organizations. *Human Relations, 59*(2), 175–194.
Johnson, M. D., Morgenson, F. P., & Hekman, D. R. (2012). Cognitive and affective identification: Exploring the links between different forms of social identification and personality with work attitudes and behavior. *Journal of Organizational Behavior, 34*(3), 1142–1167.
Kallinikos, J. (2006). The institution of bureaucracy: Administration, pluralism, democracy. *Economy and Society, 35*(4), 611–627.
Kimpakorn, N., & Tocquer, G. (2009). Employees' commitment to brands in the service sector: Luxury hotel chains in Thailand. *Journal of Brand Management, 16*, 532–544.
King, C. (2010). "One size doesn't fit all": Tourism and hospitality employees' response to internal brand management. *International Journal of Contemporary Hospitality Management, 22*(4), 517–534.
King, C., & Grace, D. (2008). Internal branding: Exploring the employee's perspective. *Journal of Brand Management, 15*, 358–372.
King, C., & Grace, D. (2012). Examining the antecedents of positive employee brand-related attitudes and behaviors. *European Journal of Marketing, 46*(3/4), 469–488.

King, C., & So, K. K. F. (2015). Enhancing hotel employees' brand understanding and brand-building behavior in China. *Journal of Hospitality & Tourism Research, 39*(4), 492–516.
King, S. (1991). Brand building in the 1990s. *Journal of Consumer Marketing, 8*(4), 43–52.
Knox, S., & Bickerton, D. (2003). The six conventions of corporate branding. *European Journal of Marketing, 37*(7/8), 998–1016.
Leijerholt, U. (2019). Public sector branding: An internal brand management perspective (Doctoral Dissertation No. 104; Studier i Företagsekonomi. Serie B, p. 86). Umeå University.
Leijerholt, U., Biedenbach, G., & Hultén, P. (2019a). Branding in the public sector: A systematic literature review and directions for future research. *Journal of Brand Management, 26*, 126–140.
Leijerholt, U., Chapleo, C., & O'Sullivan, H. (2019b). A brand within a brand: An integrated understanding of internal brand management and brand architecture in the public sector. *Journal of Brand Management, 26*, 277–290.
Leijerholt, U., Biedenbach, G., & Hultén, P. (2020). Internal brand management in the public sector: the effects of internal communication, organizational practices, and PSM on employees' brand perceptions. *Public Management Review*. https://doi.org/10.1080/14719037.2020.1834607.
Löhndorf, B., & Diamantopoulos, A. (2014). Internal branding: Social identity and social exchange perspectives on turning employees into brand champions. *Journal of Service Research, 17*(3), 310–325.
Mawdsley, H., & Lewis, D. (2017). Lean and mean: How NPM facilitates the bullying of UK employees with long-term health conditions. *Public Money & Management, 37*(5), 317–324.
McCoy, S., & Venter, P. (2016). Corporate brands as catalysts in times of change: Lessons from a south African bank. *Journal of Brand Management, 23*, 38–54.
Mitchell, C. (2002). Selling the brand inside. *Harvard Business Review, 80*(1), 99–101.
Mosley, R. W. (2007). Customer experience, organisational culture and the employer brand. *Journal of Brand Management, 15*(2), 123–134.
O'Riordan, J. (2013). *Public service motivation (State of the Public Service Series)*. Institute of Public Administration.
Perry, J. L., & Hondeghem, A. (2008). Building theory and empirical evidence about public service motivation. *International Public Management Journal, 11*(1), 3–12.
Perry, J. L., Hondeghem, A., & Wise, L. R. (2010). Revisiting the motivational bases of public service: Twenty years of research and an agenda for the future. *Public Administration Review, 70*(5), 681–690.
Piehler, R. (2018). Employees' brand understanding, brand commitment, and brand citizenship behavior: A closer look at the relationships among construct dimensions. *Journal of Brand Management, 25*(3), 217–234.
Piehler, R., King, C., Burmann, C., & Xiong, L. (2016). The importance of employee brand understanding, brand identification, and brand commitment in realizing brand citizenship behavior. *European Journal of Marketing, 50*(9/10), 1575–1601.
Punjaisri, K., & Wilson, A. (2007). The role of internal branding in the delivery of employee brand promise. *Journal of Brand Management, 15*, 57–70.
Punjaisri, K., & Wilson, A. (2011). Internal branding process: Key mechanisms, outcomes and moderating factors. *European Journal of Marketing, 45*(9/10), 1521–1537.
Punjaisri, K., Wilson, A., & Evanschitzky, H. (2009). Internal branding to influence employees' brand promise delivery: A case study in Thailand. *Journal of Service Management, 20*(5), 561–579.
Saleem, F. Z., & Iglesias, O. (2016). Mapping the domain of the fragmented field of internal branding. *Journal of Product & Brand Management, 25*(1), 43–57.
Sataøen, H. L., & Wæraas, A. (2015). Branding without unique brands: Managing similarity and difference in a public sector context. *Public Management Review, 17*(3), 443–461.

Schmidt, H. J., & Baumgarth, C. (2018). Strengthening internal brand equity with brand ambassador programs: Development and testing of a success factor model. *Journal of Brand Management, 25*(3), 250–265.
Sirianni, N. J., Bitner, M. J., & Brown, S. W. (2013). Branded service encounters: Strategically aligning employee behavior with the brand positioning. *Journal of Marketing, 77*(6), 108–123.
Sujchaphong, N., Nguyen, B., & Melewar, T. C. (2015). Internal branding in universities and the lessons learnt from the past: The significance of employee brand support and transformational leadership. *Journal of Marketing for Higher Education, 25*(2), 204–237.
Vallaster, C., & de Chernatony, L. (2005). Internationalisation of services brands: The role of leadership during the internal brand building process. *Journal of Marketing Management, 21*(1–2), 181–203.
Vallaster, C., & de Chernatony, L. (2006). Internal brand building and structuration: The role of leadership. *European Journal of Marketing, 40*(7/8), 761–784.
Vallaster, C., & Lindgreen, A. (2013). The role of social interactions in building internal corporate brands: Implications for sustainability. *Journal of World Business, 48*(3), 297–310.
Wæraas, A. (2008). Can public sector organizations be coherent corporate brands? *Marketing Theory, 8*(2), 205–221.
Wæraas, A., & Byrkjeflot, H. (2012). Public sector organizations and reputation management: Five problems. *International Public Management Journal, 15*(2), 186–206.
Wæraas, A., & Solbakk, M. N. (2009). Defining the essence of a university: Lessons from higher education branding. *Higher Education, 57*, 449–462.
Walsh, K. (1994). Marketing and public sector management. *European Journal of Marketing, 28*(3), 63–71.
Whisman, R. (2009). Internal branding: A university's most valuable intangible asset. *Journal of Product & Brand Management, 18*(5), 367–370.
Xiong, L., King, C., & Piehler, R. (2013). "That's not my job": Exploring the employee perspective in the development of brand ambassadors. *International Journal of Hospitality Management, 35*, 348–359.

Branding the Public Service of Canada: A Model Worthy of Emulation or Unfulfilled Promise?

Tim A. Mau

Introduction

Corporations have long been concerned with developing a recognizable brand, preferably one that is global in reach, as a means of creating a powerful identity that can enhance the trust and loyalty of both employees and customers and ultimately improve economic performance. Like many other private sector business principles adopted in the public sector as part of the new public management (NPM) movement, the use of brands and branding processes is now ubiquitous in the public sector. This includes both politics and the governance process more broadly where branding has been practiced with increasing regularity around the globe (Eshuis and Klijn 2012; Temporal 2015).

Branding is clearly a multifaceted phenomenon. In the private sector, branding typically involves specific goods or the corporation itself. While public goods, such as a bridge, are sometimes branded, this is usually neither as important nor as prevalent as in the private sector due to the lack of competition. Different forms of branding tend to emerge in the public sector: politicians and political parties are marketed as brands; nations, cities, and other geographic regions are subjected to "place" or "location" branding (Nimijean 2006); government as a whole and its various departments, ministries, and other public sector organizations can be branded like corporations (Marland et al. 2017); and, finally, a variety of government policies and services are often branded (Marland 2016).

Government branding is the focus of this chapter. More specifically, it explores the Canadian federal government's effort to emulate leading private sector organizations attempting to brand themselves as employers of choice. The impetus for doing so is obvious. Organizations are in the midst of a global war for talent, one for which the government is not immune (Glenn 2012). For many years, successive clerks of

T. A. Mau (✉)
Department of Political Science, University of Guelph, Guelph N1G 2W1, ON, Canada
e-mail: tmau@uoguelph.ca

S. M. Zavattaro (ed.), *Public Branding and Marketing*,
https://doi.org/10.1007/978-3-030-70505-3_8

the Privy Council, the head public servant in the Canadian federal government, identified the critical importance of recruiting and retaining as well as identifying and developing public service employees, particularly those who are the "best" and "brightest" and would comprise the leadership cadre of the future, in their annual reports to the prime minister on the status of the public service. While talk is cheap, there was a palpable shift from rhetoric to action in 2007, when the former Canada Public Service Agency[1] formally spearheaded a plan to brand the public service (CPSA 2007). Notably, Temporal (2015, p. 249) characterized this particular initiative as "the most interesting example from a branding perspective" of the many countries that have been "engaged in transformational processes to enhance the efficiency, effectiveness and image of their public service."

The purpose of this chapter is to examine the concept of branding as it applies to the public sector and, more specifically, to provide a critical evaluation the Canadian federal government's public sector branding exercise. Has Canada truly been the vanguard with respect to branding as an employer of choice? The chapter begins by outlining the nature of the human capital challenge in the federal public service, establishing the difficulty of recruiting and retaining top talent. It then examines the concepts of branding and employer of choice as they apply to the public sector, including the perceived advantages and ongoing limitations associated with such strategies. More specifically, it addresses how branding the public sector as an employer of choice can potentially overcome the talent management obstacles it faces. The final two sections of the chapter provide a critical assessment of the impetus for and implementation of the 2007 initiative to brand the Public Service of Canada. It is argued that despite good intentions and the discernible benefits derived from branding the public service, the initiative, like so many government reform efforts, has been one of unfulfilled promise. The government has yet to deliver a compelling public sector brand that has coalesced the views of current and prospective public servants and citizens alike around the name, values, and identity of the Canadian public service.

Human Resources Challenges in the Public Sectosr

As was the trend in many developed countries, Canada began to implement a number of administrative reforms in the early-to-mid-1980s that saw the gradual transformation of the public sector under the auspices of NPM to one that could be characterized as post-bureaucratic. As has been well documented elsewhere, NPM ushered in a series of administrative reforms that have had a profound impact on Canadian public administration, including, among other things, a retrenchment of the state through both expenditure and personnel reductions; the introduction of

[1] As a result of a strategic review of the federal government's HRM function in 2008, the Canada Public Service Agency was disbanded and replaced by the Office of the Chief Human Resources Officer, a branch housed within the Treasury Board Secretariat.

mission, vision, and value statements; and a new emphasis on the bottom line, competition, performance management, leadership, and service quality (Dwivedi et al. 2009; Kernaghan et al. 2000).

Of particular concern for this chapter are the implications of NPM, with its emphasis on private sector business principles, on the various human resource management (HRM) practices in the public sector. As Barker and Mau (2017, p. 253) noted, "the government shifted from a traditional model of personnel administration that was highly centralized and rule-based as per the tenets of the Weberian rational-legal bureaucracy to one that was more in line with the market-based orientation associated with the NPM movement." In many instances, these NPM-inspired policy choices served to undermine the federal government's ability to recruit and retain the competent cadre of public servants required to provide services efficiently and effectively. This is important – and problematic – because, as Lavigna and Hays (2004, p. 237) were apt to point out, "good government requires good people."

First, the desire to make the state less intrusive by reducing the size and scope of government meant there were periods of contraction in the number of public servants working in the core public administration. Downsizing, or "rightsizing" as it was often called, occurred in the mid-1980s under the Brian Mulroney government and then a decade later as part of the Program Review exercise initiated by the Jean Chrétien Liberals. Although the size of the core public administration did gradually increase during the first decade of the new millennium, after the Stephen Harper-led Conservative Party won a majority government in the 2011 general election, the public service underwent further reductions.

These personnel reductions, in combination with protracted wage and hiring freezes, meant there were fewer public servants to carry out the business of government. Those who remained in the public service tended to be older, knowledge-based employees who had to contend with increased workloads and high levels of stress (Lowe 2001). Not surprisingly, the morale of public servants waned. It was not simply that there had been cutbacks in public sector employment leading to challenging working conditions that caused discontent; public servants were disturbed these reductions occurred in the mid-1980s while the Brian Mulroney government introduced the PS 2000 renewal initiative whereby these human resources were recognized as "its most important asset" (Canada 1990, p. 16). From their perspective, there was a clear disconnect between what the government of the day was pronouncing publicly, on the one hand, and the HRM policies it was pursuing, on the other hand (McIntosh 1991).

As a result of these personnel reductions and the increased reliance on the use of temporary contract workers and public-private partnerships for service delivery, the sanctity of the public service bargain that had evolved over time had been severely compromised (Savoie 2003). In effect, the notion of a career service, so central to the evolution of the Canadian bureaucracy, was fundamentally transformed, which resulted in a reduced sense of loyalty among public servants (Lindquist 2006, p. 26).

Clearly, then, several policies pursued as part of the NPM agenda negatively affected the federal government's ability to recruit and retain public service employees. However, there were other HRM dilemmas – ones faced by most of the international community – that also posed challenges for the Canadian federal government: (1) an aging workforce, (2) a paucity of leaders within the ranks, (3) heightened competition from the private sector for top talent, and (4) negative perceptions of the public sector (Lavigna and Hays 2004, p. 237–38). In the first instance, the average age of Canadian federal public servants in 2018 was 44.6 years of age, while the average age of executives was 50.2 years. These are slight improvements from 2014 when the corresponding figures were 44.9 and 50.4 years of age, respectively. However, Canadian federal public servants are significantly older than they were in 1983 when the average age of a public servant was 39.3 years and the average age of executives was 48.1 years (TBS 2015, 2019).

With respect to the second HR dilemma, the retrenchment of public sector employment in conjunction with wage and hiring freezes, as noted above, contributed to a dearth of public sector employees who were equipped to lead. Moreover, the changing nature of public sector employment, with respect to both the shift from clerical and operational to more technical and knowledge-based employment[2] and the increased reliance on alternative service delivery, which meant those who were interested in pursuing careers that contributed to the public interest had options outside of the public service (i.e., in a non-profit entity or perhaps a private company engaged in a public-private partnership with government) to satisfy that aspiration, has resulted in even more intense competition with the private sector to recruit the best and brightest employees. The impact of these HRM dilemmas has been significant. Although tens of thousands of Canadians seek employment opportunities with the federal public service every year,[3] the desire is not simply to fill government positions but rather to attract the best and brightest, which is difficult when grappling with these myriad issues.

Jocelyne Bourgon, a former clerk of the Privy Council, was the first to acknowledge this growing HRM conundrum when she wrote of a "quiet crisis" in the fourth annual report to the prime minister on the status of the public service. She noted that 70% of the executive group members were eligible to retire within 8 years, which constituted a potential loss of significant experience and expertise. More troubling was the fact that so few people recognized the problem or were willing to do anything about it. She went on to say the public sector was going to find it "increasingly difficult to retain, motivate and attract people essential to the work of the Public Service over the coming decades" (Clerk 1997, p. 37). Part of her response was to launch *La Relève* (awakening), a public sector renewal initiative designed to overcome the failure of PS 2000.

Although this public sector reform was as flawed as its predecessor, Bourgon had nonetheless "attracted attention to this issue like no one before her and public

[2] For example, clerical positions only represented about 14% of the public service in 2008, whereas they constituted 24% in 1983; during that same time, there was a fivefold increase in the number of computer and system specialists working in the federal bureaucracy, four times as many lawyers, and three times the number of economists (Clerk 2008, p. 9, 29).

[3] Between 2000 and 2008, the federal public service recruited 12,000 to 15,000 new employees per year (Clerk 2008, p. 10).

service renewal became a critical pillar in Canada's ability to retain a leading edge, competitive governance regime" (Zussman 2010, p. 223). Notably, each successive clerk has prioritized public service renewal by acknowledging the critical importance of recruiting and retaining as well as identifying and developing public service employees. What has largely been lacking, however, is a clearly articulated plan for achieving the goal of the modernization and renewal of the public service. Nonetheless, there have been a number of statements in the various annual reports of the clerk, which hinted at the need to brand and position the public service as an employer of choice to attract and retain the best and brightest employees. This is not surprising. As Reichenberg (2002, p. 1) argued, "If governments are both to compete successfully for talented applicants and retain high performing employees, they need to brand government as an employer of choice that provides challenging work, progressive human resource policies, and opportunities for promotion and career development."

Concepts of Branding and Employer of Choice in the Public Sector

If organizations are to be successful, they ultimately must be able to attract, motivate, and retain talented people. Several factors influence these three key objectives, including external economic conditions, how the organization is perceived by potential employees (i.e., brand or the organization's reputation), the organizational mission, opportunities for growth and development, levels of compensation, the quality of supervision and leadership within the organization, and the nature of the job itself (Belcourt and Taggar 2002; Herman and Gioia 2001; Lenaghan and Eisner 2006; Ray et al. 2018). Moreover, as more Generation Y employees have entered the workforce, with a palpable disconnect between these individuals and the cultures of the organizations they joined, the retention challenges have changed. These younger employees want flexible work arrangements, mentorship, continuous learning, opportunities for community participation, and constant communication (Rajasekaran 2010, p. 30). Historically, these have not been areas where governments, with their rigid bureaucratic organizations and complete neglect of the concept of leadership (Mau 2007; Van Wart 2003), have fared particularly well. However, through time, the public sector was able to transform many of these historic limitations into organizational core competencies.

This is where the marketing management concept of branding is relevant. It is yet another example whereby the public sector, as part of NPM, has adopted concepts and practices widely embraced (often, but not always, with good effect) in the private sector. According to leading researchers, brands are "symbolic construct[s] that [consist]…of a name, term, sign, symbol or design, or a combination of these, intended to identify a phenomenon and differentiate it from different phenomena by adding a particular meaning to it" (Eshuis and Klijn 2012, p. 6). Brands, therefore, are not the products themselves; rather they are specific kinds of symbols giving meaning and value to something in the minds of consumers. They promote identification and differentiation, often using a sign, design, or name. Brands are unique and have a personality built on trust and relationships with consumers (Wæraas 2008, p. 207).

Although people typically think of products being branded or various global corporate brands, such as Disney, BMW, or Apple, other forms of branding are identifiable, including processes (i.e., services, policies, or decision-making processes), people, and places. The public sector has not relied on the branding products all that much due to the lack of competition for the limited range of goods (e.g., passports, roads, and bridges) it produces. However, these other forms of branding are now ubiquitous in the public sector.

Marland (2016) provides a detailed account of how the Canadian federal government has used branding broadly as a strategy to centralize the control over its communications by ensuring simplicity and consistency. The focus of this chapter, however, is much narrower in scope. More specifically, it examines employer branding, which emerged in the 1990s as "the hottest strategy in employment" for responding to the talent acquisition and retention problem (cited in Tanwar and Prasad 2016, p. 188). Through time, employer branding, which aimed to establish the employer identity of an organization, became synonymous with the quest for the employer of choice designation. It remains a popular approach. A 2018 Conference Board report identifies building a compelling employer brand and/or developing a strong employee value proposition as one of the top organizational strategies to attract, retain, and engage employees (Ray et al. 2018, p. 13).

It should be noted, however, that employer of choice or ideal employer "is basically a self-proclaimed achievement" (Lenaghan and Eisner 2006, p. 99). As such, organizations will have greater credibility in making such claims if the popular press recognizes them as being one of the best companies for which to work. Nevertheless, the efficacy of the strategy is obvious. By establishing a brand as an employer of choice, the organization strives to increase its competitive advantage in terms of attracting and retaining its employees. The hope is that in doing so the organization can foster both greater loyalty and productivity from its employees.

An Organisation for Economic Co-operation and Development (OECD) report, prepared for the Competitive Public Sector Employment Project (Äijälä 2001), identified four areas where governments have introduced reforms to become an employer of choice. First, governments must build trust and address the widespread negative perceptions of the public service as an employer, which can be accomplished by improving the image of public servants and better explaining the nature and methods of work in public sector organizations to prospective employees. Second, governments need to reform HRM systems by bolstering career development and career planning; developing managerial and leadership capacity, particularly through the adoption of leadership competency models; and fostering the mobility of employees between the public and private sectors. Third, they need to create better working conditions. This includes, but is not limited to, developing new wage policies or pay systems. Other non-material incentive systems, including flexible work arrangements, flatter hierarchies, and high standards for technical equipment, are also required. Finally, they need to enhance professionalism in the public service through the provision of ongoing staff training and development, especially induction training or onboarding for new employees.

Employer branding involves three steps (Tanwar and Prasad 2016). First, the employer must establish the employee value proposition, which is "the psychological contract or promise between employee and employer, defining the tangible and intangible deal that organizations are willing to provide in exchange for employee effort, commitment and performance" (Ray et al. 2018, p. 2). Relevant aspects would include pay and benefits, interesting and challenging work, flexible work arrangements, and opportunities for training and development. Second, the employer must use the value proposition externally to market the employer brand to prospective employees. Finally, the employer must undertake internal marketing to develop an engaged and committed workforce.

Branding as an Employer of Choice: Benefits, Risks, and Limitations

Public sector branding, Temporal (2015, p. 8) argues, "is here to stay." He offers several reasons why all public sector organizations must engage in the process of branding, the first of which is their need for both differentiation and attraction. Furthermore, they have recognized brands as strategic assets with tangible benefits, be it increased tourism and investment, greater international political influence, or the attraction and retention of talented human resources. Finally, like it or not, both citizens and stakeholder groups are formulating positive and negative images of the public sector; public sector organizations, therefore, should manage and control those images to ensure the positive ones prevail.

Several perceived benefits are associated with branding. This includes promoting recognition for the organization's products and services and setting it apart from the competition. Branding also allows for the retention of messages and provides motivation and direction for staff as well as greater commitment and loyalty. A clear brand strategy can help the organization and its employees stay focused on its mission and vision. Most importantly, a strong brand provides value to the organization in that higher prices can be charged for goods and services (Eshuis and Klijn 2012; Temporal 2015).

Most of these benefits are also relevant to the public sector, but there are other reasons why branding in that realm is considered advantageous. As Temporal (2015, p. 35) argues, "the civil service brand image has a major impact on that of the nation, as it is the nexus between the country, its citizens, the private sector, and international relations of any kind." Therefore, it is difficult to have a positive image of a country if its public service is not efficient, effective, and ethical. The Canadian government outlined why a branding strategy for the public service was necessary and important:

- ...to strengthen and promote Canada's foundational values of freedom, democracy, human rights and the rule of law;
- ...to ensure that the Public Service meets the evolving needs of Canadians and Canadian society with excellence in policy development and advice, and professional service delivery...;
- to respond to public expectations of: more accountability, better management of tax dollars and improved core public services;
- to build employees' trust and pride, and encourage them to act as ambassadors for their organization, its values, products and services;
- ...[to] provide a strong foundation and a canvas for more focused, departmental sub-brands, while influencing the culture at large...;
- to support recruitment and retention strategies... (CPSA 2007, p. 5–6).

Therefore, while branding the public sector has broader strategic benefits, many of the advantages of branding directly relate to HRM issues, namely, recruitment, retention, and motivation.

Focusing specifically on the strategy of branding as an employer of choice, the rewards are significant. First, fewer marketing and recruitment expenses are incurred because potential employees will want to work for the organization, which, in turn, will produce a waiting list of high-caliber people. Second, productivity will be enhanced because employees will be more competent and effective in the performance of their duties. Third, loyalty to the organization will be strengthened and turnover reduced, which will contribute to better quality and service outcomes. Fourth, employers and employees will have less stress and more fun. Fifth, given the continuity of personnel, it is easier to engage in long-term planning. Finally, the organization becomes more attractive to both investors and customers (Herman and Gioia 2001; Love and Singh 2011).

On the face of it then, it is difficult to envision why public sector branding as an employer of choice is anything but a positive approach to addressing the human resources challenge it faces. Nonetheless, caution is warranted. The first issue is whether a large, complex organization like the public sector can clearly and consistently articulate a corporate brand with a single identity when it comprises hundreds of different organizations of varying purposes and structures. In the words of one scholar, "public organizations are by nature inconsistent and complex entities that are difficult to include under one single identity definition" (Wæraas 2008, p. 217). Moreover, corporate branding requires uniqueness and differentiation, while public sector institutions have a need for legitimacy and the equal provision of services (Sataoen and Wæraas 2015).

Eshuis and Klijn (2012) identify other risks associated with branding. One relates to the potential tension existing between the many different audiences – citizens, employees, potential employees, departments and agencies, and other stakeholders – for whom the brand is supposed to be relevant. It may be the case that the brand speaks to one group and not another. More pointedly, there is a legitimate concern that branding can be a misleading representation of reality, resulting in manipulation (or outright propaganda). Moreover, as other scholars have highlighted, there is little empirical evidence to demonstrate how branding as a governance strategy is effective (Karens et al. 2016, p. 486).

Some of the factors often derailing government reform initiatives, such as insufficient human or financial resources, a lack of political support, or negative reactions by key stakeholder groups, can also negatively impact attempts to engage in public sector branding (CPSA 2007, p. 21–22). Significant investments of time, human capital, and money are required throughout the branding process, but they are particularly important for producing the necessary support materials during the implementation phase. Without these resources, the branding exercise is unlikely to succeed. Furthermore, the political and administrative support required to accomplish public sector branding will simply not materialize if there is widespread cynicism regarding the branding initiative among current and future employees, citizens, and the media. Even if there is support for the branding initiative among the political executive at the outset, it can often dissipate as government priorities shift or when the election of a new government ushers in a new agenda altogether. Notably, senior administrative leaders must also drive the initiative given branding is a strategic undertaking spanning the entire public service.

The importance of political and administrative leadership is not to be underestimated. To some extent, employer of choice, like many other HRM and administrative reform concepts, has become a hackneyed phrase. As some pundits note, many organizations overtly profess a desire to become an employer of choice without understanding the importance of that objective. Herman and Gioia (2001), p. 63) argue, "it is more than just a buzz phrase; it is representative of a whole new design of corporate culture." From the outset, therefore, establishing the public sector as an employer of choice is a formidable objective to achieve because engendering cultural change is perhaps the most difficult challenge organizational leaders face. Effecting cultural change is a long, arduous process requiring all employees to embrace and internalize the new organizational reality. The next section of the chapter examines the impetus for the Canadian government's 2007 initiative to brand the public service, revealing that the strong, sustained political and administrative leadership required to ensure its success was simply not forthcoming.

Leading the Charge? The Impetus for Branding the Public Service

As noted above, the successful implementation of a branding initiative requires leadership. Ultimately, those with the formal position, access to resources, and motivation will have the greatest ability to influence others while fostering a new culture to support the employer of choice brand (Herman and Gioia 2001). In the context of the public sector, there are two dimensions to leadership, political and administrative. In the Canadian context, assessments of that leadership are decidedly mixed.

Reflecting on the transition from one Liberal prime minister to another, one commentator wrote of the need for more robust political leadership with respect to establishing the next generation of employees in the public sector (Malloy 2004). The Martin government (2004–2006), however, was short-lived and was consequently unable to implement measures to transform the public service. The arrival of the new Conservative government in 2006, despite being a minority, seemingly provided the political leadership necessary to renew the public sector. In his first year in power, for example, Prime Minister Stephen Harper established an Advisory Committee on the Public Service to advise both him and the clerk on its renewal and future development. Harper was quoted as saying in the press release announcing this new initiative that the public service had served Canadians well and the government was committed to ensuring it remained strong (Canada, Prime Minister Harper 2006). That was certainly encouraging. So, too, were the signals he gave in a couple of speeches that year where he stated, "effective government requires effective public servants" (cited in Clerk 2007, p. 13) and "…being a public servant is not merely an honorable profession, but also vitally important to the well-being of our country" (cited in Clerk 2006, p. 1).

It appears, therefore, the requisite political leadership existed to engage in a corporate branding exercise to position the federal government as an employer of choice. Remarkably, the press release announcing the creation of the Advisory Committee made explicit references to branding. In the first instance, it was to assert that "the 'Public Service brand' is probably less clear and perhaps less positive in the public's mind than in decades past." As such, one of the issues the Advisory Committee was mandated to provide advice on related to "branding the Public Service as a trusted and innovative institution of national importance" (Canada, Prime Minister Harper 2006). It was not surprising, then, in its first annual report the Advisory Committee concluded the public service needed a strong, positive brand to help market the public service as "an attractive employment option for talented Canadians" (Clerk 2007, Annex 3, p. 9).

As vital as political leadership may be for the successful creation and implementation of a public sector branding strategy, the need for administrative leadership is equally important. Kevin Lynch, clerk of the Privy Council at the time, provided a sound articulation of the role that he and his successors could play: "The role of the clerk, as the most senior federal public servant, is to define major objectives, to set out benchmarks against which we can measure our progress, and to drive a continuing process of change and renewal that will take us well into this new century" (Clerk 2007, p. 1). This is exactly the type of leadership required if the employer of choice designation were to become a reality.

Lynch went on to note that to recruit successfully and remain a vital national institution that contributes to Canada's competitive advantage, the public service had to be branded as a career of choice in the minds of talented young Canadians. He then outlined the factors deemed to be essential for attracting and retaining employees – meaningful work, strong leadership, opportunities for learning and development, appropriate compensation, career advancement, and policies and practices that contribute to a flexible workplace – suggesting these were all benefits that the public service could offer employees (Clerk 2007, p. 21). His stated strategy over the medium term was to articulate and refine a public sector brand that could be adapted to meet the needs and circumstances of various public sector organizations.

While Kevin Lynch is credited with having launched and prioritized the renewal of the public service built around the development of a public service brand, his predecessors were also cognizant of the need for action to reinvigorate the institution. For example, even during a period of retrenchment in the public sector, Glen Shortliffe wrote of his commitment to the "renewal and rejuvenation" of the public service, noting that an exemption from the external hiring freeze had been granted to recruit "a limited number of top-quality graduates" (Clerk 1994, p. 2). Moreover, Mel Cappe was the first clerk to make explicit reference to the need for becoming an employer of choice in his inaugural report to the prime minister on the status of the public service, a theme he reiterated in his second report (Clerk 2000, p. 7; Clerk 2001). In his view, the public service could have an exceptional workplace and become an employer of choice by focusing on recruitment, workplace well-being and retention, and the promotion of learning and development for all employees. Ultimately, he identified the need for the public service to become a learning organization, reiterating Jocelyne Bourgon's (Clerk 1997) earlier appeal to do so.

Alex Himelfarb, former clerk of the Privy Council, perhaps came the closest to providing a vision of the Canadian public service. He wrote: "We can justifiably aspire to setting the world standard for a professional public service" (Clerk 2003, p. 9). This required certain competencies, including financial management, HRM, information and communications management, leadership development, professional development, and service delivery, all of which were to be infused throughout all public sector organizations at all levels. He went on to say, "Our professionalism should be our 'brand' and our voice on the global stage" (Clerk 2003, p. 10).

In his next report, Himelfarb restated his vision for creating a public service "that is nothing less than the finest...in the world" (Clerk 2004, p. 8) and set out the ideals to which all public servants should aspire. In his view, the Canadian public service should have the following characteristics: high levels of integrity and public trust; be focused on people, and comprising employees who are passionate about their work; use creativity and innovation to constantly transform; serve the public interest by achieving results and value for money; possess strong policy leadership; open and transparent; representative of the country's diversity and linguistic duality; and ultimately a vital national institution where all public servants are leaders.

In short, all four previous clerks who served prior to Kevin Lynch articulated in their own way the need to develop a strong public service. The most perceptible difference is Lynch appeared to be more purposeful and strategic in initiating his renewal plan. At that time, the Public Service of Canada did not have an overall and integrated corporate brand, although, with the establishment of the Federal Identity Program (FIP) in 1970, there has been an attempt to achieve a "common corporate identity of the federal government in Canada and on the world stage" (Marland 2016, p. 246). The intent was not to have a public service branding strategy replace the FIP but rather to complement the visual elements already in place. Central to his efforts was tasking the Canada Public Service Agency to develop a brand framework by March 2008.

The next section of the chapter outlines the year-long process spearheaded by that entity to develop a public service brand. It was a comprehensive, inclusive process that in many respects is deserving of the high regard it has been afforded (Temporal 2015). However, it is also clear the promise of the branding exercise has gone unfulfilled. More importantly, whether the initiative is "a work in progress," as has been suggested (Marland 2016, p. 248; see also PCO 2014, p. 25), is debatable. There is little tangible evidence to indicate a master brand for the public service has been or will be articulated.

Framework for Branding the Public Service of Canada

In 2007, with the requisite political and administrative leadership in place, Kevin Lynch tasked the Canada Public Service Agency to lead the process of developing and implementing an overall and integrated corporate brand for the public service. The intent was this master brand would be flexible enough to encapsulate the many sub-brands that already existed for various public sector organizations, for example, Canada Post, the Department of National Defense, or the CBC (Canada's public broadcast network). The rationale for the corporate brand was straightforward:

> It is about getting Canadians in general to understand and appreciate that the Public Service excels at doing its job well and is essential to the well-being of Canada and Canadians. This way, people will want to work for the Public Service, as we position it as the employer of choice, which will also attract new talent and make employees proud to be part of the Public Service (CPSA 2007, p. 3).

Although the Canada Public Service Agency assumed responsibility for the initiative, three separate support committees aided in this process. The first was the Brand Working Committee with director-general-level employees from across various agencies and departments. Second, a Steering Committee comprising assistant deputy ministers and senior officials from across the public sector was created. The Expert Advisory Group, which included external representatives from academia, the private sector, and the media to act as a sounding board, was the third committee. Moreover, to ensure full representation and get full support for the initiative, union officials were also involved at the committee level.

Ensuring there was broad representation from all the key stakeholders in these support groups was important to legitimize the process. After all, this exercise had to establish an overall umbrella brand that would address and appeal to distinct audiences: current employees, recruits, parliamentarians, and Canadians as a whole. Each of these audiences can be segmented further and, in many cases, would require a different but complementary approach. For example, the category "prospective employees" includes students, mid-career professionals, new graduates, executive members, and members of the equity groups. Current employees include individual public servants, managers, unions, APEX (Association of Professional Executives), and functional communities. When referring to Canadians as a whole, this includes citizens, key influencers from business, communities and academia, and the media.

Several key considerations were required for each audience. For example, with respect to public sector recruits, the employee value proposition for recent graduates would be distinct from the one for mid-career professionals. What was clearly stipulated was one of the primary purposes of branding was to assist with recruitment efforts. Moreover, the public service brand "must differentiate the Public Service from other employers in order to become an employer of choice" (CPSA 2007, p. 13). For current employees, the objective of the brand was to promote a culture of pride and engagement; for parliamentarians, it was to obtain their understanding, engagement, and trust. With respect to Canadians, the challenge was to reinforce the brand promise through every point of contact they have with the public service. This is what makes corporate branding more complicated than product branding; in effect, all employees – in this case, public servants – must become "brand managers" (Wæraas 2008, p. 207). The value proposition had to be comprehensive by capturing culture, leadership, policy, programs and services, people strategies, institution, system, and results.

The branding exercise was planned through five phases, spanning from the fall of 2007 to the spring of 2008, the first of which involved internal analysis and intelligence gathering and saw the creation of the various committees. Phase 2 was dedicated to research, conducting employee focus groups and stakeholder research, conducting a gap analysis, and confirming the branding objectives. Strategic planning was the focus of the third phase. During this period, the brand platform, promise, and character would be established, while both the brand identity and the implementation framework were developed and tested. Phase 4 was dedicated to developing and finalizing the brand strategy as well as the internal and external implementation plan. The final phase was implementation and evaluation.

Within these five phases, several branding steps were undertaken, the first of which was to define the public service vision. The mission was identified as "serve the public and the public interest," while the vision stated, "The Public Service of Canada is a world-class institution attracting, motivating and engaging highly skilled people to service the public and the public interest" (CPSA 2007, p. 33). Underpinning the mission and vision were the democratic, professional, ethical, and people values of the public service. The brand character, positioning, and promise had yet to be identified in the framework document. Other steps in the branding process included the identification of the key audiences; establishing objectives for each one; developing a multi-year strategy; developing the brand identity, messages, and actions; and, finally, implementing then evaluating the brand.

A great deal of thought, planning, and preparation went into the early phases of the branding strategy; this included consultations with key groups. It also involved the distribution of real-life stories profiling the various jobs and associated responsibilities of public servants (called "A Day in the Life"); public service success stories, including a link to the top 10 reasons to join the public service (https://www.canada.ca/en/privy-council/corporate/clerk/public-service-success-stories.html);

and interesting data about the public service ("Did You Know?"). Navigating the federal government website will also lead to "Proudly Serving Canadians," recently created and launched by the Treasury Board Secretariat, which provides a series of short videos and stories that demonstrate the diversity and rewards associated with public sector employment (TBS 2016, p. 5). The government also initiated off-campus recruitment events to target university graduates more effectively.

All these efforts are noteworthy and are vast improvements in terms of marketing public service careers more effectively. However, as the next section of the chapter establishes, this process failed to produce an entrenched public service master brand. Even though five federal public sector organizations were among Canada's Top 100 Employers in 2018 and seven different entities were on Forbes' List of Canada's Best Employers,[4] it is a stretch to suggest that the federal public service as a whole is an employer of choice.

Unfulfilled Promise: Whither the Public Service Brand?

It is virtually impossible not to feel disheartened by the public sector branding exercise. Despite all the good intentions and support of both political and administrative leaders; a comprehensive, well-designed process; and the buy-in from various internal and external communities and stakeholder groups, there is presently no discernible master brand for the public service. Citing an internal Public Works and Government Services Canada document, Marland (2016, p. 248) claims that "serving Canadians with integrity and distinction" was positioned as the public service master brand. However, if there is little or no external recognition of that brand, then clearly it is wanting.

As the previous section of this chapter outlined, the government developed a thorough framework for developing a public service brand, but there was no obvious indication of the outcome of the process, certainly not to those audiences external to government, which undermines the effort to brand the Public Service of Canada as an employer of choice. After all, if one of the central objectives of the brand were to facilitate the recruitment of talented Canadians to work in the public service, then presumably the brand would need to be both easily recognizable and accepted by citizens. This does not appear to be the case. Ironically, this branding initiative seems to have suffered from one of the pitfalls identified as part of the "lessons learned" portion of the framework document: "Branding exercises generally fail with the 'actions' portion of implementation. They also seem to lose

[4]The Canada Revenue Agency, Department of Canadian Heritage, Treasury Board of Canada Secretariat, and National Energy Board were on the first list, while Parks Canada; Fisheries and Oceans Canada; Immigration, Refugees and Citizenship Canada; the Department of Justice; Environment and Climate Change Canada; Statistics Canada; and the Canada Border Services Agency all made the Forbes list (Clerk 2018, p. 23).

steam when the brand champion leaves or is no longer engaged in the process" (CPSA 2007, p. 32).

The branding exercise appears to have been derailed during the implementation phase, which may have been a product of shifting political and administrative priorities. According to the framework document (CPSA 2007, p. 28–29), the development and finalization of the branding strategy was to have occurred in February 2008, with a testing of the brand promise to follow the next month and the development of implementation and evaluation plans in April. A soft launch of the brand was earmarked for June 2008, with further development of the brand strategy from June to October. Implementation and evaluation of the brand was slated for October 2008 and onward.

It is reasonable to expect this branding strategy would have figured prominently in the annual reports of the clerk, especially the 2008 and 2009 reports when Kevin Lynch was still the head of the public service. However, the only mention Lynch makes to the branding initiative in the 15th annual report, when much of the work would have been completed, was to state: "…we will follow up on the advice of the Advisory Committee on Senior Level Retention and Compensation…that defining and communicating the public service brand – who we are and what we do – is critical to successful recruitment and retention" (Clerk 2008, p. 16). The next annual report is not any more enlightening. Lynch makes reference to his commitment to strengthen the public service brand and mentions the development of a user-friendly website for job seekers, a project that was "moving forward, though proceeding slower than we might have hoped." He also highlighted career fair pilot projects at four universities, which included a branding element by "showing graduating students that Canada's public service is a committed and interesting employer, offering a wide variety of attractive careers" (Clerk 2009, p. 6). One of his objectives for the upcoming year was to implement a revised career fair "to support effective post-secondary recruitment with clear branding of the public service" (Clerk 2009, p. 17).

Given the financial and human resources expended in the planning and formulation of the branding strategy, the lack of detail about the progress of the initiative in the annual reports of the clerk is surprising. Even reports produced by the Canada Public Service Agency, which had responsibility for driving the branding exercise, were rather uninformative. For example, the department's 2008–2009 Report on Plans and Priorities simply stated efforts were underway to develop a branding strategy for the public service (CPSA 2009b, p. 30), while the Departmental Performance Report merely acknowledged branding the public service as an employer of choice was a priority and referred to the four career fair pilots that had been organized (CPSA 2009a, p. 11).

While Kevin Lynch's support for the branding initiative arguably waned during his tenure as clerk, it seems to have completely dissipated under the leadership of his three successors, none of whom even mentioned branding in their annual reports. Wayne Wouters, Janice Charette, and Michael Wernick followed Kevin Lynch as clerk, and they all continued to emphasize the need for recruitment and renewal but not in the context of branding the public sector as an employer of choice. Equally troubling were the lack of signals from the Prime Minister's Advisory Committee on the Public Service that the public sector master brand remained important. In its sixth report, the Advisory Committee did intimate the public service was a career of choice for young Canadians, but the only way it would remain so was to make the public service more productive, leaner, and more relevant (Clerk 2012, Annex B, p. 3). Two years later, the Advisory Committee made the startling suggestion that efforts should be made for "'re-branding' the public service career for young people as an opportunity to develop transferable skills for the private and not-for-profit sectors as well" (Clerk 2014, Annex C, p. 13). Neither of the two subsequent annual reports of the Advisory Committee mentions branding the public service, and the committee has not met since 2016 (Personal communication, May 18, 2018).

It was rather baffling to recommend a rebranding occur when there was no indication a public service master brand had ever been implemented.[5] A plausible explanation for this curious situation is that, politically, the tide was shifting. The global economic recession of 2008 struck when the branding exercise was to be finalized and implemented and it proved to be all-consuming. Moreover, after years of incurring deficits in response to the financial crisis, the Conservative government was eager to demonstrate it could be a sound steward of the public purse. During the campaign and after winning a majority in the 2011 general election, therefore, the Conservative government resurrected the politics of fiscal austerity. The first throne speech of the new government underscored this shift: "Our Government will also support the efforts of the public service to modernize the way it works…" (Governor General 2011, p.15). While seemingly a subtle shift in phraseology, using the term "modernize" instead of "renew" implied making the public service smaller and leaner. That is exactly what happened. From March 2012 to March 2013, the population of the federal public service decreased by 5.5%, which represented a contraction of nearly 17,000 employees (Clerk 2014, Annex B). Therefore, given the public sector was reeling from budget and personnel cuts, positioning the government as an employer of choice was problematic.

With austerity in vogue once more, Wayne Wouters launched yet another initiative to renew the public service. Announced in his annual report (Clerk 2013, p. 15), *Blueprint 2020* was designed to engage all employees in refining the public service vision and developing a "concrete blueprint for action." The desire was to ensure the federal public service is "a world-class Public Service equipped to serve Canada and Canadians now and into the future" (PCO 2013, p. 4). *Destination 2020*, a follow-up document highlighting ideas and actions taken to realize that vision, actually referred to branding. Oddly, 7 years after the process for developing a public sector master brand was launched, it was only characterized as a "work in progress" and not a *fait accompli*. The document further noted, "…employee engagement will continue with the goal to further refine the strategy and address this priority area" (PCO 2014, p. 25). Two new actions were identified to help the public service reach its destination: (1) an engagement process with public servants to define and com-

[5] A freedom of information request submitted in the spring of 2018 to the Privy Council Office (A-2018-00074) failed to produce any documentation confirming the adoption of a public sector master brand.

municate the public service brand and (2) a government landing page to profile what public servants do and to promote employment opportunities.

Other signals pointing to the failure of the master branding strategy are evident. Writing about the precarious state of the federal public service, Zussman (2010, p. 237, emphasis in original) wrote: "…further work is needed to rethink the value proposition for new employees. Specifically, why should someone work for the government today *other than for the generous benefits that come with the job*?". Ng and Sears (2015) similarly suggest there is a lack of awareness among citizens of public sector employment opportunities and the virtues of working therein. In an earlier study of nearly 20,000 Canadian university students by Ng and Gosset (2013), only 28.9% of respondents stated they would choose a public sector career over employment in the private sector. About one-third of respondents identified the Government of Canada as an employer of choice. These numbers reveal that much work remains to be done by the government to brand the Public Service of Canada as an employer of choice.

Moreover, there have been a number of references to the need for "branding" or "re-branding" in the public sector. Take, for example, a Conference Board of Canada study where human resource leaders from major public sector organizations participated in a roundtable discussion on their human capital challenges.[6] A commonly expressed concern was in recent years it had become more challenging to attract talented public sector employees and rarely were public sector organizations regarded as employers of choice. As such, there was a perceived need "to rebrand and communicate the value of embarking on a career in the public sector" (Cooper and Jackson 2017, p. 19). They concluded, "The challenge is to integrate, into the brand, the value of public service as a vocation, or a calling, in contrast with private sector employment" (Cooper and Jackson 2017, p. 34).

Another Canadian think tank also identified the need to rebrand the public service for the next generation of employees (Public Policy Forum 2017). This study, which relied on several roundtable discussions and in-depth interviews with students and young professionals countrywide, focused on identifying improvements to public sector recruitment and retention practices. These results suggest the federal government is not yet an employer of choice; moreover, participants articulated the need for government to modernize its recruitment and retention strategies. Several suggestions were offered for accomplishing this objective, including having more on-campus recruitment events; expanding recruitment drives to target students from non-traditional disciplines like business, science, and engineering; and allowing applicants to more easily track their application at various points in the process by modernizing the online recruitment system. Most significantly, the government needs to do a better job of both better explaining the breadth and diversity of public sector careers and providing a compelling rationale as to why a public sector career is worth pursuing.

[6] Seventeen public sector HR leaders participated in the roundtable, of which three were from the federal government and another two were from federal government agencies. The remaining participants worked for provincial and municipal public sector organizations.

Conclusion

The period from 1945 until the early 1970s has been described as the "golden age for bureaucracy." During that time, the government had no problem recruiting the best and brightest university graduates, since, as Savoie (2008, p. 53) notes, "the goal of many young university graduates in the immediate postwar years was to work for the common good, and the state provided the most promising avenue to do so." Public service was a vocation or calling, and the majority of public servants were willing to place duty over self for a place in the village. By the 1970s, however, interventionist governments fell out of favor as did the bureaucrats, increasingly viewed with suspicion and disdain by politicians, media, and the public, who enabled the welfare state to flourish. During a few decades, with the demise of Keynesianism and the rise of NPM, the value of the public service waned; it was subject to a barrage of criticism and the morale of public servants plummeted.

Therefore, although attracting the most talented Canadians to the ranks of the public service remained as important as it was in an earlier era, recruiting and retaining these individuals became increasingly challenging. As the war for talent emerged in the 1990s, this problem was only exacerbated. In response, as it had done so frequently in the past with the ascendancy of NPM, the federal government looked to prevailing trends in the private sector and embraced the concept of branding as an employer of choice.

The purpose of this chapter was to examine the Canadian federal government's 2007 initiative to create a master brand for the public service. All the critical success factors appear to have been in place. High-level political and administrative support for the branding exercise was evident from the outset. The requisite research was conducted to understand how key audiences perceived the public sector and rigorous planning was involved in designing the brand strategy. Staff from across the public sector and other key stakeholders were engaged early in the branding process and at other key decision points to create legitimacy. Furthermore, the government has clearly taken steps to articulate more effectively the diversity of careers available in the public service and why talented Canadians, recently graduated from university or mid-career, would choose to join its ranks.

However, more than a decade after the branding initiative was launched, it is apparent that no master public service brand was ever established, and it remains highly unlikely that one will be developed any time soon. Although there have been occasional passing references during the past several years to the fact that branding the public service is an ongoing exercise, there does not appear to be any real commitment by either the political executive or the administrative leaders in the federal bureaucracy at this point in time to follow-through with this branding campaign. Until such a time that a new champion for branding the Canadian public sector as an employer of choice emerges, nothing more than lip service will be paid to achieving this objective. Therefore, despite the early promise of a well-crafted branding initiative, it is difficult to draw any conclusion other than the Canadian government squandered the opportunity to develop a strong master brand for the public service as an employer of choice "where ambitious, talented and conscientious people can work collectively for the public good" (Public Policy Forum 2017, p. 13).

In the end, however, the failure of this initiative may have been inevitable. While public administration has been able to embrace a number of corporate practices quite successfully, such as the adoption of mission, vision, and value statements in the 1990s, other attempts to incorporate private sector business principles as part of the NPM reform agenda, despite their widespread and ongoing use, have been much less effective. This would include the use of pay-for-performance and leadership competency frameworks. As this case study demonstrated, attempts to undertake corporate or whole-of-government branding in the public sector are equally problematic. This is not to suggest the concept of branding is inapplicable to the public sector. Clearly, there are a number of instances where branding has been successfully applied in the public sector be it to attract residents and tourists or promote economic development (Zavattaro 2014) to ensure consistent communications (Marland 2016) or to brand specific policies or garner international political influence (Temporal 2015). Moreover, as noted in the chapter, several Canadian public sector organizations have been able to brand themselves successfully as employers of choice by garnering inclusion on some of the lists of best employers.

Branding the entire public sector as an employer of choice, however, is probably unrealistic. Even the largest multinational companies, such as Apple with 137,000 full-time employees in some 25 countries around the world, simply do not have the same diversity and complexity that is characteristic of the Public Service of Canada with hundreds of public sector organizations of dozens of varying types. Given the public sector is not a single entity, characterized by a common organizational culture and governed by a uniform set of policies, it seems inconceivable how a single public sector master brand can be established to position it as an employer of choice. A seemingly more sensible strategy for any government to pursue would be to continue with a decentralized approach whereby individual public sector organizations can emphasize their specific features that make them desirable places to work.

References

Äijälä, K. (2001). *Public sector—An employer of choice? Report on the competitive public employer project*. Paris: OECD.

Barker, P., & Mau, T. A. (2017). *Public Administration in Canada* (2nd ed.). Toronto: Nelson.

Belcourt, M., & Taggar, S. (2002). *Making government the best place to work*. Toronto: IPAC.

Canada. (1990). *PS 2000: The renewal of the Public Service of Canada—A synopsis*. Ottawa: Minister of Supply and Services Canada.

Canada. (2006, November 21). *Prime Minister Harper Establishes Advisory Committee on the Public Service*. (News Release). Available online at https://www.canada.ca/en/news/archive/2006/11/prime-minister-harper-establishes-advisory-committee-public-service.html.

Canada. Canada Public Service Agency (CPSA). (2007). *A framework for developing a brand for the Public Service of Canada*. Ottawa: CPSA.

Canada. Canada Public Service Agency (CPSA). (2009a). *2008–09 departmental performance report*. Ottawa: Her Majesty the Queen in Right of Canada.

Canada. Canada Public Service Agency (CPSA). (2009b). *2008–09 a report on plans and priorities*. Ottawa: Her Majesty the Queen in Right of Canada.

Canada. Clerk of the Privy Council. (1994). *Public service 2000: Second annual report to the Prime Minister on the Public Service of Canada*. Ottawa: Her Majesty the Queen in Right of Canada.

Canada. Clerk of the Privy Council. (1997). *Fourth annual report to the Prime Minister on the Public Service of Canada*. Ottawa: Her Majesty the Queen in Right of Canada.

Canada. Clerk of the Privy Council. (2000). *Seventh annual report to the Prime Minister on the Public Service of Canada*. Ottawa: PCO.

Canada. Clerk of the Privy Council. (2001). *Eighth annual report to the Prime Minister on the Public Service of Canada*. Ottawa: PCO.

Canada. Clerk of the Privy Council. (2003). *Tenth annual report to the Prime Minister on the Public Service of Canada*. Ottawa: PCO.

Canada. Clerk of the Privy Council. (2004). *Eleventh annual report to the Prime Minister on the Public Service of Canada*. Ottawa: PCO.

Canada. Clerk of the Privy Council. (2006). *Thirteenth annual report to the Prime Minister on the Public Service of Canada*. Ottawa: Her Majesty the Queen in Right of Canada.

Canada. Clerk of the Privy Council. (2007). *Fourteenth annual report to the Prime Minister on the Public Service of Canada*. Ottawa: Her Majesty the Queen in Right of Canada.

Canada. Clerk of the Privy Council. (2008). *Fifteenth annual report to the Prime Minister on the Public Service of Canada*. Ottawa: Her Majesty the Queen in Right of Canada.

Canada. Clerk of the Privy Council. (2009). *Sixteenth annual report to the Prime Minister on the Public Service of Canada*. Ottawa: Her Majesty the Queen in Right of Canada.

Canada. Clerk of the Privy Council. (2012). *Nineteenth annual report to the Prime Minister on the Public Service of Canada*. Ottawa: PCO.

Canada. Clerk of the Privy Council. (2013). *Twentieth annual report to the Prime Minister on the Public Service of Canada*. Ottawa: PCO.

Canada. Clerk of the Privy Council. (2014). *Twenty-first annual report to the Prime Minister on the Public Service of Canada*. Ottawa: PCO.

Canada. Clerk of the Privy Council. (2018). *Twenty-fifth annual report to the Prime Minister on the Public Service of Canada*. Ottawa: PCO.

Canada. Governor General. (2011, June 3). Here for all Canadians: Stability, prosperity and security. Speech from the throne to open the first session of the forty-first parliament of Canada. (Ottawa: Her Majesty the Queen in Right of Canada).

Canada. Privy Council Office (PCO). (2013). *Blueprint 2020. Getting started—Getting your views*. Ottawa: Her Majesty the Queen in Right of Canada.

Canada. Privy Council Office (PCO). (2014). *Destination 2020*. Ottawa: Her Majesty the Queen in Right of Canada.

Canada. Treasury Board of Canada Secretariat (TBS). (2015). *Demographic snapshot of Canada's Public Service, 2014*. Ottawa: Her Majesty the Queen in Right of Canada.

Canada. Treasury Board of Canada Secretariat (TBS). (2016). *Treasury Board secretariat public service renewal results, 2016*. Ottawa: TBS.

Canada. Treasury Board of Canada Secretariat (TBS). (2019). *Demographic snapshot of Canada's Public Service, 2018*. Ottawa: Her Majesty the Queen in Right of Canada.

Cooper, J., & Jackson, S. (2017). *Rethinking public sector HR*. Ottawa: Conference Board of Canada.

Dwivedi, O. P., Mau, T. A., & Sheldrick, B. (Eds.). (2009). *The evolving physiology of government: Canadian Public Administration in Transition*. Ottawa: Ottawa University Press.

Eshuis, J., & Klijn, E. (2012). *Branding in governance and public management*. New York: Routledge.

Glenn, T. (2012). The state of talent management in Canada's public sector. *Canadian Public Administration, 55*(1), 25–51.

Herman, R., & Gioia, J. (2001). Helping your organization become an employer of choice. *Employment Relations Today*, *28*(2), 63–78.

Karens, R., Eshuis, J., & Klijn, E. (2016). The impact of public branding: An experimental study on the effects of branding policy on citizen trust. *Public Administration Review, 76*(3), 486–494.

Kernaghan, K., Marson, B., & Borins, S. (2000). *The new public organization*. Toronto: IPAC.
Lavigna, R., & Hays, S. (2004). Recruitment and selection of public workers. *Public Personnel Management, 33*(3), 237–252.
Lenaghan, J. A., & Eisner, A. B. (2006). Employers of choice and competitive advantage. *Journal of Organizational Culture, Communications and Conflict, 10*(1), 99–109.
Lindquist, E. (2006). *A critical moment: Capturing and conveying the evolution of the Canadian Public Service*. Ottawa: Canada School of Public Service.
Love, L. F., & Singh, P. (2011). Workplace branding. *Journal of Business and Psychology, 26*(2), 175–181.
Lowe, G. (2001). *Employer of choice? Workplace innovation in government*. Ottawa: Canadian Policy Research Network.
Malloy, J. (2004). The next generation? Recruitment and renewal in the federal public service. In G. B. Doern (Ed.), *How Ottawa spends, 2004–2005: Mandate change in the Paul Martin era* (pp. 277–295). Montreal: McGill-Queen's University Press.
Marland, A. (2016). *Brand command: Canadian politics and democracy in the age of message control*. Vancouver: UBC Press.
Marland, A., Lewis, J. P., & Flanagan, T. (2017). Governance in the age of digital media and branding. *Governance, 30*(1), 125–141.
Mau, T. A. (2007). Public sector leadership development: The Canadian model considered. *Leadership Review, 1*(4), 267–293.
McIntosh, R. (1991). Public service 2000: The employee perspective. *Canadian Public Administration, 34*(3), 503–511.
Ng, E. S. W., & Gosset, C. W. (2013). Career choice in Canadian public service. *Public Personnel Management, 42*(3), 337–358.
Ng, E. S. W., & Sears, G. J. (2015). Toward representative bureaucracy. *Review of Public Personnel Administration, 35*(4), 367–385.
Nimijean, R. (2006). The politics of branding in Canada. *Revista Mexicana de Estudios Canadienses, 11*, 67–85.
Public Policy Forum. (2017). *Building a dynamic future: The next generation of public service talent*. Ottawa: Public Policy Forum.
Rajasekaran, V. (2010). *The road to retention: Youth perspectives on transforming organizations into choice employers*. Ottawa: Public Policy Forum.
Ray, R., et al. (2018). *DNA of engagement 2018: Moments that matter throughout the employee life cycle*. New York: Conference Board.
Reichenberg, N. (2002). Branding the government as an employer of choice. In *International recruitment and selection strategies*. Washington, DC: IPMA-HR.
Sataoen, H. L., & Wæraas, A. (2015). Branding without unique brands: Managing similarity and difference in a public sector context. *Public Management Review, 17*(3), 443–461.
Savoie, D. (2003). *Breaking the bargain: Public servants, ministers, and parliament*. Toronto: University of Toronto Press.
Savoie, D. (2008). *Court government and the collapse of accountability in Canada and the United Kingdom*. Toronto: IPAC.
Tanwar, K., & Prasad, A. (2016). Exploring the relationship between employer branding and employee retention. *Global Business Review, 17*(3), 186–206.
Temporal, P. (2015). *Branding for the public sector*. Chichester: John Wiley and Sons.
Van Wart, M. (2003). Public sector leadership theory: An assessment. *Public Administration Review, 63*(2), 214–228.
Wæraas, A. (2008). Can public sector organizations be coherent corporate brands? *Marketing Theory, 8*(2), 205–221.
Zavattaro, S. (2014). *Place branding through the phases of the image*. New York: Palgrave Macmillan.
Zussman, D. (2010). "The Precarious State of the Federal Public Service: Prospects for Renewal" in G.B. Doern and C. Stoney, eds. How Ottawa Spends 2010–2011: Recession, Realignment and the New Deficit Era. (Montreal: McGill-Queen's University Press), 219–240.

Branding for Morale

M. Blair Thomas

The Spanish Media Broke the Influenza Story Wrong in 1918

The chapter begins in a home office 16 miles north of the Florida-Georgia line in 2020, while the world is facing its most threatening global pandemic since 1918. Across the news and social media landscapes are reminders that our world is in a feverish tailspin with Italy, China, and the United States competing for the unwanted title of "most lives lost" during the 2020 novel coronavirus outbreak. March Madness has taken on a whole new meaning, with the cancellation of collegiate and professional sports alike and millions glued to daily press briefings. Commercial flights are as spacious as chartered ones given a societal fear of flying, and toilet paper demand is at an all-time high. What "Spanish flu" was to people in the Roaring '20s, "coronavirus" or "COVID-19" is to 2020. If one wants to make light of a terrible situation, it certainly will not be those affected by the disease nor those responsible for branding initiatives in Spain or with the Corona beer brand.

Despite rendezvouses with different viral threats in recent memory (SARS, Avian Bird, West Nile, Ebola), one has to go back to 1918 for a time that shares similarities with contemporary challenges. While we are not currently in the midst of an intercontinental war, the world was dealing with its first when the Spanish flu virus was responsible for 600,000 deaths in the United States and 21 million lost in total globally (Zeitz 2020). None of the three most recently recognized hypotheses place the ground zero in Spain, but rather Fort Riley (Kansas), France, or China. When King Alfonso XIII of Spain become ill with the virus in May 1918, the coverage of the virus increased (Andrews 2020). Coincidentally, the Spanish media was the first to cover the virus extensively given their status as a country that retained its neutrality as battles raged on around them. This status protected them from the wartime censors of their contemporaries.

M. B. Thomas (✉)
Department of Political Science, Valdosta State University, Valdosta, GA, USA
e-mail: michaelbthomas@valdosta.edu

S. M. Zavattaro (ed.), *Public Branding and Marketing*,
https://doi.org/10.1007/978-3-030-70505-3_9

Unfortunately, in their pursuit to cover the outbreak was an absent mindedness that the coverage possibly needed to be branded, as well as the flu itself. The lapse in foresight allowed the American and Allied media to tag the virus with the nickname "Spanish" flu to keep wartime morale up. A century later, US President Donald Trump dubbed COVID-19 the "Chinese Virus" on social media, taking a different route than his international contemporaries by tagging an individual country with "responsibility" for the virus. How pandemics are viewed is just one element of countless others where branding (private by nature of the press influence and public by nature of the government influence) impacted public servant morale without the general public's explicit awareness of the activity happening.

When discussing branding, there are several ways to approach the concept. This chapter examines branding from the purview of internal and external morale. Using survey results and follow-up interviews from a sample or Florida city managers and relevant local officials with knowledge of the practices, the basis of this contribution is to discuss the idea of morale and its place in global branding. While the chapter maintains a domestic flavor, the chapter discusses concepts that carry a global application.

Defining Concepts

Before progressing further, it is worth establishing some definitions. Although branding and morale are the topics most covered, social marketing and marketing are as well; therefore, all four of them are defined in the context of the chapter. This chapter uses Kanter's (1977) definition on morale as an "attitudinal response to work conditions that has an impact on the behavior of individuals within the organization." Branding is a conscious effort put forth by public officials to establish a brand on published documentation and relevant materials (Anholt 2007; Fay and Zavattaro 2016; Keller 1993; Thomas et al. 2020). Marketing is a purposeful strategy used to influence stakeholder perceptions, and social marketing is a concerted effort made to educate and change behaviors based on visual images and narratives (Fay and Zavattaro 2016; Thomas et al. 2020). Social marketing is the application of private marketing concepts designed to influence the voluntary behavior of citizens to improve personal welfare and the overall community in which people reside (Kotler and Zaltman 1971).

As such, another relevant definition to consider is strategic management. Consistent with Bryson (2011, n.p.), strategic management is "the appropriate and reasonable integration of strategic planning and implementation across an organization (or other entity) in an ongoing way to enhance the fulfillment of its mission, meeting of mandates, continuous learning, and sustained creation of public value."

One of the best aspects of branding in theory or practice is the space encourages tapping into creative energy like few other professions allow. In some ways, these individuals live the mantra that *The New York Times* best-selling author Elizabeth Gilbert wrote in 2015: "If you're alive, you're a creative person." While the energy driving branding is about forming ideas, those ideas are merely vessels for something greater: connections. For example, think about the five purchases you have made of $75 or more. Where did you buy the product? Where did you go? If you went online, what website did you purchase from? What type of device did you use? In this exercise, were you able to avoid thinking about branding in any sort of way? If you were, you might be the first person to pull off this feat.

While that may have been a private sector-based exercise, if you have ever purchased airplane tickets, then you have crossed over into the realm of public branding unintentionally (place branding, specifically). Depending on where you are choosing to travel, the city or region has a connection to you in some way. If it is your hometown, then you may have personal memories that draw you back. If it is somewhere new, perhaps the area has intrigued you in some way. Maybe you are captivated by the nightlife, the potential to find your inner self inside a centuries-old temple or for a new adventure and thrills that provide you with stories to last a lifetime. Competition to be part of your future memories is just one reason why private sector entities have long engaged in branding. It is easy to understand how public sector entities such as local and regional governments engage in these activities as well.

Place branding focuses on particular geographical spots including countries, provinces/states, and cities, highlighting focused physical and emotional elements that give the area a specific meaning (Eshuis and Klijn 2012). At the state and level municipal level, several efforts come to mind in this effort, highlighted by the state of New York and the Las Vegas Convention and Visitors Authority. The brainchild of Milton Glaser, the **I ♥ NY** campaign of 1977, came about as a concerted effort by the state of New York as the New York State Department of Economic Development (known today as the Empire State Development (ESD)) sought ways to attract citizens to the region as New York City (and as a result the entire state) was hurt by a reputation that the nation's most populated municipality had developed into an "asphalt jungle and ungovernable city" (Greenberg 2008; Eshuis & Klijn 2012). On any item that can fit the wording on its surface to names of episodes of iconic television shows (season 4, episode 18 of *Sex and the City* in 2002, e.g.), Glaser's work is iconic, and the city benefits from it to this day.

It has been almost 20 years since firm R&R Partners worked with the Las Vegas Convention and Visitors Authority to create the infamous "What Happens Here, Stays Here" slogan in 2003. It was famously buoyed by then First Lady Laura Bush's coy response: "Jay, what happens in Vegas stays in Vegas," to then-host Jay Leno's inquiry on *The Tonight Show* on what she and President George W. Bush did during a 2004 campaign stop in Las Vegas and later kept in vogue by popular ads and constant references in *The Hangover* movies and other pop cultural references. To this day, once a typeface is used or the first couple of words of a certain phrase are spoken, many will see an image of New York or in their mind's eye or mouth the end of "in Vegas, stays in Vegas." Feelings also accompany these actions as well, a key point that will be touched on later in this chapter.

While New York and Las Vegas are two of America's hotbeds when it comes to branding and imagery, Florida also captures the imagination across an entire spectrum of ages. Butterbeer is served and Dole whips fly as Harry Potter and Mickey Mouse headline the protagonists used by Universal Studios Orlando and Walt

Disney World, respectively, to plant, mold, and then capture the imaginations of children everywhere. Buoyed by the sights and sounds reserved for only fantasies, thrill rides and pyrotechnic shows set imaginations ablaze as Orlando is the place in which dreams can come true. For others, Miami comes to the forefront with its legacy in pop culture in the past 40 years – from Don Johnson and Phillip Michael Thomas fighting crime in their 1984–1989 television hit "Miami Vice" to the 2002 blockbuster video game "Grand Theft Auto: Vice City" loosely based on the city. On July 8, 2010, National Basketball Association (NBA) star LeBron James famously declared that he was "taking his talents to South Beach," while the cast of MTV's "Jersey Shore" raided the city's club scene that same year and once again in the context of their reboot "Jersey Shore: Family Vacation" in 2018. Few places conjure up public branding like the state of Florida. Evidence can be found in the numbers, as the Sunshine State has received an increase in tourists for eight consecutive years with 126.1 million visitors in 2018 (Santana 2019). Given the imagery and the emotions tied with it disseminated across the world, it is appropriate that the state serves as the case for this chapter.

Data Sources

Data for this study was accessed from two sources: an administered survey distributed in June 2017 to city managers and other related public officials who are tasked with public communications in Florida's cities with responses representing approximately 40 cities and seven follow-up interviews with respondents that wanted to discuss activities further. The survey covered public activities such as branding, marketing, social marketing, and strategic management and also included several questions on internal and external morale. It was distributed via the Tallahassee-based Florida League of Cities whose goals are to "shape policy, share the advantages of cooperative action, and exchange ideas and experiences" (Florida League of Cities 2016). The second source of data is from follow-up interviews with seven survey participants who wanted to engage further in discussing survey-related topics. The questions were written to provide more information into how public agencies use public activities within their municipalities, how success is determined, and what other practitioners as well as scholars should know further about the concepts. Survey responses and commentary derived from the questionnaire and interviews that focused on internal and external morale serve as the basis for this chapter.

Morale

An emerging area of public branding that practitioners are wanting to learn more about is morale. Specifically, this chapter investigates how a sample of city managers and communication officials in cities approached the concept of branding morale. Research on morale traces back at least a century to Hall's (1920) work on the concept. In the 1930s and 1940s, morale was viewed as a misused or understudied term (Watson 1942). Baehr and Renck (1958) characterized the research on morale into three approaches: personal determinants of needs (highlighted by McGregor's work), a hierarchy of needs (led by Maslow's (1946) seminal work on human motivation), and the third that placed importance on someone's interactions with others in the context of a professional group and the pattern of group activity driven by Mayo (1933) and later Lewin (1951). Scholars such as Gordon (1955) tried to answer related questions such as "How can employee morale be measured satisfactorily? How is morale in my organization? How can morale be improved?" Baynes (1967) features terms such as "mind" and "spirit" with other elements in the fold such as "courage" and "self-discipline" as ways to describe morale.

In an organizational sense, Doherty (1988) posited that "low" psychological morale seemed to indicate a person sees themselves as an individual lacking power or consequentially is important socially within an organization. Others have argued morale is associated with individual and group elements (Child 1941; Gal 1986). Perhaps the most appropriate way to think about it from a macrolevel perspective is to refer to Johnsrud et al.'s (2000) citation of Kanter's (1977) definition discussed earlier in the chapter as an "attitudinal response to work conditions that has an impact on the behavior of individuals within the organization."

When diving deeper into morale, the concept can be viewed two different ways in the context of the public sector: as internal morale (how do public employees feel about a public organization or city within the confines of their organization) and external morale (outside of the organization and relates to the citizenry as a whole and how they feel about the city). In terms of discussing morale, scholars have examined the topic in various forms, including internal communication and organizational climate (Settle et al. 2013), the relationship between pay and morale (Leavitt 1996), the link between satisfaction and morale (Benge and Hickey 1984), a connection between organizational culture and morale (Arunchand and Ramanathan 2013), and public service motivation and internal morale (Pandey et al. 2008).

No matter where you are in your career, inevitably you will find yourself having some sort of feelings toward your business card. At your first professional job, perhaps just having a business card produced excitement or a sense of pride. As you progressed, maybe you enjoyed (or hated) seeing the job title you held at the time printed out formally. Perhaps it is the company that you represent now that harkens similar feelings. As someone that grew up in Portland, Oregon, countless friends loved seeing a Nike "swoosh" emblazed across their card. Meanwhile, others were proud to see "City of Portland," "Oregon Zoo," or "Portland State University" on theirs, three public organizations held in high regard in the area. The pride that someone feels about an organization, their place within it, and the activities associated with the entity can all serve as influential factors in addition to the ones mentioned in prior studies to internal and external morale.

Survey and Interview Findings

Internal Morale In terms of improving internal morale, 20% of respondents to the survey reported their city developed branding mechanisms ("IM + Mechanism" in Fig. 1) for that specific purpose ("IM" in Fig. 1). When questioned further about whether internal morale or external morale ("EM" in Fig. 1) was focused on more heavily, respondents noted they were discussed about the same amount. Within municipalities that lacked an internal morale mechanism, 43.8% claimed they are unaware of the extent to which morale was discussed in the workplace.

From these responses, while municipalities are engaging in branding, survey respondents indicated that due to the acknowledgment that marketing and branding are recognized as activities new to the public sector, there was an apprehension public officials face in their application of these activities. One official in a Miami suburb supported this sentiment:

> Marketing and public relations actually became an office three years ago. The city manager established this office. Prior to that there was no marketing or branding. There was no real strategy, or marketing and promotion efforts done in the city prior to the creation of this office.

In another interview response, an official who worked in one Orlando suburb provided insight on the importance of balancing the dynamics that exist within a municipality's government, given the practices of marketing and branding have a pronounced role in a community:

> Some department directors were super excited and couldn't wait to get rid of their inconsistent crap. They wanted to get on board. Then there was some, that you know with change, you know managing change is a whole process within its self, so we understood that there was going to be some sense of sensitivity because there was a lot of pride associated with some of the stuff we have been doing in the past, we have a lot of employees that have been with us for decades and just love what they do, take pride in what they do, and so we were sensitive that we were going to be changing some of that.

Marketing and branding can influence both internal and external morale. From an internal perspective, good can come from departments collaborating on a project

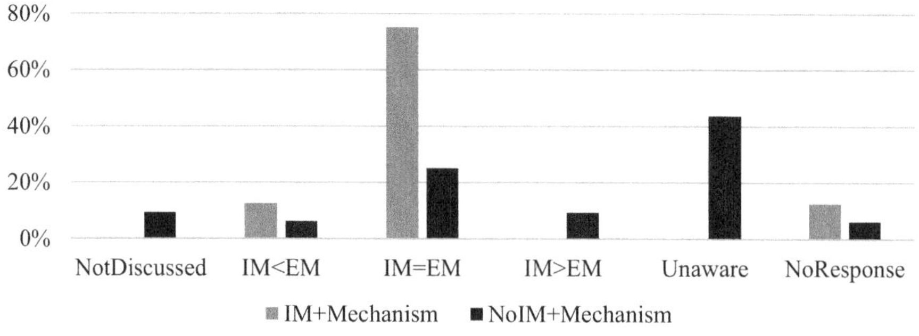

Fig. 1 Morale discussion regarding branding initiatives (by percentage)

that reflects the best in what the city has to offer. Whether it is an external marketing or branding project where a city looks to highlight the best of what a municipality has to offer, a choice to seek buy-in can only boost internal morale if the decisions made support the comments or suggestions needed to achieve buy-in. This feeling ultimately could have lasting effects far exceeding the life of a campaign. If we take the business card example and expand on it, someone can feel a sense of increased pride upon looking at it when their organization works toward a common goal, especially if the organization already has a good public reputation. There is research that supports the idea that organizational reputation has larger effects. Wæraas and Dahle (2019, p. 278) report that reputation can serve as a significantly influential factor in the way an organization "guid[es] strategies, practices and control systems." As a result, a project intended to increase external morale ultimately is likely to provide internal morale benefits if executed correctly. This is one reason why acknowledging and understanding morale is important.

Social Media and Morale Morale can surface in how social media are used. Several cities are active in their use among elected officials. While some communication departments or city managers will shoot a directive to elected officials to participate on social media apps (Twitter, for instance), elected officials in two suburbs took up using the mechanism to spread word of municipal branding initiatives while boosting the morale of the city as a result. An official in a Miami suburb described the usage: "It [has] been organic (elected officials using hashtags), I think it's extremely helpful and important when the faces of the community are our elected officials, and when they are engaged using social media, people just start to follow." Public organizations can witness an increase in external morale when people see the faces of their community leaders in broader spaces and when there is a feeling of access to those leaders.

For example, if Orlando Mayor Buddy Dyer were to engage on Twitter at a microlevel (per tweet) or a macrolevel (overall usage of Twitter), one can theorize that if a tweet received negative coverage or was viewed as misconstrued by the city or the public, then both internal and external morale would fluctuate, likely in a negative direction. One would expect the opposite to happen logically if it were viewed in a positive light. With social media's presence in our daily lives combined with how inexpensive it is for individuals and organizations alike, along with how it has become a key way to reach constituents, it is logical that using social media to improve various facets of an organization, such as morale, would be a route for public agencies to take and a practice to engage in.

Branding, Marketing, and Social Marketing In looking at how public officials see their municipalities using branding, marketing, or social marketing to increase internal morale, at least one in five believed their cities used all the three activities for that purpose. In the survey, 32.5% of cities used social marketing for increasing internal morale. In terms of marketing, while the numbers are not as high as social marketing, they are worth noting as 27.5% reported using marketing to increase internal morale, and one in four (25%) would use branding for those purposes. What this illustrates is public administrators are considering engaging in the behavior. As branding becomes a more socially accepted practice in the public sector, one would expect to see these numbers rise along with morale via the other two mechanisms as well. Given the rise of literature in strategic planning in public administration over the course of the past three decades (most recently by Bryson and George 2020), it is reasonable to believe that branding will become a formal part of planning moving forward. Consequently, any discussions to increase internal or external morale with branding (or marketing for that matter) would also become formalized as a result.

Internal Training Programs Training programs are another avenue that cities are exploring to improve employee morale. Fifteen percent of the sample stated their municipalities had them and that their communications departments or human resource/relations agencies were responsible for their functioning. Half of those municipalities thought branding, marketing, and social marketing would be used to improve employee morale in the future. What this shows is that leaders in different cities are confirming that there is value in examining morale through yet another avenue.

Individuals were asked about the types of benefits that city managers saw when using public branding mechanisms (e.g., programs, initiatives, etc.) for the intention of improving organizational internal morale. An average of just over three (3.3) benefits was discussed. These benefits are listed with their frequency in Fig. 2. Respondents were provided with several options for the benefits witnessed, including ("IncreasedwithinOrg"), enhanced communication between administrators ("BetterAdminComm"), enhanced communication between management and administrators within the organization ("BetterCommAdmin&Man"), emergence of pride within an organization ("EmergingOrgPride"), improved communication between administrators and citizens ("BetterCitizenComm"), and "other." Over 80% (83.3) reported an increase in internal morale within the organization.

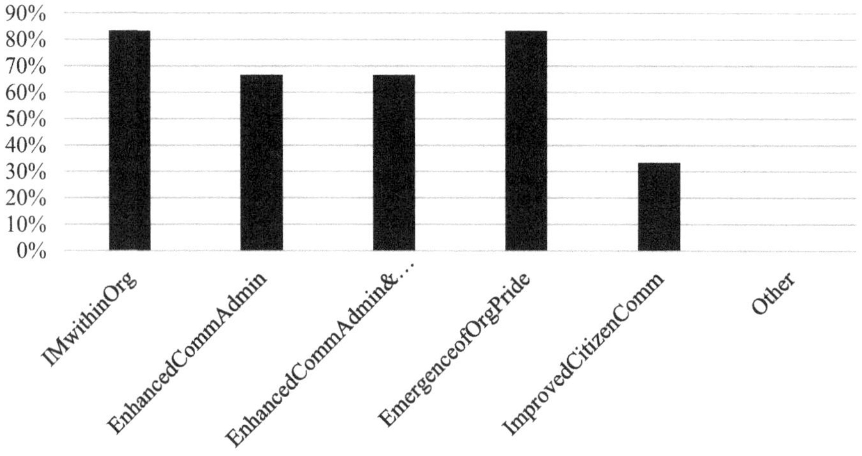

Fig. 2 Benefits experienced by cities that use branding to improve internal morale (by percentage)

In municipalities where branding initiatives were not used specifically to improve internal morale, five categories were provided (and listed in Fig. 1) that described the level of internal and external morale discussion among public officials within their cities. These included IM not discussed ("NotDiscussed"), internal morale is less discussed than external morale ("IM < EM"), internal morale is discussed just as often ("IM = EM"), internal morale is discussed more than external morale ("IM > EM"), and the respondent is "unaware of discussion" within the organization.

The proportion of individuals that responded is reflected in each category in Figs. 2 and 3. The following options provided to the public officials included increased morale within the organization ("IncreasedwithinOrg"), enhanced communication between administrators ("BetterAdminComm"), enhanced communication between management and administrators within the organization ("BetterCommAdmin&Man"), emergence of pride within an organization ("EmergingOrgPride"), improved communication between administrators and citizens ("BetterCitizenComm"), and "other." What is worth noting is that in places where internal morale is talked about more often than external morale, there are better interorganizational dialogue within the departments of the agency and an increased level of pride associated within the organization. One-third of respondents did not witness and benefits within their organization, which is an illustration that organizational benefits were not a primary reason behind branding implementation. Conversely, what this shows is that in this sample, public organizations where internal morale was viewed as a priority, benefits were captured.

Formal Strategic Planning Although there is a litany of benefits gained with benefits associated with branding, an element that influences how they are captured is based largely on the way things are implemented within the context of formal stra-

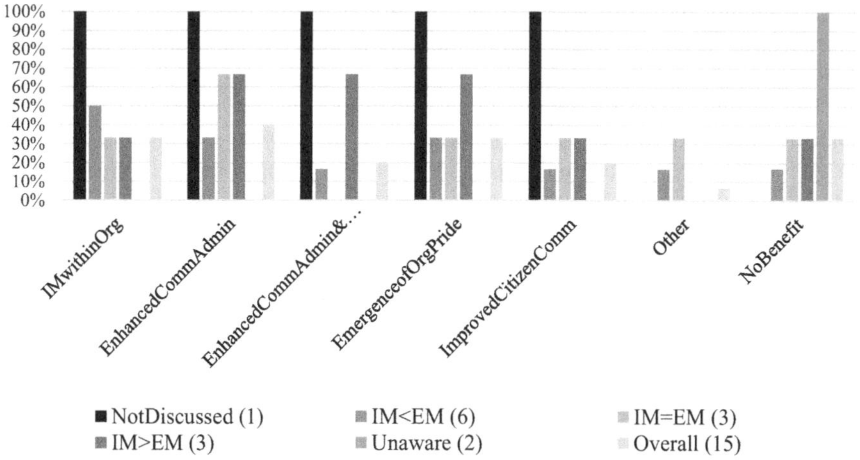

Fig. 3 Benefits experienced by cities that do not use branding to improve internal morale (by percentage)

tegic planning. In the confines of a municipality where one person is in charge with few administrators serving under them, the desire or necessity to gain internal buy-in is hardly an issue. In those instances, a town clerk or an individual charged with like duties that deal with branding relies on the support of their superior, such as a city manager or mayor. In these cities, a city manager of a rural town located within the central part of the state notes internal buy-in carries less importance given that individuals in her position are often self-reliant and have the freedom to do as they wish: "It's pretty much me pushing things forward. I do not need approval."

Concluding Thoughts and Future Research

At a time when public organizations are starting to embrace the practice of branding, this chapter serves as an exploration of how morale can enhance a city's brand from an internal and external perspective. Morale is an element that can be branded as a separate entity or embedded as an exercise in social media usage among a municipality itself or by a public official, different types of internal training, and other creative opportunities where communication can be improved and as a result morale can be improved. In this sample of Florida's public officials, there are two general observations that stand out. First, when morale is discussed among cities, public officials tend to discuss internal morale rather than external morale. Second, municipalities that utilize branding for internal morale see an increase in the phenomenon and the emergence of organizational pride, which parallels some of the findings reported by Waeraas and Dahle (2019). In these municipalities, there is an increase in communication across multiple departments. As cities grow, the ability to engage in quality dialogue increases in importance as citizens look to public officials for services. To see these types of benefits in the future however, city managers, communication officers, and related officials must be willing to start the conversation of how to weave morale building in their current branding plans or in their future plans.

This chapter addresses morale in an exploratory manner given its sample size and thus does not allow for the opportunity to make wide-ranging assumptions. What the chapter does accomplish is to serve as a launching point for future research. Morale is a concept studied in private sector practice, but not in public sector entities. Understanding the significance of internal morale on human resource motivation in a public setting is an understudied element worth investigating. How morale is approached by city managers is a macrolevel concept that could be explored by scholars as well. Investigating how morale is impacted in different forms (through partnerships, e.g.) and what cities can do to improve it internally is a phenomenon worth studying. For example, Nike recently partnered with the New York City Department of Parks and Recreation on an apparel deal that was announced in November 2019 and launched at the start of 2020. While it is unclear why the city entered into the agreement, seeing the department's logo across the world likely would have a tangible impact on their employees (whether good or bad), given

Nike's impact on pop culture. As branding continues to emerge as an activity that leading public administrators can no longer ignore, it is worth a long-term observation of how cities address morale among their peers and with their citizenry.

References

Andrews, E. (2020, March 27). Why was it called the 'Spanish Flu?' History.com. Retrieved from: https://www.history.com/news/why-was-it-called-the-spanish-flu.

Anholt, S. (2007). *Competitive identity: The new brand management for nations, cities, and regions*. New York: Palgrave Macmillan.

Arunchand, C., & Ramanathan, H. (2013). Organizational culture and employee morale: A public sector enterprise experience. *Journal of Strategic Human Resource Management, 2*(1), 1.

Baehr, M., & Renck, R. (1958). The definition and measurement of employee morale. *Administrative Science Quarterly*, *3*(2), 157–184.

Baynes, J. (1967). *Morale: A study of men and courage*. New York: Praeger.

Benge, E., & Hickey, J. (1984). *Morale and motivation: How to measure morale and increase productivity*. New York: Franklin and Watts.

Bryson, J. (2011). *Strategic planning for public and nonprofit organizations: A guide to strengthening and sustaining organizational achievement* (Vol. 1). Hoboken: John Wiley & Sons.

Bryson, J., & George, B. (2020). Strategic management in public administration. In B. G. Peters & I. Thynne (Eds.), *The Oxford encyclopedia of public administration*. New York: Oxford University Press, forthcoming.

Child, I. (1941). Morale: A bibliographical review. *Psychological Bulletin, 38*(6), 393.

Doherty, J. (1988). Psychological morale: its conceptualisation and measurement. The Doherty Inventory of Psychological Morale (DIPM). *Educational Studies, 14*(1), 65–75.

Eshuis, J., & Klijn, E. (2012). *Branding in governance and public management*. New York: Routledge.

Fay, D., & Zavattaro, S. (2016). Branding and isomorphism: The case of higher education. *Public Administration Review, 76*(5), 805–815.

Florida League of Cities. (2016). About us. Retrieved from https://www.floridaleagueofcities.com/about-pages/about

Gal, R. (1986). Unit morale: From a theoretical puzzle to an empirical illustration: An Israeli example 1. *Journal of Applied Social Psychology, 16*(6), 549–564.

Gilbert, E. (2015). *Big magic: Creative living beyond fear*. New York: Penguin.

Gordon, O. (1955). A factor analysis of human needs and industrial morale 1. *Personnel Psychology, 8*(1), 1–18.

Greenberg, M. (2008). *Branding New York: How a city in crisis was sold to the world*. New York: Routledge.

Hall, G. (1920). *Morale, the supreme standard of life and conduct*. New York: D. Appleton.

Johnsrud, L., Heck, R., & Rosser, V. (2000). Morale matters: Midlevel administrators and their intent to leave. *The Journal of Higher Education, 71*(1), 34–59.

Kanter, R. (1977). *Men and women of the corporation*. New York: Basic Books.

Keller, K. (1993). Conceptualizing, measuring, and managing customer-based brand equity. *The Journal of Marketing, 57*(1), 1–22.

Kotler, P., & Zaltman, G. (1971). Social marketing: An approach to planned social change. *Journal of Marketing, 35*(3), 3–12.

Leavitt, W. (1996). High pay and low morale: Can high pay, excellent benefits, job security, and low job satisfaction coexist in a public agency? *Public Personnel Management, 25*(3), 333–341.

Lewin, K. (1951). *Field theory in social science: Selected theoretical papers*. New York: Cartwright.

Mayo, G. (1933). *The human problems of an industrial society*. New York: Macmillan.
Pandey, S., Wright, B., & Moynihan, D. (2008). Public service motivation and interpersonal citizenship behavior in public organizations: Testing a preliminary model. *International Public Management Journal, 11*(1), 89–108.
Santana, M. (2019, February 20). Visit Florida reports record 126.1 million visitors to state in 2018. Retrieved from: https://www.orlandosentinel.com/business/tourism/os-bz-visit-floridatourism-record-20190220-story.html
Settle, Q., Telg, R., Carter, H., & Irani, T. (2013). Internal communication and morale in a natural resources public organization. *Journal of Applied Communications, 97*(3), 19–32.
Thomas, M., Fay, D., & Berry, F. (2020). Strategically marketing Florida's cities: An exploratory study into how cities engage in public marketing. *American Review of Public Administration, 50*(3), 275–285.
Wæraas, A., & Dahle, D. (2019). When reputation management is people management: Implications for employee voice. *European Management Journal, 38*, 277–287.
Watson, G. (1942). Civilian morale: Second yearbook of the society for the psychological study of social issues. Boston: Houghton Mifflin.
Zeitz, J. (2020, March 17). Rampant Lies, Fake Cures and Not Enough Beds: What the Spanish Flu Debacle Can Teach Us About Coronavirus. Retrieved from: https://www.politico.com/news/magazine/2020/03/17/spanish-flu-lessons-coronavirus-133888

Brand on the Run? Marketization, Market Position, and Branding in Upper Secondary Schools

Dag Yngve Dahle

What is the relationship between the level of marketization, market position, and branding efforts in public sector schools? These issues are explored in this chapter, using data from public sector upper secondary schools in two areas in Norway. After introducing New Public Management ideas in the public sector in the 1990s, deregulation has been the norm in the public sector in general and in public sector education in particular in Norway. Authorities in the Norwegian capital of Oslo have taken a neoliberal approach to public education (Hovdenak and Stray 2015), and have introduced a dominant market logic in the field of upper secondary education.

Specifically, they have marketized upper secondary education through the introduction of free choice of schools, test-based accountability, and per capita funding where funds are tied to the student (Haugen 2020). Thus, public sector schools try to appear attractive to compete for students, especially high-performing students. Branding is a possible tool for schools to attract good students. Unlike Oslo, authorities in the neighboring areas of Follo and Romerike have not chosen the same level of marketization of public sector upper secondary schools. In both areas, which belong to a different county than Oslo, there is only partly free choice of schools, and authorities have introduced little test-based accountability and per capita funding. In this context, Follo and Romerike represent an interesting contrast to Oslo.

Until now branding in schools has received scant attention from scholars, and possible links between marketization, market position, and branding remain rather unexplored. Thus, responding to calls for research into "the potential benefits and possible shortcomings of using branding principles in public organizations"

D. Y. Dahle (✉)
Østfold University College, Halden, Norway

School of Economics and Business, Ås, Norway
e-mail: dag.y.dahle@hiof.no

S. M. Zavattaro (ed.), *Public Branding and Marketing*,
https://doi.org/10.1007/978-3-030-70505-3_10

(Leijerholt et al. 2019, p. 133), the present chapter examines these issues by providing clues to the following guiding questions:

1. *How do schools at different popularity levels engage in branding efforts of their schools?*
2. *To which extent does this unfold differently in a highly marketized area like Oslo than in the less marketized areas surrounding the city?*

The present study contributes to existing literature in several ways. It contributes to scholarship within public sector branding by highlighting market position as a motor for branding strategies, ultimately leading to differences in branding strength. The study also highlights how institutional arrangements, namely free choice of schools, per capita funding, and test-based accountability, act as drivers of organizations' market position. Another contribution is to shed light onto how these institutional arrangements run parallel with a move toward market logics as the dominating institutional logics within public sector upper secondary education. A core contribution is to unveil how school executives choose to brand their schools in a marketized setting, providing a rare insight into the day and life of public sector schools exposed to market or market-like settings. Relatedly, the study shows how branding strategies appear and unfold in two settings compared, one highly marketized and one little marketized setting, providing empirical data on differences in branding strategies relating to the level of marketization.

Before presenting the theoretical framework and methodological approach, the research context, in this case public sector upper secondary schools in Oslo, Follo, and Romerike, is laid out in detail, including how public sector education in the city is part of a Nordic tradition and way of organizing education at this level. In addition, certain relevant demographical features of Oslo as a city are presented.

Research Context

Education in a Divided City

The research setting for the present study is public sector upper secondary schools in the Norwegian capital of Oslo and its neighboring areas Follo and Romerike. Oslo represents an interesting setting for research on upper secondary schools, as it is a rather divided city with pronounced divisions along economic, social, and ethnic dimensions. Between 1860 and 1900, much of the wealthy, predominantly white upper middle-class and prosperous elite settled in the leafy western parts of the city (Myhre 2017), while lower middle-class and working-class people lived in the more industrialized, suburban, and, later on, multiethnic eastern districts.

While there are some working-class areas in the west, and areas peopled by the well-off in the east, including areas transformed by gentrification, the city is to this day characterized by a prominent and, even, increasing east/west divide (Ljunggren and Andersen 2015). The divide applies to fortune (Wiborg 2017), income (Andersen and Ljunggren 2014), health (Elstad 2017), crime rate (Sætre et al. 2018), and immigrant share (Wessel 2017), materializing as more favorable life chances for those living in the city's affluent neighborhoods (Toft and Ljunggren 2016).

Of particular interest is the work by Hansen (2017) on classed recruitment to Oslo's upper secondary schools. Using data from a 10-year period, she finds that recruitment to upper secondary schools has become less *geographically* segregated; students from different districts, for example, in the east and west, increasingly go to the same schools. However, segregated recruitment based on *social background* and *class position* has not been reduced in the same period. Students from privileged backgrounds mainly choose schools in the western and central parts of the city, while a majority of students from non-privileged backgrounds seek schools in the eastern districts. Concurrently, the privileged schools based on popularity are mainly situated in the west and central parts of the city, while the marginalized schools are found in the eastern districts. Hansen (2017, p. 16) describes how "students from privileged backgrounds to a greater extent than students from non-privileged backgrounds will perceive enrolling at an unpopular east-end school as a form of social degradation."

As a contrast, the surrounding areas of Follo and Romerike have a notable more equal distribution of wealth and social resources than Oslo, and they are much less divided along ethnic dimensions. While there is some inequality in these areas, like in most other areas, there is not the same wedge-like divide between the affluent and the less affluent as in Oslo.

The Nordic Education Model

Historically, the Norwegian education system belongs to "the Nordic model of education" developed in the rebuilding era after World War II, a period in which universally available education for all children was regarded as the primary motor for reducing social and economic gaps in the population and as a catalyst for social mobility. Certain values were building bricks, namely, "social justice, equity, equal opportunities, inclusion, nation building, and democratic participation for all students, regardless of social and cultural background and abilities" (Imsen et al. 2017, p. 568). Today, most high schools in Norway are publicly owned and run, and, "even though the number of private schools has risen, the private school market is still strongly regulated" (Haugen 2020, p. 2). However, similar to the education systems in many western countries, the education system in Norway has been exposed to New Public Management ideas and neoliberal management reforms and policies since the first part of the 1990s.

Unlike neighboring Sweden, Norway has not introduced a voucher system (Imsen et al. 2017), but has introduced neoliberal elements into the systems for student admission, school funding, and education quality measures, resulting in marketization of the upper secondary school field through a governmentally planned and supervised quasi-market (Glennerster 1991). Quasi-markets in the public sector are not designed to maximize profit, but often to increase efficiency, create transparency, and boost consumer choice. At the high school level, government in Norway has not made such measures mandatory for municipalities through a nationwide policy, but has provided it as an option.

The Oslo School Market

Local authorities in Oslo have embraced the option fully and have implemented several of the elements. Free choice of schools has been introduced as the mechanism governing student admissions. Through this system, students can apply to and be accepted by any public high school within the city, as only student grades from primary and lower secondary schools determine whether a student is accepted by upper secondary schools or not, and schools cannot reject students through *cream skimming* (Haugen 2020, p. 5). If the grades are not good enough for admission to the desired school, then students are guaranteed a place in another school with available places.

Free choice of schools is combined with per capita funding, a funding principle where the amount of public funding schools receive depends on incentives like the number of students they admit, including the amount of students with special needs and students from minority backgrounds. Funding also depends on the number of students who graduate from the respective schools. The combination of free choice of schools and per capita funding is viewed as a catalyst for competition, as "competition as a device is effective only when market recruitment is directly related to funding and the survival or well-being of individual organizations" (Ball 2008, p. 54). In other words, money follows the student in a system where students are free to choose whichever school they want.

In addition, schools in the city are economically and budgetarily self-governed and must themselves bear the financial consequences of attracting few applicants or having a low graduation rate. To control if school results are good enough, an extensive test schedule based on standardized testing has been introduced. Until recently, full school results have been published on a website[1] open to everyone, making comparison between schools easy. However, school-specific results are no longer published, making direct comparison difficult. School principals are held accountable through contracts "where evaluation criteria and financial incentives are connected to the results" (Haugen 2020, p. 7). Summed up, the public sector upper secondary schools in Oslo are governed through a multiple whammy of marketization: a whole package consisting of free choice of schools, per capita funding, decentralization, standardized testing, accountability, and, to some extent, comparison are the building bricks of the upper secondary school market in the economically, socially, and ethnically divided city of Oslo.

[1] Results are published at www.minosloskole.no.

High Schools in Follo and Romerike

In the more economically, socially, and ethnically uniform areas of Follo and Romerike, the upper secondary school fields are less marketized compared to those in Oslo. Unlike Oslo, these areas do not allow for completely free choice of schools. Instead, they have chosen a zoned approach where students can only apply to schools in a number of municipalities close to their home.[2] Within the application zone, admission is based on student grades. For Follo, this means students can apply to nine schools,[3] including two schools in a neighboring area to the south. Although nine schools may provide students with the choice between several schools, the schools are for the most part situated far apart – it is more than 50 kilometers (31 miles) between the northernmost and the southernmost school. Some of the schools are situated within a 10-kilometer (6.2 miles) radius, but transportation is much less frequent and is more time-consuming than in Oslo, making it difficult to commute to schools. There is no metro system or tram network, with infrequent buses and local trains as the sole options. Hence, Follo does not have a functioning school market, and there is little real competition between these schools.

Much of the same applies to Romerike. This area is divided into two application areas: one for the northern part (with five schools) and one for the southern part (with ten schools).[4] In the southern area, a few schools lie within a 10-kilometer radius, but again transportation is relatively time-consuming. Due to great distances between the other schools, most students chose a school close to home. Like Follo, this is not a well-functioning market, and choice of school is not really free.

Relatedly, unlike Oslo, funding of upper secondary schools in these areas is not based on the per capita principle where money follows the student. Instead, schools are given a base funding independent of the number of applicants and graduation rate. Another crucial difference from Oslo is a much more relaxed testing schedule. Schools here take part in the yearly national tests, but are not subjected to a battery of extra tests as is the case in Oslo. As follows, there are few results to compare, which virtually eliminates direct comparison of results between schools. Accountability based on evaluations and test results seems to be of less concern here than in Oslo. Summed up, there is a huge divide in marketization between Oslo and Follo to the south and Romerike to the north, with the latter areas being little marketized compared to Oslo.

[2] An exception is if they want to apply for a specialization that none of the schools in their zone offer, in which case they may apply to schools in the whole county.

[3] Eight of the nine schools offer Specialization in General Studies.

[4] 12 of these 15 schools offer Specialization in General Studies.

Theoretical Framework

According to Kornberger (2010), we live in a "brand society" where organizational branding is key. Organizational branding, defined as the "systematic effort to develop and present the organization as one unified brand" (Christensen et al. 2008, p. 64), is the culmination of a change in organizational identity creation: 1) The identity is increasingly tied to outsiders' view of the organization, and 2) the brand becomes the channel for actual identity construction. Hence, organizations of today operate successfully if they are able to "put themselves out there, to convey who they are, what they do, and what they stand for" (Fombrun et al. 2004, p. 95; Wæraas 2019).

Organizational identity is increasingly regarded as a flexible, changing, and unstable notion. Contrary to prior assumptions, "the apparent durability of identity is somewhat illusory," and organizational identity "is actually relatively dynamic" and "potentially precarious and unstable" (Gioia et al. 2000, p. 64). This paves the way for building, maintaining, and protecting organizational identity through branding.

Organizational branding has two dimensions: the internal dimension involves management of employees' commitment to and support of the desired brand (Miles and Mangold 2004; Wæraas and Dahle 2020), whereas the external dimension involves visual and verbal communication to establish emotional bonds with external stakeholders (Leijerholt et al. 2019). In the present chapter, the external dimension of organizational branding is in focus.

School Branding

Public sector branding has received relatively little scholarly attention, and results are contradictory. In their review, Leijerholt et al. (2019) considered all papers published before and during 2016 and identified only 92 relevant papers all in all. To avoid conceptual misunderstandings, some conceptual clarification is offered: *Public sector brands* is conceptualized as "public organizations, public services, or public policies that among stakeholders are associated with unique and attractive values, meanings, and characteristics," while *public sector branding* involves the "process of systematically creating such associations in the minds of stakeholders" (Dahle and Wæraas 2020, p. 3). Some similarities with corporate marketing exist, but whereas marketing includes establishing awareness and a will to purchase among customers, branding forms identities, values, and meanings.

Branding in public sector schools is even more uncharted waters. Several studies explore branding of private or independent schools (Cheng et al. 2016; Polat et al. 2010; Trivitt and Wolf 2011), but few focus on public sector schools. Lubienski (2007) studied public sector schools, but highlighted promotion materials only. Gewirtz et al. (1995), on the other hand, focused on visual techniques for public sector schools branding.

Exploring branding of public sector high schools in New York City, DiMartino and Jessen (2016) found that partner organizations increasingly incorporated their names into school names, logos, symbols, and promotion material, which was effective to attract prospective students and their parents. Cucchiara (2008) followed a re-branding initiative of public sector elementary schools in downtown Philadelphia, where schools were branded as urban elite schools for upper middle-class children by downplaying race and class. This proved effective, as it attracted kids from privileged families, but at the same time it raised the entry bar for qualified but less privileged children. Dahle and Wæraas (2020) found that marketization of upper secondary schools in Oslo set up inescapable incentives for school branding, and that both privileged and less privileged schools demonstrated branding efforts.

School Markets and Institutional Logics

In the present study, the upper secondary school market is viewed as an *organizational field* comprised of "organizations that, in the aggregate, constitute a recognized area of institutional life: key suppliers, resource and product consumers, regulatory agencies, and other organizations that produce similar services and products" (DiMaggio and Powell 1983, p. 148). Institutional logics theory is used as a framework for examining and explaining the relationship between marketization of upper secondary schools, responses to market pressure, and school branding. As follows, branding efforts as a response to market exposure are understood in relation to the core mechanisms of institutional logics.

According to Alford and Friedland (1985) and Friedland and Alford (1991), institutional logics set up broad belief systems to shape the cognition, behavior, and communication of actors in organizational fields (DiMaggio 1979), leading to "supra-organizational patterns of human activity." Consequently, the construct of institutional logics is defined as "the socially constructed, historical patterns of material practices, assumptions, values, beliefs, and rules by which individuals produce and reproduce their material subsistence, organize time and space, and provide meaning to their social reality" (Thornton and Ocasio 1999, p. 804). Importantly, the prevailing institutional logic not only provides unwritten rules for action and interaction; it functions as a guide for how to "interpret the organizational reality" (Thornton and Ocasio 1999, p. 804).

Both the behavior-, the interaction-, the cognition-, and the interpretation-shaping functions of institutional logics are seen as embedded in central institutions in human societies, namely capitalism, the bureaucratic state, democracy, family, and truth (incorporating religion and science) (Friedland and Alford 1991). These societal institutions "provide a distinct set of often contradictory logics that form the bases of political conflicts" (Thornton and Ocasio 1999, p. 805). The set of societal institutions embedded with institutional logics was revised by Thornton and Ocasio (1999) and Thornton et al. (2012) to include the market, the corporation, the organization, the professions, the state, the family, the industry, religion, inter-organizational networks, and geographic communities.

Because governing of public secondary schools in the city has been reformed according to neoliberal principles, the field of upper secondary schools is seen as a market with a prevailing institutional logic rooted in economic principles (Ertimur and Coskuner-Balli 2015). Simultaneously, the teaching professions may inhibit another institutional logic based on principles other than economic ones. This professional logic, which for educators is rooted in "common values and beliefs about quality standards" and pedagogical methods in schooling of children and teenagers, can facilitate "normative isomorphism" (Hattke et al. 2016, p. 238) valid in the institutional field of upper secondary education. As the municipal Oslo school administration operates "within state legal frameworks" (Helgøy and Homme 2016, p. 53) and is organized as a bureaucracy, institutional logics tied to the bureaucratic state apply to upper secondary schools in the city. Thus, "bureaucratic logics, based on process control, democratic participation, and state intervention" (Hattke et al. 2016, p. 238), probably is in action as well as market and professional logics. However, the influx of market logic has to some extent taken place on the expense of educators' professional logic.

Methodology

Data Collection

The present study builds on the study of branding in Oslo city high schools by Dahle and Wæraas (2020) and Dahle (2020), in which official admission statistics provided by the Oslo municipal administration and necessary admission grade levels for the Specialization in General Studies for each school were presented. This served as a proxy for the schools' popularity and thereby market position.

In this study, the same methodology is used, but with new high schools admission statistics for 2019/2020 and a sample of 15 schools in Oslo, and an additional 6 schools in Follo, and 9 schools in Romerike, 30 schools in total. Among the schools offering Specialization in General Studies, the sample represented 2/3 of the schools in Oslo, 6/7 of the schools in Follo, and 3/4 of the schools in Romerike. By purposive sampling (Silverman 2013), schools with different admission levels were identified. In Oslo, five schools were on the lowest admission level (10 to 37.4 admission points necessary for admission) and were viewed as marginalized. Five schools were on a medium admission level (37.5 to 44.9 points), while five schools were on the highest admission level (45 to 60 points) and were viewed as privileged. Correspondingly, in Follo and Romerike, five schools were on the highest level, six were on the medium level, and four schools were on the lowest level.[5]

[5] Due to lower admission levels in Romerike, the level for the privileged group of schools in Romerike was set to 43 points and above.

Branding efforts, particularly how school management chose to brand their schools, were studied through examination of the web sites and social media profiles of the schools. Specifically, on the web sites for each school, the front page, the "About Us" section, and the under-sections for "Our profile," "Strategic plan," "How we work," and/or "History"[6] were studied. For social media sites, the front page and the "About" section were studied. The analysis was based on the level of distinctiveness of the branding (Fombrun et al. 2004), specifically "emphasizing the distinctive characteristics that make it different from its competitors" (Kapferer 2008, p. 96). All statements were translated from Norwegian to English by the author.

Analysis

An in-depth analysis of textual branding efforts by the 30 sampled schools was carried out by performing a content analysis of branding statements from the schools' websites and Facebook and Instagram profiles. Branding statements were analyzed by content analysis with thematic coding (Kuckartz 2014), using the software Provalis QDA Miner. Content analysis is a research method rooted in "systematic examination of communicative material" (Mayring 2000, p. 266), in this case text. It is based on classification of codes, including "breaking down the text into single units of analysis, and oriented to a system of categories" (Mayring 2000, p. 267). During the first data-driven step of coding, first-order codes were assigned to branding statements. As a result of this process, 36 codes were identified and assigned to statements.

In the second step of coding, axial coding (Corbin and Strauss 1990) was applied to reduce the codes into second-order categories. For example, "common characteristics," "good enough," and "for all" (first-order) were coded as "generic" (second-order), while "better than others," "top results," and "university affiliation" (first-order) were coded as "differentiating" (second-order). The procedure led to three second-order categories, "Generic," "middle," and "differentiating," which reflect the extent of how branding differentiates one organization or product from another in a positive way. The full coding scheme with categories is shown in Table 1.

Results

Dahle and Wæraas (2020) unveiled that marketization of upper secondary education led to schools at different admission levels engaging in branding efforts, in order to cope with the competition for students and, thus, public funding. Thus, both privileged and marginalized schools actively try to improve their brand. The present analysis nuances and expands on these findings.

[6] Not all school web pages had a section for "History."

Table 1 Codes and categories extracted from the schools' branding efforts

Codes	Categories
Better than others	Differentiating branding
Top results	
Ambitious	
Many applicants	
University collaboration	
Business cooperation	
STEM focus	
Famous alumni	
Awards received	
History	
Pride	
Progressive	
Specializations	
Talent programs	
Both generic and differentiating	Middle-level branding
Both supportive and challenging	
Competent staff	
Modern buildings	
Up-to-date equipment	
Extra-curricular activities	
Unleash potential	
Forward-looking	
Special vibe	
Curiosity and exploration	
Common characteristics	Generic branding
Good enough	
Graduation focus	
Reduce absence	
Student attachment	
Extra help to students	
Student voice	
For all	
Mastery	
Tolerance	
Supportive	
Safety	

Differentiation in Oslo

The study confirms the findings by Dahle (2020), who shows how schools in Oslo generally used unique logos, slogans, and values commonly regarded as positive in their branding efforts, and that marginalized schools demonstrated a less differentiating and more generic branding than privileged schools (Dahle 2020). The more thorough examination of branding statements in the present study reveals that the marginalized schools' branding efforts were generally reliant on generic characteristics which arguably did little to improve the schools' brand. This ran along three dimensions: (1) empty phrases ("open," "inspiring," "united"), (2) supportive features ("caring," "safe," "help"), and (3) non-elitist ("reduce absence," "all students graduate," "low threshold").

In contrast, privileged schools generally relied on unique, non-generic, differentiating characteristics with a less supportive and more elitist approach, which was meant to brand them as prestigious, high-quality institutions ("proud academic traditions," "leading," "good results") with celebrated alumni, including Nobel Prize winners, a famous mathematician, an internationally renowned painter, and a former prime minister and now Secretary General of NATO. Consequently, the highly popular, privileged schools were able to depict themselves as more attractive than their marginalized counterparts.

Generic in Follo/Romerike

The analysis unveiled that branding efforts took on a different character and were on another level in the surrounding areas of Follo and Romerike than in Oslo. As there were no notable differences between the branding in the areas of Follo and Romerike, these 2 areas and their 15 schools were viewed and analyzed together. Only 2 of 15 schools used unique logos both on the website and on social media profiles, which is far less than the 10/15 ratio Oslo. Eight other schools used a logo on only one of the two platforms, signalling an inconsistent use of logos. Most schools had vision statements, but only seven schools used slogans or slogan-like statements. In contrast, almost all the schools in Oslo used slogans or slogan-like statements.

While all Oslo schools sought to brand their schools in one way or another, school branding seemed to be of less concern in the surrounding areas. Put simply: schools in Follo and Romerike generally engaged in both less-eager and less-differentiating branding efforts than their counterparts in Oslo. This came to view through the analysis of branding statements by the privileged schools in these areas. While the Oslo schools bragged about a prestigious history, top results, and famous alumni, the schools in Follo and Romerike took on a less boastful approach. None brought forward top results on a national level or claimed their school was highly ranked, like one school in Oslo, which declared that it was "among the country's leading schools."

Instead, a privileged school in Follo stated it was "the best school in the municipality" before adding it had "the world's best students." Another school proclaimed it had "knowledgeable teachers and students with high ambitions," and they specialized "in Specialization in General Studies." No mentions of well-known alumni were found, but this may partly be because these schools have fewer famous alumni to boast about. The contrast to Oslo and the mention of Nobel Prize winners, pioneering academics, and state leaders is however huge. Few mentions of prestigious history were found. The closest came one school in Romerike, established 136 years ago, of which the school only informed that it was "the oldest upper secondary school in the area" and that the school was regarded as "impressive and modern for its time." As part of this modest approach, these schools used less differentiating and, to some extent, generic characteristics in their efforts to positively stand out from other schools in positive ways. One school was concerned with "the joy of learning," while another acknowledged that students want to "be seen and respected."

Figure 1, which illustrates the frequency of differentiating characteristics with an elitist approach, like good results, prestigious alumni, traditions, history, awards, STEM focus, and university collaboration combined, clearly shows how privileged schools used a more differentiating branding than less privileged schools and that this finding is more pronounced in the more marketized upper secondary school field in Oslo than in the surrounding areas. Other schools pointed to non-elitist and supportive measures seen in the communication by marginalized schools in Oslo, exemplified by statements like "all students can learn," "inclusive," and "90 percent (of the students) shall graduate."

Some schools at the *middle* level also pointed to differentiating characteristics in their branding efforts ("university school," "participates in international programs like Erasmus"), but generally this was accompanied by supportive and non-elitist communication. This was well illustrated by one school, which stated that "students

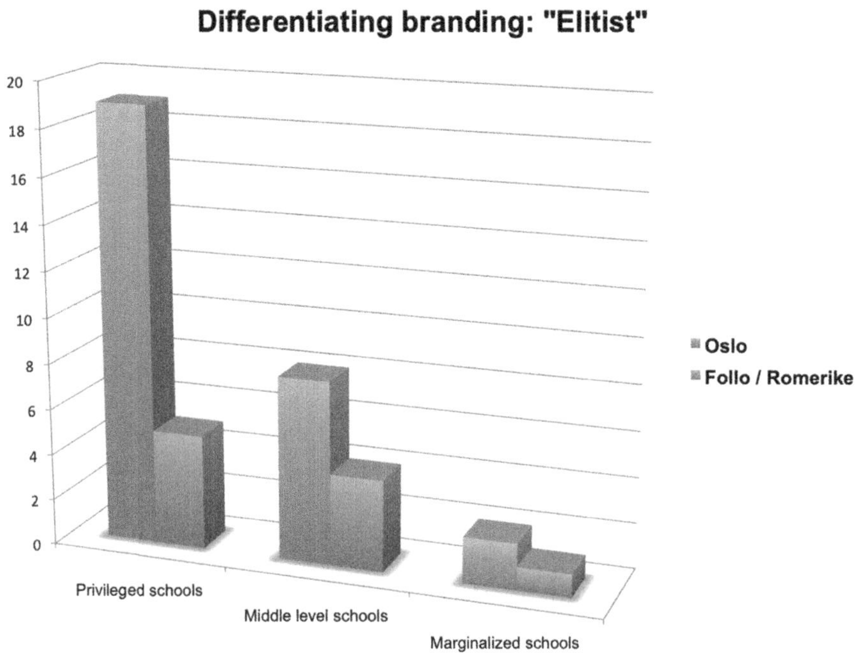

Fig. 1 The use of differentiating branding by schools in different market positions in Oslo and Follo/Romerike

feel [included]," but also that the school is "demanding." In Follo and Romerike, there were similarities between the branding efforts by schools at the middle and the privileged level. As such, the differences between schools at each level in Follo and Romerike appear to be relatively minor compared to those in Oslo.

At the marginalized level, schools relied upon more or less generic characteristics bordering on empty phrases ("one step beyond," "mastery of life"), partly based on non-elitist and supportive values ("complete and graduate"; "open, tolerant, and inclusive"; "trust"; "respect"). Interestingly, this was the case even for privileged schools in Follo and Romerike. A summation of the frequency of supportive and non-elitist characteristics with an inclusive, "for all" approach, like tolerance, mastery, safety, diversity, care, solidarity, well-being, inclusion, belonging, success for all students, and extra help, is shown in Fig. 2.

In the marketized Oslo school market, it is mainly the marginalized schools which demonstrate this type of supportive branding, while the privileged schools hardly use it at all. This, however, does not apply to schools in the less marketized Follo and Romerike, where this type of branding is used by schools at all levels, and the frequency is even a little higher in the privileged than in the marginalized schools.

Summed up, analysis of the social media and website communication by privileged schools in Follo, Romerike, and Oslo leads to three core findings: 1) Schools in Follo and Romerike seem to be less concerned about branding than their counterparts in Oslo; 2) branding by the Oslo schools has a distinctly more differentiating character, based on elitism, prestige, and quality, than schools in the two

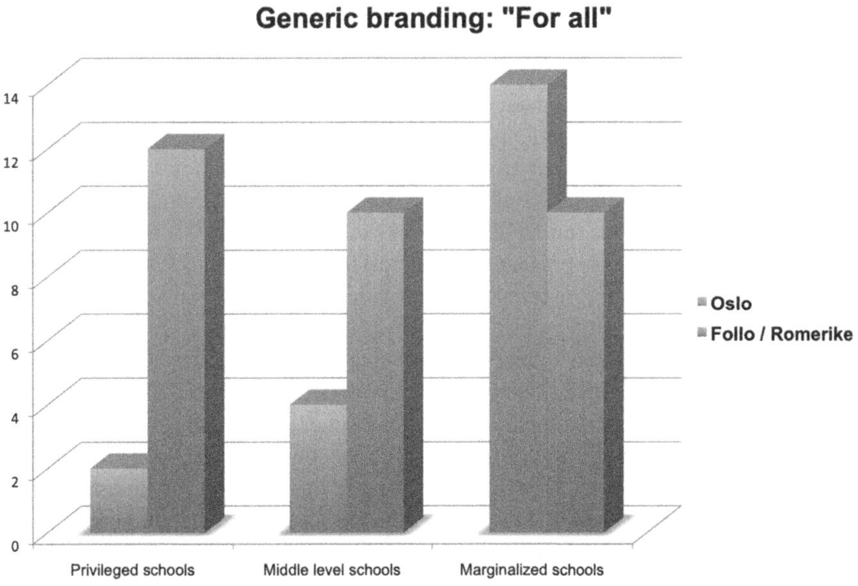

Fig. 2 The use of generic branding with supportive features by schools in different market positions in Oslo and Follo/Romerike

neighboring areas; and 3) branding efforts by schools in Follo and Romerike do not mirror the pronounced difference in branding strength between privileged and marginalized schools in Oslo, as there are only minor differences in branding strength between schools on different levels in these two areas. An overview of branding characteristics and example statements reflecting differentiating, middle-level, and generic branding by schools at the three popularity levels in Oslo and Follo/Romerike is presented in Tables 2 and 3.

Discussion

This study has a dual purpose. Firstly, the study explores the extent of branding efforts by upper secondary schools at different levels of popularity, measured by grades necessary for admission. Secondly, it explores possible differences in school branding efforts in the highly marketized school field in Oslo and the less marketized neighboring areas of Follo and Romerike, providing a comparative perspective.

Theoretical Implications

The findings are understood in light of institutional logics theory. Widespread branding efforts in the most marketized areas, in this case Oslo in contrast to the surrounding areas, can be seen as empirical support of the existence of broad belief systems within upper secondary schools in Oslo. With the implementation of neoliberal policies, the upper secondary school field in Oslo took on market qualities as schools had to compete for students and, thereby, funding. Thus, a market logic was introduced in tandem with institutional arrangements like free choice of schools and per capita funding. This has continued to shape patterns of human activity, for example, branding efforts initiated by school executives in Oslo. Simultaneously, a market logic has been less present in Follo and Romerike due to little school marketization in these areas compared to Oslo, which again may explain why schools in these areas demonstrate less branding than many schools in the capital.

As the institutional arrangement of free choice of schools affects all public upper secondary schools in Oslo, it is an inescapable feature of a marketized school sector. This implies that the dominating market logic is rather evenly diffused in schools on different popularity levels. Still, privileged schools demonstrate stronger branding than marginalized schools. There may be several explanations for this. It may be that the market logic is in fact *not* evenly diffused, but is more prominent in privileged schools, or that school executives in privileged schools more effectively see to that branding concerns are infused in the organization and its members. Alternatively, it may be that marginalized schools are hybridized in the sense that different logics are active within them, for example, that professional logics play a greater role in

Table 2 Statements representing different differentiating levels used by schools at three popularity levels in Oslo and Follo/Romerike

	Oslo	Follo/Romerike
	Differentiating branding	*Middle-level branding*
Privileged schools	"Proud academic traditions" "The mathematician Niels Henrik Abel's school" "Nobel prize winner Trygve Haavelmo went to school here" "Among the country's leading schools" "Good results for many years" "Cooperates with a nearby research institution" "An ambitious and creative school environment" "At this school a young Edvard Munch was a student" "Proud" "We want to make a difference" "Among the highest number of applicants in Oslo" "Roots as far back as 1153 A.D."	"The best school in the municipality - the world's best students" "A diverse learning community" "We strive to see the potential in every student" "All students can learn" "The joy of learning" "To be seen and respected" "A university school" "Knowledgeable teachers and students with high ambitions" "We specialize in specialization in general studies" "By 2023 90 percent shall graduate, and achieve better grades" "Open, tolerant, and inclusive"
	Middle-level branding	*Middle-level/generic branding*
Middle-level schools	"Traditional and forward-looking" "Oslo's second oldest secondary school" "A diverse school with university collaboration" "Oslo's best school for entrepreneurship" "An inventive school" "High ambitions for *all* students" "Against all odds" "100 years of history" "Students graduate with good exam results" "Well-being leads to better learning" "Go as far as you want" "Open, safe, pleasant, ambitious"	"Always towards new goals" "Forward-looking, exploring, and student-engaging" "A university school" "Students shall reach their maximum potential in a safe environment" "Participates in international programs like Erasmus" "Open and tolerant" "Students feel included, but also that the school is demanding" "Stimulate curiosity, a need for exploration, and enthusiasm for creativity" "Inclusive learning"
	Generic branding	*Generic branding*
Marginalized schools	"Inspiring learning environment" "Proud, varied and united" "Safety and well-being for all students" "Everyone fits in" "Low threshold, high tolerance" "Reduce absence" "Complete and graduate" "An approved diploma at graduation" "Sees your talents" "Speak up, be proud" "Nobody should feel lonely" "We provide a whole team around each student" "We strive for our students to complete their studies and graduate"	"One step beyond" "Safe and innovative" "Mastery of life" "Sustainability" "Flexibility" "Open, tolerant, and inclusive" "Trust, respect, and appreciation" "Students shall complete and graduate" "Inspiring learning environment" "Inclusive learning environment" "Screening of students' skills in English, Norwegian, and mathematics" "We offer personal tutors if necessary"

Table 3 Characteristics of the branding efforts by all the schools in the study, as identified in the coding and categorization process

	Oslo		Follo/Romerike	
	School #	Branding characteristics	School #	Branding characteristics
Privileged schools	1	Logo, Latin slogan, prestigious alumni, history, academic traditions, results, graduation rate	16	Logo, slogan, community, joy of learning, inspiration, care, history
	2	Logo, slogan, awards, modernity, ambition, results, make a difference, professionality, competent teachers	17	Logo, IB class, diversity, well-being, ambitions for all students, university collaboration, internationalization
	3	Logo, slogan, results, nationally leading, research and university collaboration, industry collaboration, STEM focus	18	Logo, Latin slogans, professionalism, IB class, ambition, competent teachers, progressiveness, inclusion, safety, respect
	4	Slogan, student environment, diversity, results, awards, history, traditions, well-being	19	Logo, mastery, graduation focus, tolerance, inclusion, ambitions for all students, university collaboration, STEM focus
	5	Logo, vibrance, creativity, progressiveness, student environment, university collaboration	20	Graduation focus, safety, hard work, feedback to students, diversity, student environment
Middle-level schools	6	Logo, slogan, ambition, inclusion, progressiveness, internationalization, location	21	Logo, university collaboration, internationalization
	7	Logo, slogan, STEM focus, internationalization, best in Oslo, student popularity, awards	22	Logo, slogan, forward-looking, mastery, graduation focus, student voice, inclusion
	8	Logo, slogan, history, results, forward-looking, nationally leading, awards	23	Slogan, community, inclusion, hard work, mastery, graduation focus, tolerance, trust, competent teachers
	9	Logo, student popularity, university collaboration, diversity, progressiveness, ambitions for all students, teamwork	24	Logo, slogan, professionalism, openness, creativity, progressiveness, solidarity, industry and university collaboration
	10	Logo, slogan, sports profile, talent nurturing, student environment, inclusion	25	Mastery, flexibility, graduation focus, openness, tolerance

(continued)

Table 3 (continued)

	Oslo		Follo/Romerike	
	School #	Branding characteristics	School #	Branding characteristics
Marginalized schools	11	Logo, slogan, internationalization, university collaboration, graduation focus, mastery, safety	26	Logo, slogan, inclusion, ambitions for all students, internationalization
	12	Slogan, history, diversity, tolerance, safety, well-being, graduation focus	27	Mastery, flexibility, graduation focus, openness, tolerance, award
	13	Logo, slogan, IB class, student environment, belonging, student voice, success for all students, safety, well-being	28	Logo, slogan, safety, innovation, progressiveness, mastery, sustainability
	14	Logo, safety, inspiring, competent teachers, graduation focus	29	Logo, safety, inspiration, trust, inclusion, STEM focus, internationalization
	15	Logo, slogan, diversity, student environment, graduation focus, safety, well-being	30	Forward-looking, mastery, inclusion, learning environment, screening of students, extra teaching if necessary

marginalized schools than in privileged schools. Or it may be that the differentiating qualities of the schools' branding are in fact products of the schools' market position. As popular and privileged schools seem to have more to boast of in terms of unique and differentiating "selling points" than their less privileged fellow schools, they enter the competition ahead of the marginalized schools. Marginalized schools find themselves in a situation where they have fewer differentiating features to use in their branding efforts. They do not have the outstanding results of the other schools, they lack a prestigious history, and they find little famous alumni to brag about. As a result, they struggle to brand their schools, and, thus, they are not able to present themselves as so attractive as the privileged schools are able to do.

Practical Implications

Marketization of the upper secondary school field in Oslo has more or less been a deliberate and planned policy by both conservative and social democratic local governments, not an outcome of political compromises or the by-product of other political decisions. Seen from the decision-makers' side, a policy infused with a market logic has effectively led to market behavior by schools and their executives. The package of marketization measures introduced in Oslo, with free choice of schools combined with per capita funding as cornerstones, seems to be quite effective for governing the school field as a quasi-market. It functions as a quasi-market where schools find it hard not to act according to market rules, including trying to bolster their attractivity through branding. The opposite is the case in areas where policy makers have not tried to set up a quasi-market, in this case the less marketized areas surrounding Oslo. Here schools demonstrate less eager branding efforts and weaker branding than their counterparts in the capital city.

In Oslo, the local government led by a market-friendly right-wing coalition has expressed few concerns about a possible growing divide between privileged and marginalized schools in the city. The current left-wing coalition in power have signaled that they will adjust and soften free choice of schools to reduce segregation and decrease the divide between schools. So far this has not been done, although publication of all school results has been altered to avoid direct comparison of test results. The policy makers will continue to have to deal with the effects of free choice of schools combined with per capita funding in a city as socially, economically, and ethnically divided as Oslo. Many of the high-performing schools, as identified by standardized test results, are popular. Many of them are situated in the affluent districts in the western city center and the western districts. In a divided city, free choice of schools may lead to these schools becoming even more popular, at the expense of lower-performing and less popular schools in the eastern parts of the city. Such a development is probably being accentuated by the link to per capita funding. A stronger and probably more effective branding by the privileged schools, as unveiled by the present research, may accentuate the development even further.

As a core practical implication, practitioners should take care when implementing school marketization in cities strongly divided along economic, social, and ethnic dimensions, as it is likely that it will set in motion a spiral toward a prominent division between A- and B-level players among both schools, students, and teachers. On the other hand, it may boost both privileged schools and students alike and lift some of them to elite status, which probably will be a desirable turn-out for some policy makers.

Limitations and Avenues for Future Research

The present study, like other studies, has limitations. One limitation is the level of branding is not as high in public sector upper secondary schools in Norway as in similar schools in the Anglo-American and Asian world, mainly due to schools being reliant on limited public budgets alone. As these schools in Oslo demonstrate a low to middle level of branding (Dahle and Wæraas 2020), they may not be as representative for public sector secondary schools as desired, yielding slightly different results than elsewhere. On the other hand, Oslo as a zone of school experimentation represents an exciting setting for a study on school marketization and branding in light of institutional logics. Another limitation is that the study relies on content analysis of branding statements alone. While school principals in Oslo spoke about branding in the study by Dahle and Wæraas (2020), their thoughts and rationalizations behind their schools' *actual branding efforts* remain undisclosed. The same applies to their colleagues in Follo and Romerike. Exploring this through interviews could have added to the analysis of branding statements in positive ways, something which represents an avenue for further research.

A third limitation is that the introduction of a market logic is not seen in comparison with a point in time before the market logic was introduced. Such a temporal dimension could have bolstered the design of the study, providing an interesting comparative perspective. This too represents an avenue for further research, although it would have to rely on existing data from the time before marketization was introduced to the upper secondary school market in Oslo. Alternatively, a before and after -study could be designed in tandem with the possible case of a planned introduction of marketization measures. A fourth limitation is that the question of institutional logics cannot directly be identified through analysis of branding statements. If this had been the case, it could have bolstered the link between the empirical and the theoretical part of the study. In itself this represents an avenue for future research, that is, to examine empirically the shift from a professional to a market logic.

References

Alford, R. R., & Friedland, R. (1985). *Powers of theory: Capitalism, the state, and democracy*. Cambridge: Cambridge University Press.

Andersen, P. L., & Ljunggren, J. (2014). Gylne ghettoer. Inntektselitens bostedssegregering i Oslo, 1980–2005. In O. Korsnes, M. Hansen, & J. Hjellbrekke (Eds.), *Elite og klasse i et egalitært samfunn* (pp. 126–143). Oslo: Universitetsforlaget.

Ball, S. J. (2008). *The education debate*. Bristol: Policy Press.

Cheng, A., Trivitt, J. R., & Wolf, P. J. (2016). School choice and the branding of Milwaukee private schools. *Social Science Quarterly, 97*(2), 362–375.

Christensen, L. T., Morsing, M., & Cheney, G. (2008). *Corporate communications: Convention, complexity and critique*. London: SAGE.

Corbin, J. M., & Strauss, A. (1990). Grounded theory research: Procedures, canons, and evaluative criteria. *Qualitative Sociology, 13*(1), 3–21.

Cucchiara, M. (2008). Re-branding urban schools: Urban revitalization, social status, and marketing public schools to the upper middle class. *Journal of Education Policy, 23*(2), 165–179.

Dahle, D. Y. (2020). *Marks of distinction: Branding responses to market logics in schools*. Working Paper 6/2020. Norwegian University of Life Sciences, Aas, Norway.

Dahle, D. Y., & Wæraas, A. (2020). Silence from the brands: Message control, brand ambassadorship, and the public interest. *International Journal of Organization Theory & Behavior, 23*(3), 259–280.

DiMaggio, P. (1979). Review Essay: On Pierre Bourdieu. *American Journal of Sociology, 84*(6), 1460–1474.

DiMaggio, P., & Powell, W. W. (1983). The iron cage revisited: Collective rationality and institutional isomorphism in organizational fields. *American Sociological Review, 48*(2), 147–160.

DiMartino, C., & Jessen, S. B. (2016). School brand management: The policies, practices, and perceptions of branding and marketing in New York City's public high schools. *Urban Education, 51*(5), 447–475.

Elstad, J. (2017). Helseulikhetenes by. In J. Ljunggren (Ed.), *Oslo – ulikhetenes by* (pp. 171–187). Oslo: Cappelen Damm Akademisk.

Ertimur, B., & Coskuner-Balli, G. (2015). Navigating the institutional logics of markets: Implications for strategic brand management. *Journal of Marketing, 79*(2), 40–61.

Fombrun, C. J., Van Riel, C. B., & Van Riel, C. (2004). *Fame & fortune: How successful companies build winning reputations*. Upper Saddle River, NJ: FT Prentice Hall.

Friedland, R., & Alford, R. R. (1991). Bringing society back in: Symbols, practices, and institutional contradictions. In P. J. DiMaggio & W. W. Powell (Eds.), *The new institutionalism in organizational analysis* (pp. 232–267). Chicago: University of Chicago Press.

Gewirtz, S., Ball, S. J., & Bowe, R. (1995). *Markets, choice, and equity in education*. Buckingham: Open University Press.

Gioia, D. A., Schultz, M., & Corley, K. G. (2000). Organizational identity, image, and adaptive instability. *Academy of Management Review, 25*(1), 63–81.

Glennerster, H. (1991). Quasi-markets for education? *The Economic Journal, 101*(408), 1268–1276.

Hansen, M. (2017). Segregering og ulikhet i Oslo-skolen. In J. Ljunggren (Ed.), *Oslo – ulikhetenes by* (pp. 257–275). Oslo: Cappelen Damm Akademisk.

Hattke, F., Vogel, R., & Woiwode, H. (2016). When professional and organizational logics collide: Balancing invisible and visible colleges in institutional complexity. In J. Frost, F. Hattke, & M. Reilhen (Eds.), *Multi-level governance in universities: Strategy, structure, control* (pp. 235–256). Cham: Springer.

Haugen, C. R. (2020). Teachers' experiences of school choice from 'marginalised' and 'privileged' public schools in Oslo. *Journal of Education Policy, 35*(1), 68–94.

Helgøy, I., & Homme, A. (2016). Educational reforms and marketization in Norway – A challenge to the tradition of the social democratic, inclusive school? *Research in Comparative and International Education, 11*(1), 52–68.

Hovdenak, S. S., & Stray, J. H. (2015). *Hva skjer med skolen?: en kunnskapssosiologisk analyse av norsk utdanningspolitikk fra 1990-tallet og frem til i dag*. Bergen: Fagbokforlaget.

Imsen, G., Blossing, U., & Moos, L. (2017). Reshaping the Nordic education model in an era of efficiency. Changes in the comprehensive school project in Denmark, Norway, and Sweden since the millennium. *Scandinavian Journal of Educational Research, 61*(5), 568–583.

Kapferer, J.-N. (2008). *The new strategic brand management: Creating and sustaining brand equity long term*. London: Kogan Page.

Kornberger, M. (2010). *Brand society: How brands transform management and lifestyle*. Cambridge: Cambridge University Press.

Kuckartz, U. (2014). *Qualitative text analysis: A guide to methods, practice and using software*. London: SAGE.

Leijerholt, U., Biedenbach, G., & Hultén, P. (2019). Branding in the public sector: A systematic literature review and directions for future research. *Journal of Brand Management, 26*(2), 126–140.

Ljunggren, J., & Andersen, P. L. (2015). Vertical and horizontal segregation: Spatial class divisions in Oslo, 1970–2003. *International Journal of Urban and Regional Research, 39*(2), 305–322.

Lubienski, C. (2007). Marketing schools: Consumer goods and competitive incentives for consumer information. *Education and Urban Society, 40*(1), 118–141.

Mayring, P. (2000). Qualitative content analysis. *Forum Qualitative Social Research/Forum Qualitative Sozialforschung, 1*(2), 159–176.

Miles, S. J., & Mangold, G. (2004). A conceptualization of the employee branding process. *Journal of Relationship Marketing, 3*(2–3), 65–87.

Myhre, J. E. (2017). Oslos historie som delt by. In J. Ljunggren (Ed.), *Oslo – ulikhetenes by* (pp. 31–56). Oslo: Cappelen Damm Akademisk.

Polat, S., Abat, E., & Tezyurek, S. (2010). The perceived corporate image of private secondary schools by student's and parent's views. *European Journal of Educational Studies, 2*(2), 65–76.

Sætre, M., Hofseth, C., & Kjenn, B. (2018). *Trender i kriminalitet 2018–2021: Digitale og glokale utfordringer.[rapport politiet]*. Oslo: Oslo politidistrikt.
Silverman, D. (2013). *Doing qualitative research: A practical handbook*. London: SAGE.
Thornton, P. H., & Ocasio, W. (1999). Institutional logics and the historical contingency of power in organizations: Executive succession in the higher education publishing industry, 1958–1990. *American Journal of Sociology, 105*(3), 801–843.
Thornton, P. H., Ocasio, W., & Lounsbury, M. (2012). *The institutional logics perspective: A new approach to culture, structure and process*. Oxford: Oxford University Press.
Toft, M., & Ljunggren, J. (2016). Geographies of class advantage: The influence of adolescent neighborhoods in Oslo. *Urban Studies, 53*(14), 2939–2955.
Trivitt, J. R., & Wolf, P. J. (2011). School choice and the branding of Catholic schools. *Education Finance and Policy, 6*(2), 202–245.
Wæraas, A. (2019). They put themselves out there: A longitudinal study of organizational expressiveness. *Corporate Reputation Review, 23*, 267–279.
Wæraas, A., & Dahle, D. Y. (2020). When reputation management is people management: Implications for employee voice. *European Management Journal, 38*(2), 277–287.
Wessel, T. (2017). Det todelte oslo – etniske minoriteter i øst og vest. In J. Ljunggren (Ed.), *Oslo – ulikhetenes by* (pp. 81–104). Oslo: Cappelen Damm Akademisk.
Wiborg, Ø. (2017). Hvor er formuene i Oslo? Gylne ghettoer og deres utvikling over tid. In J. Ljunggren (Ed.), *Oslo – ulikhetenes by* (pp. 105–120). Oslo: Cappelen Damm Akademisk.

Part III
Concepts and Cases: Place Branding in Action

For this part of the book, chapters focus on what I am calling branding in action. The chapters reflect real-world applications of place branding and the sometimes-unintended consequences associated with such initiatives. In the scholarly literature about place, political, and public branding, there are countless case studies of how people or organizations go about implementing branding plans. The chapters herein add to that knowledge by bringing to light successes and challenges associated with branding strategies.

One reason place branding and marketing remain so difficult is because there is scant evidence the practices "work." This is because there is a real question surrounding what it means to define a brand strategy as working. As noted throughout the book, brands sometimes lack conceptual clarity given the myriad stakeholders involved in the creation, implementation, and (hopefully) evaluation of the brand identity. While there is isomorphic pressure to establish a brand strategy to "keep up with the Joneses" (Fay and Zavattaro 2016), there is sometimes little evidence of a return on investment for places and nations (Anholt 2008). The problem is branding campaigns predicated only on symbolic actions without strategy or substance (Anholt 2008).

Recent research in the field is offering suggestions for taking place branding practice and scholarship to the next level. One direction is the integration of spatial planning into the process and outcomes. For instance, Grenni et al. (2020) examine how spatial planning can be tied to a region's cultural narratives to create a more robust branding initiative. In urban planning, as more attention is paid toward collaborative process involving myriad stakeholders in spatial allocation, the same happens when spatial awareness is incorporated into branding processes that tap into a place's existing native narratives instead of creating new narratives with few ties to the place (Grenni et al. 2020).

Similarly, van Assche et al. (2020) note the integration of spatial planning in place branding is novel because the two fields often run parallel in both research and practice. The authors note this is because those engaged in a branding exercise might not think directly about how the brand supports (or not) local businesses, job creation, infrastructure, and recruitment and retention (van Assche et al. 2020). The key is to understand both spatial planning and branding as key governance issues requiring time, attention, and integration to paint a more holistic picture of the place and what its space can offer stakeholder groups. This is an open challenge when place branding is put into action.

Another relatively new (about a decade or so) avenue is the explicit incorporation and study of narrative tactics in brand development and communication. This narrative turn recognizes the social construction of reality inherent in brand development and promotion (marketing). As Hansen (2010) articulates, narratives about a place come from diverse sources – locals, business owners, media outlets, etc. Some might be promoting a brand image different from what the place itself is trying to promote via its own government-created (or organization-created) narratives. Narratives are powerful ways of organizing our world, and Hansen (2010) notes even in a socially constructed environment it does not mean "everything goes" (p. 271). Instead, for narratives to "work," there needs to be a meaningful connection to the place for some stakeholder group. Creating narratives from nothing rarely succeeds.

When it comes to understanding narratives, one might look to the public policy literature to understand how shaping a narrative can help (or hurt) its outcomes. Jones and McBeth (2010) argue a narrative has several parts: a story with a time-based sequence of events (a beginning and end), a plot with dramatic moments and characters, and a morally driven ending. They tie their elements to neuroscience, indicating "narrative cognition may be fundamental to meaningful human existence" (p. 330), thus making narratives a powerful component of socially constructing reality. If place branding and marketing experts want to employ a narrative strategy, knowing the elements of narratives and how they work is a good first step to making those stories even stronger.

Another new way these narratives are emerging is through social media, specifically using the tools for digital storytelling. In this volume, Rudolf and Waeraas indicate how cities throughout three European countries use social media to communicate brand stories. As social media continue to change, with new tools seemingly coming out almost every month, this is an emerging trend to watch – especially at the intersection of narratives and brand promotion.

Hudak (2019) notes the digital technologies and platforms mean "residents are progressively sharing more than just their advice for local attractions and flavorful cuisine" to instead focus on "spaces for residents to address community issues, advocate for social change, and preserve local history and culture" (p. 97). These tools allow residents themselves to become brand ambassadors, as usually residents might only be asked to participate in a survey or a focus group about the place brand. Digital storytelling allows them to create their own place narratives apart from what is often the tightly controlled central messaging strategy. In this way, brand co-creation is both easy and a bit scary, as someone who has a negative opinion and narrative can just as easily commandeer the space. In this volume, Minkman et al show us what happens when narratives come together to undermine a branding strategy. So, brand and marketing experts should account for the power of digital storytelling going forward.

A final trend seems to be getting a clearer picture of how to evaluate a brand's success beyond economic measures. Most measures of success depend on easily countable things, such as hotel stays, restaurant receipts, or airline travel to and from a place. These measures are usually tied with destination and tourism branding, but what happens when places are not big tourism hubs? Then how does one know branding efforts are working to achieve stated ends? One way to examine this is through public values, or measuring how a place adds value to someone's life (Herezniak and Anders-Morawska 2020). Some public values can include inclusivity, communicative action, democratic values, openness, transparency (Jørgensen and Bozeman 2007). Place brand creators, especially those associated with government agencies, often do not measure these kinds of abstract concepts because they lack time, money, and human resources (Zavattaro 2016).

Much of the success measures come from those implementing the brand (Hanna and Rowley 2013; Herezniak and Anders-Morawska 2015; Zavattaro et al. 2013). A trend is and should continue to be user perception and connection – brand image and emotional affect. Again, this is not an easy task given measuring emotions falls more in line with psychology, but if brands are about creating connections, then there should be a move more toward the public values instead of only economic ones (Herezniak and Anders-Morawska 2020). For instance, Zenker and Rütter (2014) found citizen satisfaction is directly related to positive citizenship behaviors (such as being a good neighbor) and place attachment. While not easy to study, the findings indicate the human side of branding efforts are just as important as the economic ones.

The chapters in this part all highlight the varying ways place brands are implemented. One take away from chapters in this part, as well as some throughout the book, is the proclivity to implement a brand as a problem-solving tool. This could make sense if one sees place branding as a governance strategy, as sometimes governance is about problem mitigation. It does not make sense if there is only some pretty images and pictures meant to convince someone a singular point of view is correct. While challenging, the best branding campaigns are those created co-collectively and implemented co-productively (Kavartzis & Hatch, 2013).

References

Anholt, S. (2008). Place branding: Is it marketing or isn't it? *Place Branding and Public Diplomacy, 4*, 1–6.

Fay, D. L., & Zavattaro, S. M. (2016). Branding and isomorphism: The case of higher education. *Public Administration Review, 76*(5), 805–815.

Grenni, S., Horlings, L. G., & Soini, K. (2020). Linking spatial planning practices and place branding strategies through cultural narratives in places. *European Planning Studies, 28*(7), 1355–1374.

Hanna, S., & Rowley, J. (2013). Place brand practitioners' perspectives on the management and evaluation of the brand experience. *Town Planning Review, 84*(4), 495–515.

Hansen, R. H. (2010). The narrative nature of place branding. *Place Branding and Public Diplomacy, 6*, 268–279.
Herezniak, M., & Anders-Morawska, J. (2020). Public value-drive place branding. The way forward? *Place Branding and Public Diplomacy*. https://doi.org/10.1057/s41254-020-00185-0
Herezniak, M., & Anders-Morawska, J. (2015), City brand strategy evaluation: in search of effectiveness indicators. *Journal of Place Management and Development*, 8(3), 187-205.
Hudak, K. C. (2019). Resident stories and digital storytelling for participatory place branding. *Place Branding and Public Diplomacy, 15*, 97–108.
Jones, M. D., & McBeth, M. K. (2010). A narrative policy framework: Clear enough to be wrong? *Policy Studies Journal, 38*(2), 329–353.
Jørgensen, T. B., & Bozeman, B. (2007). Public values: An inventory. *Administration & Society, 39*(3), 354–381.
Kavaratzis, M., & Hatch, M. J. (2013). The dynamics of place brands: An identity-based approach to place branding theory. *Marketing Theory, 13*(1), 68–96.
Van Assche, K., Beunen, R., & Oliveira, E. (2020). Spatial planning and place branding: Rethinking relations and synergies. *European Planning Studies, 28*(7), 1274–1290.
Zavattaro, S. M. (2016). Exploring managerial perceptions of place brand associations in the US Deep South. *International Journal of Culture, Tourism and Hospitality Research, 10*(1), 53–69.
Zavattaro, S. M., Daspit, J. J., & Adams, F. G. (2015). Assessing managerial methods for evaluating place brand equity: A qualitative investigation. *Tourism Management, 47*, 11–21.
Zenker, S., & Rütter, N. (2014). Is satisfaction the key? The role of citizen satisfaction, place attachment, and place brand attitude on positive citizenship behavior. *Cities, 38*, 11–17.

#thatswhywestallis Shows the World How Good West Allis Feels

Jenny Kosek

When its namesake factory, the Allis Chalmers Corporation, closed in the 1980s, the City of West Allis, Wisconsin, fell into a difficult period in its history. Since the early 1900s, the Allis Chalmers factory had employed thousands of the City's residents, providing well-paying, secure jobs that contributed to West Allis becoming a close-knit and desirable community in which to live. But when the Allis Chalmers factory – and so many large manufacturing facilities like it in the City – shuttered its doors during the latter half of the twentieth century, unemployment and poverty levels rose. West Allis soon became perceived as being high-crime and rundown. By the mid-1990s, the City had earned the undesirable nickname "Dirty 'Stallis'," and local media seemed committed to furthering the City's negative reputation by seizing on disparaging stories at every turn.

But West Allis began rapidly transforming in the new century. With Mayor Dan Devine's election in 2008 and the appointment of City Administrator Rebecca Grill in 2015, a strong leadership team was in place to steer the City of West Allis in a bold new direction. An intensive community collaboration led to the creation of the City of West Allis Five-Year Strategic Plan, which was adopted by the West Allis Common Council in 2017. The number one goal in the Five-Year Plan is "Brand/Image/Destination." With this, the City of West Allis sealed its commitment to change the public's perception about the City and ensure residents, business owners, neighboring communities, and beyond knew that rumors of West Allis' demise were greatly exaggerated.

J. Kosek (✉)
Communications Strategist, City of West Allis, West Allis, WI, USA
e-mail: joksek@westalliswi.gov

S. M. Zavattaro (ed.), *Public Branding and Marketing*,
https://doi.org/10.1007/978-3-030-70505-3_11

Building the Team

As the Five-Year Strategic Plan was being developed, City leadership recognized a key part of changing the public's perception of the City was through positive communications and public relations. To that end, the City created its Communications Department in 2016. Led by 20-plus-year veteran employee Jonathan Matte, additional positions were created to develop a fully functioning in-house marketing team to execute brand management, social media management, public relations, print communications, and website management. By 2018, the department was fully staffed with a full-time videographer, graphic designer, events coordinator, and communications strategist, in addition to two interns and a full-time staff member dedicated to operating a full-service in-house print shop.

As the Communications Department took shape, the City issued a request for proposals (RFP) to contract with a professional marketing agency to begin to develop a new brand and marketing strategy for the City. Mayor Devine and City Administrator Grill knew a new logo was not enough to overcome the strongly held, negative public opinions toward West Allis and recognized that the City's partner through this RFP needed to be committed to a multi-year project that would include strategic actions to reposition the City in the public eye in order for it to attract businesses, new residents, and tourists.

After reviewing over a dozen responses to the RFP, Savage Solutions, LLC., a Milwaukee-based marketing firm, was chosen to lead the City of West Allis rebranding and marketing strategy project. The Savage team was an ideal match for the City. Staffed by a team of young creatives and experienced in municipal marketing, the City was confident that Savage could deliver the engaging, modern elements needed to move the City forward and execute an effective and measurable marketing strategy that would make an impact.

Strategy and Goals

Savage Solutions began its work by conducting focus groups. These conversations included City leaders and officials, City employees, residents, business owners, members of the local school district, and leaders of local non-profits and organizations, all from diverse backgrounds and demographics. In between focus groups, the Savage team spent time in the City of West Allis; visiting restaurants, shops, and businesses; and getting a feel for what the community was about and what made it strong.

The Savage team quickly learned immense pride ran through the community. Focus group feedback noted how many people had lived in the community for their entire lives. Participants noted they loved the people in the City, reporting that everyone was kind and welcoming and that West Allis was an unpretentious place to be. People were proud of the City's manufacturing history and excited to see younger people moving to the area. Focus group respondents spoke enthusiastically about the City's dining scene and abundance of unique restaurants and were excited by the diversity of the community. But they also acknowledged the challenges West Allis faced: an aging housing stock, lack of amenities to retain younger residents, and overcoming the burden of its negative perception.

The Savage team worked closely with a core City team comprised of the mayor, city administrator, director of communications, and communications strategist. Combining guidance from this small internal committee with the information gathered through the focus groups, the Savage team got to work creating a new brand and corresponding campaign that would speak to the energy, passion, and pride within the West Allis community.

#thatswhywestallis

It was clear to Savage Solutions that West Allis residents were proud of the City's past and excited about its future. This led their design team to create a logo that acknowledged the City's long history by including its "established 1906" date, but that was youthful, fun, and bright, to support the new developments, residents, and opportunities in the City's future. The circle represents the City's ever-changing, ongoing commitment to the community's spirit and strength, and the circle is open to speak to inclusion and unity among residents.

The orange, turquoise, and gray color palette reinforces the youthful energy of the community while upholding the City's proud manufacturing past.

The scripted font felt genuine and sincere, with a bit of nostalgic attitude to further embrace the City's past. The upward slant speaks to the forward motion of the community and its drive to keep growing and developing.

One thing focus group participants were eager to share with Savage Solutions was their "why": why they loved West Allis and why they were raising their families here, buying homes here, starting businesses here, or moving businesses here. The focus group participants were excited to tell the Savage team about all of the reasons they were passionate about the community; Savage leveraged that enthusiasm as they developed the multi-year marketing plan and thatswhywestallis.com website, which invites visitors to tell their personal West Allis story.

Brand Launch

The brand debuted to the public on Saturday, October 13, 2018, with a big kickoff event at West Allis Farmers Market, a City destination. T-shirts with the new logo were given away, and all local major media channels attended the event to announce "West Allis' New Look." Simultaneously, a vigorous ad campaign launched that included print, radio, and digital ads to drive people to the thatswhywestallis.com website to learn about the new brand and the opportunities and energy in the West

Allis community. A unique @thatswhywestallis Instagram account was also created to promote the brand and drive users to the website.

With the new brand launched to the public, a comprehensive marketing campaign that included social media ads, digital ads, and traditional advertising was initiated to draw attention to the positive changes occurring in West Allis and to drive visitors to learn more at thatswhywestallis.com.

Achievement

Savage Solutions compiled numerous metrics in the fourth quarter of 2018 relating to the brand launch and corresponding launch marketing campaign:

- Unique users to thatswhywestallis.com: 12,298.
- Page views: 29,403.
- Instagram impressions: 187,361.
- Digital advertising impressions (via social media and Google ads): 1,629,962.
- Traditional advertising impressions (television, radio, outdoor, display network, print): 9,107,105.
- Total media impressions: 10,737,067.

These metrics continued to grow in 2019 as the 3-year marketing plan entered its first full year of execution. Television, radio, outdoor advertisement, social media, digital advertising, and print ads were leveraged, earning more than 90 million impressions.

With more eyes and more interest being drawn to West Allis, the benefits to the community of this campaign are also taking hold. Three large-scale redevelopment projects are proceeding in the City, which will reinvigorate formerly empty or underutilized land parcels as they become mixed-used developments offering residential and commercial opportunities. Inquiries to the Mayor and Department of Development from businesses interested in opening or relocating to West Allis have increased. Incoming messages through the City's social media channels have also changed tone since 2017, from being largely negative in sentiment to largely positive.

In addition, #thatswhywestallis has become a rallying cry for City pride, so much so that local media – who previously spoke negatively of the City of West Allis – have begun using it to promote the positive changes going on in the community.

Budget

No residential tax dollars were spent on this project, which is funded by tax revenue the city earns from Wisconsin local room tax laws. These laws require that a percentage of hotel room taxes collected within a municipality be used by the municipality to promote tourism promotion and development. There are guidelines that define what constitutes acceptable tourism spending. This project met those requirements, and room tax revenue funds paid for fees associated with this work.

The first phase of this project included extensive research by the Savage Solutions team, creative time developing the logo and brand identity, and time spent planning the rollout of the new brand. This phase totaled $47,500 and spanned over 6 months of work.

The second phase cost $20,000 and allowed Savage Solutions to create a 3-year marketing plan to promote West Allis' business and tourism opportunities. $100,000 was provided from the City of West Allis Tourism Commission budget, funded by the local room tax monies, to support the launch campaign, including digital and traditional advertising and promotional item purchases. Savage Solutions coordinated and executed the public advertising components; the City's in-house Communications team created and coordinated production of promotional items, including t-shirts, tote bags, mugs, etc., and also created graphics and copy to promote the new brand on the City's digital platforms.

Ongoing media spends average $100,000 per year, and 2019 media impressions totaled over 90 million impressions. As the 3-year marketing plan continues, the future looks bright for West Allis indeed.

Backfiring Boomerang Brands in Jakarta: The Case of the Great Garuda

Ellen Minkman, Vidar Stevens, and Roanne van Voorst

Branding as an Emotional Governance Strategy

All through the world, governments are busy with policy actions for adaptation to climate change (Pielke et al. 2007). In fact, adaptation has become part of the discourse of global warming and is now widely recognized as a necessary response to the threat posed by the expected climatic changes. Climate adaptation – i.e., the ability or potential of a system to respond successfully to climate variability and change (Nielsen and Reenberg 2010) – as a phenomenon is not new. What is new, however, is a sense of urgency has entered the scene, where governments have engaged in a race against time to understand how adaptation can be facilitated, supported, and ultimately sustained in societies at risk from climate change impacts (Coulthard 2008, p. 479). Recent studies about local climate adaptation practices have indicated there are various barriers that render adaptation difficult as a response to climate change. Specifically, these studies have shown the adaptive capacity of humans depends not only on economic or technological factors but is also influenced by social norms, values, and rules (see, e.g., Adger et al. 2005). One of the current challenges in climate adaptation research is, therefore, to gain a better

E. Minkman (✉)
Faculty of Technology, Policy and Management, Delft University of Technology, Delft, The Netherlands
e-mail: e.minkman@tudelft.nl

V. Stevens
Mulier Institute, Utrecht, The Netherlands
e-mail: v.stevens@mulierinstituut.nl

R. van Voorst
University of Amsterdam, Amsterdam, The Netherlands

Dutch Future Society, Amsterdam, The Netherlands
e-mail: r.s.vanvoorst@uva.nl

S. M. Zavattaro (ed.), *Public Branding and Marketing*,
https://doi.org/10.1007/978-3-030-70505-3_12

understanding in how *social* norms and values can be influenced to ensure the adaptive capacity of a society toward issues of climate adaptation increases.

A potential public management strategy to influence norms and values of people, or at least to trigger positive associations about a policy or service, is branding (Eshuis and Klijn 2012). Specifically, Kavaratzis (2004, p. 53) explains how branding involves selecting and associating attributes to add value to a basic policy or service; it involves deliberate attempts to influence how citizens interpret or evaluate a product or service. In a similar vein, Eshuis and Klijn (2012, p. 19) define a brand as a symbolic construct consisting of a name, term, sign, symbol, or design or a combination of these created deliberately to identify a phenomenon (e.g., a product, service, policy, place, etc.) and differentiate it from similar phenomena by adding particular meaning to it. As such, branding as a strategy can be used to connect a phenomenon with particular words, images, emotions, or ideas to facilitate the development of particular webs of associations "inside" the brain (Eshuis and Klijn 2012). In the case of climate adaptation, this would imply the more positive associations about a climate adaptation policy plan are created through a branding campaign, the higher the likelihood that people will support the policy instead of resisting it.

The literature on branding in the public sector is quite diverse. So far, there have been studies where scholars looked at practices of process branding (Eshuis and Klijn 2012, pp. 21–22), branding organizations (Balmer and Greyser 2006), branding people – or human branding – and branding places or city branding (Eshuis et al. 2014; Lucarelli and Olof Berg 2011). To our knowledge, limited focus has been on how branding can be used as an instrument to foster support for policies, like climate adaptation policies. Hence, we examine in this chapter a deliberate attempt of governments to influence people's ideas and opinions about a policy plan by forging emotional and psychological associations. We take a critical perspective on branding, as like any other governance strategy it is highly uncertain if a brand reaps the intended impact in terms of generating support or legitimacy. In the worst-case scenario, a government's branding efforts have a completely contrary effect and do the opposite from what was intended: it generates resistance. In such a situation, the branding campaign functions as a boomerang and "brings the launcher, or those standing close by, a sharp knock on the head" (David 2001, p. 1). By choosing a critical approach, we are able to discuss the limitations of branding as a governance strategy to create awareness for a particular problem situation, increase the support for a preferred policy solution, and bind people (creating brand loyalty through establishing emotional bonds) to a policy plan.

This chapter offers a telling case of how branding was used as a governance strategy to increase support for the Dutch delta approach as a policy solution to combat rapid land subsidence in the city of Jakarta, Indonesia. The Dutch delta approach consists of a series of measures, like polder creation, water management, a water sanitation program, and social protection structures. At a certain point, however, the branding campaign, which used a symbol of the national bird the Garuda, backfired, and the brand became subject of an opposition campaign to destabilize the proposed Dutch delta approach. Hence, we show in this chapter up to what point branding is a useful instrument to enhance the societal support for climate adaptation policies and what, simultaneously, the limits of branding policies are. Based on an analysis of relevant (policy) documents, 55 interviews, and 160 respondents who participated in anthropological fieldwork, we aim to provide an answer to the following research question: why did in the Jakarta case the branding campaign backfire? We discuss in subsequent order: the theoretical section, the method section, the case, the empirics, and the final reflections that follow from the chapter.

What Explains a Backfiring Boomerang Brand?

Branding is a promising, upcoming governance strategy (Lucarelli and Brorström 2013) and refers to the deliberate attempt of governments to influence people's ideas and opinions about, for example, a policy plan, policy solution, or collaboration, by forging particular emotional and psychological associations (Eshuis and Klijn 2012). Inter alia, the goal of a branding campaign can be to create awareness for a particular problem situation and increase the support for a preferred policy solution. Simultaneously, public brands can be used to bind people (creating brand loyalty through establishing emotional bonds), activate people, and communicate with media. Hence, a well-managed branding campaign holds the promise to gain more legitimacy and support for a governmental intervention in society.

Nevertheless, as van Buuren and Warner (2014, p. 3) show, it is highly uncertain that a brand reaps the intended impact in terms of generating support or legitimacy. In the worst-case scenario, a city government's branding efforts have a completely contrary effect and do the opposite from what was intended: it generates resistance. In such a situation, the branding campaign functions as a boomerang and "brings the launcher, or those standing close by, a sharp knock on the head" (David 2001, p. 1).

Following the metaphor of the boomerang, we can distinguish (following the work of van Buuren and Warner 2014; Liotta 2002) three possible impacts when launching a public branding campaign:

1. Expert handling: throwing the boomerang up in a graceful arc and catching it upon return. This is the desirable boomerang effect. The branding message is sent into the public arena and is then reflected back at the thrower as positive, reinforcing feedback.
2. Inexpert handling: the boomerang comes back, but the thrower is "clobbered" by it. In this way, the branding campaign backfires and has unintended side effects.
3. The boomerang goes astray and is lost. As Martens and Wolf (2009) describe it, the brand is deliberately set in motion, but it landed differently than expected by the throwers. The boomerang can fly off into different territory – for example, an environmental discourse can become enmeshed with economics (van Buuren and Warner 2014, p. 3).

Within the Jakarta case, the boomerang sent out by the Dutch developers' team backfired. As such, the branding campaign can be qualified as an "inexpert handling" of the branding effort. In the branding literature so far, no clear causal pathways to why a branding campaign fails have been found. We, however, believe that two (interrelated) causal pathways can be distilled by combining different insights from the literature: (1) a negative brand translation and interpretation and (2) mismanagement of the branding effort.

A Backfiring Boomerang Brand Due to a Mismatch in Brand Translation

To elucidate, an important element of the backfiring boomerang dynamics is the reinterpretation of the original brand label. Normally, when a brand message (i.e., the positive and emotional associations that government officials add to a marketing campaign) is sent out, the message is a combination of a specific project demarcation and a specific project interpretation. For example, the futuristic buildings in Rotterdam (i.e., project demarcation) give a sense of innovativeness and a "can-do" mentality to the city (i.e., project interpretation). However, if the recipients of a brand label create another demarcation-interpretation combination of the brand label and use this new interpretation in their communication and marketing efforts, the original brand can get an opposite meaning and effect.

Hence, to understand the negative backfiring boomerang branding dynamics, it is important to reconstruct the "brand translation" process of the receivers of the branding efforts. So far, literature has devoted little attention to this "brand translation" process by recipients, although there are some leads that can give us a sense of the "brand experience" of a recipient. Hommes et al. (2008), among others, argue people's deep core beliefs are hard to change or to influence by, among others things, a branding campaign. At the most, you can change their view about possible policy actions to provide more effective or efficient solutions to a problem. Hence, if a branding campaign creates an image that goes against the deep core beliefs of a person (e.g., the use of a national symbol for solidarity for marketing purposes), this person is likely to reject the image or resist it.

Although the notion of "deep core beliefs" explains why people can reject a branding message, it does not yet give an answer to why a branding campaign backfires or how "heavy opposition" is mobilized to ensure a branding campaign backfires. We expect such negative campaigning emerges due to increased negative media attention for the launched branding campaign in combination with a negative bandwagon effect. The term bandwagon effect denotes a phenomenon of public opinion impinging upon itself (sometimes also referred to as "contagion effect"), implying that in their political preferences and positions people will join what they perceive as existing or expected majorities or dominant positions in society (Schmitt-Beck 2015). Hence, if the (negative) reinterpretation of the demarcation-interpretation combination of the original brand label gains momentum in the media and society,

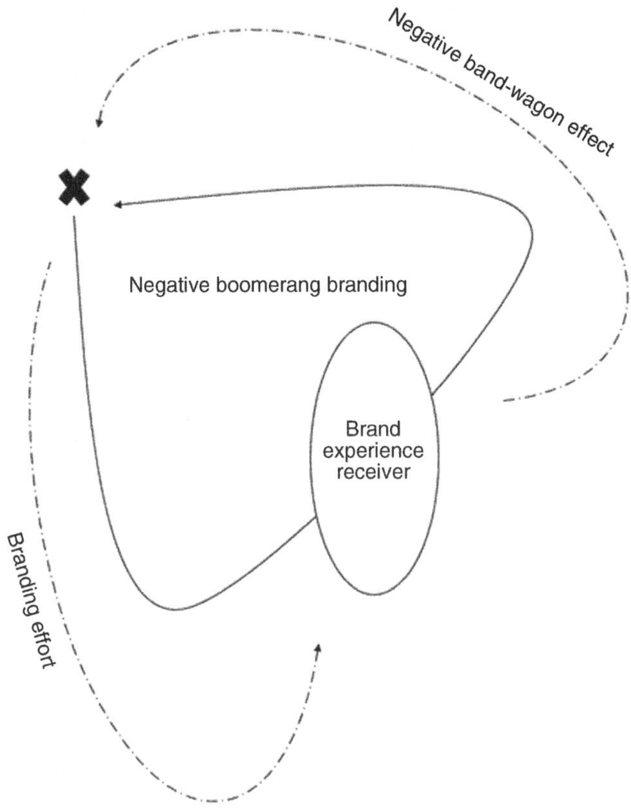

Fig. 1 Conceptual model of a backfiring boomerang brand

a branding campaign is more likely to backfire and run aground. The described relationship between a negative boomerang branding effort and the bandwagon effect is visualized in Fig. 1. This model will be used as *first causal pathway hypothesis* to examine why in the Jakarta case the branding effort of the Dutch developers' team backfired.

A Backfiring Boomerang Brand Due to Wrong Brand Management

Nevertheless, whether a branding campaign backfires or not, a large part also depends on how the branding campaign is managed. In recent years, we have seen a rise in the total number of brand management models in different streams of literature, for example, the strategic place brand management model of Hanna and Rowley (2011), the relational network brand model (Hankinson 2004), city image communication model (Kavaratzis 2004), a model of destination branding

(Cai 2002), destination branding process model (Laws et al. 2002), the 7A destination branding model (Baker 2012), and the city brand management model (Gaggiotti et al. 2008). A common point of criticism against these models, however, is they merely describe the consecutive phases of a "successful" brand management campaign instead of truly capturing the micro-level management dynamics in the interactions between brand communicator and receiver (Stevens 2019). Nevertheless, one of the few scholars who devoted particular scholarly attention to the micro-level management interactions between brand communicator and receiver is Kerr (2009).

Based on an inductive research approach, Kerr (2009, p. 127) identified five brand management activities managers need to perform to bring a brand message across to a target audience: activating, energizing, concepting, expanding, and re-energizing. Activating entails "the identification of the need and potential for a branding campaign" (Kerr 2009, p. 94). Often people who can be regarded as brand activists or brand champions, i.e., people who are seeding the idea for a branding campaign in their (working or managing) communities, begin this process. In most cases, the idea for propagating a branding campaign starts when other managing initiatives failed to work or there is an immense need for changing problem perceptions (Kerr 2009). Energizing is the act of "gaining support, resources and a management structure for the brand strategy" (ibid, p. 100). Concepting involves the practices of designing, adopting, and planning the place brand strategy. Expanding, subsequently, entails "taking the brand to the target audience and community" (ibid, p. 117). Lastly, re-energizing involves the idea of nurturing the branding campaign and (iteratively) ensuring that the goals of the campaign are achieved and the brand message reaches the target audience (Kerr 2009).

Kerr (2009) does not assume a hierarchy among the five brand management activities. Nevertheless, to some extent, there still seems to be a consecutive order between the various management activities, e.g., a brand first needs to be activated before energizing activities can take place. In addition, although managers of branding processes are called upon to perform multiple activities, Kerr (2009) shows how the salience of these management activities may vary with the emerging circumstances in the branding process. In this chapter, the focus is on whether the brand manager in the empirical case performed all brand management activities proposed by Kerr and, if so, what specific micro-level management strategies were utilized to perform these activities, for what reasons, and with what effects. In this way, this chapter empirically drills down into the management variable, which allows to explore – *as a second causal pathway hypothesis* – why, despite various management activities, the branding campaign of the Dutch developers' team backfired.

Methods

For this study, a single case study approach was used. We acknowledge the inherent limitations of using single case studies for extrapolating findings, as was described by George and Bennett (2005). Yet, we also agree with Flyvbjerg (2011, p. 305) when he argues a case study can further scientific development by the force of example. To this end, we do not pretend that the observations in the chapter are fully generalizable or highly theoretical. However, we are convinced the chapter's reflections are helpful in substantiating the development of theory on the impact of branding on gaining citizens' support for policy plans, like climate adaptation policies[1] in the Jakarta case.

For the data collection, we drew on a detailed process mapping based on an analysis of documents and a series of interviews (Koppenjan and Klijn 2004, pp. 136–137). The document analysis included policy documents, meeting minutes, position papers, annual reports, evaluations, and implementation plans of the stakeholders involved. The relevance of the documents was determined by making a selection based on whether the information in the documents said something new or extra about the context in which the Dutch developers' team tried to promote their solution for Jakarta's flooding issue, the peculiarities of the policy-making process, the various activities of the brand manager, the positions of the actors with regard to the problem situation and possible solutions, and the morphology of the actor constellation. The document analysis was complete once we reached data saturation.

Subsequently, the 55 interviews with 39 individuals helped to gain more insight into the opinions, objections, ideas, attitudes, and experiences of the stakeholders with regard to the use and value of the branding campaign to foster citizens' support for the policy solutions proposed by the Dutch developers' team to avoid Jakarta from sinking any further. The interviews typically lasted for 1–1.5 hours. Each interview was recorded and transcribed. We promised our respondents anonymity. Therefore, we numbered the interviews and used the phrase "respondent (number)" to report quotes from the interviews in this chapter. When relevant, the nationality or function type of the respondent is provided as well. We triangulated the interview data by comparing the interview responses to each other and to the document analysis findings. If we ran into inconsistencies, we followed up with respondents to ask for clarification. The interview data was collection in two rounds of interviews in May 2017 and February 2018.[2] Most Indonesian respondents are part of the decision-making elite, including high-level government officials and senior advisors of ministers and the Indonesian president.

Furthermore, we used data collected during fieldwork for 1 year (2010–2011) with four short follow-up visits between 2012 and 2016, in neighborhoods along the Ciliwung River in central Jakarta and areas around the North Bay. These areas have been inhabited for several decades, but the communities have been living with the constant threat of floods. Typical anthropological and ethnographic methods were employed during fieldwork. The third author lived in the research area for an

[1] Although land subsidence is considered the "root cause" of flooding by the Dutch consortium, climate change is expected to accelerate the flooding issue: "Based on extrapolation of current trends, such as land subsidence, rising seawater level, climate change and rapid urban growth the flooding problems are expected to become critical over the next decades for the northern part of Jakarta that now accommodates over 4 million people." – JCDS Atlas (2011, p. 3).

[2] See Minkman, Letitre, and Van Buuren (2019) and Minkman (in writing) for an extensive method description of, respectively, the data collection in 2016–2017 and in 2018.

extended period so that daily activities in and around the kampongs could be experienced first-hand and to get a sense of how the announcements of new flood measures impacted upon the people. Overall, 130 residents participated in the research, as did 30 activists and NGO employees active in the areas.

Eventually, we used the dimensions (deep core beliefs and bandwagon effect argument and the brand management strategies) of the two casual pathways, presented in the theoretical section, as sensitizing concepts to reflect on the process mapping. This allowed us to gain a notion whether during the policy-making process the brand manager was able to influence public opinion with the help of the branding effort and, if not, why the brand manager was unable to do so. Finally, we interviewed two Dutch consortium members to cross-validate our research findings. The results of our data analysis are presented in the following section.

Empirical Results

In this section, we reconstruct the branding campaign and the events that led to its backfiring. Doing so, we use the five managerial activities following Kerr (2009) as sensitizing concepts and describe the branding campaign and the counter-brand that was created along the way. We start with the brand strategy of activating, which, according to Kerr (2009), is about "seeding the idea" for a branding campaign.

Activating: Need and Potential for Brand Campaign

Like other urban deltas, Jakarta is suffering from rapid land subsidence (Syvitski et al. 2009; Wesselink 2016). Especially, North Jakarta is sinking below sea level at staggering rates of over 10 cm/year (Bucx et al. 2014). The resulting flood risk is threefold: coastal, fluvial (rivers can easily flood the low-lying city), and pluvial (precipitation has nowhere to go) (Abidin et al. 2011). Experts disagree about the cause of rapid land subsidence, but the most commonly accepted explanation blames the often-illegal groundwater extractions by industries and public and private buildings in the city. Other explanations include natural soil compaction or compaction following urban development and tectonic movements (Abidin et al. 2011). In the meantime, Jakarta continues to sink at rates up to 25 cm/year.

In February 2007, a King Tide (an exceptionally high tide) surprised Jakarta and took 80 lives while also causing severe economic losses. Following this flooding, the Indonesian government requested advice from the Dutch government. They initiated a process where a consortium of Dutch water experts could share their knowledge with Indonesian officials. This exchange evolved through time into formulating a strategic plan. The strategic plan included four major strategic alternatives to protect Jakarta from tidal flooding. In 2013, a preferred solution was formulated based on these four alternatives. Characteristic for the strategic plan was the construction of a large offshore sea dike, which would protect the city of Jakarta. The then-governor of Jakarta started communicating about this sea dike as "the Giant Sea Wall."[3] As the Dutch consortium anticipated negative associations with a "giant wall," they decided to create a brand with more positive associations about the sea dike. This led to the birth of the "Great Garuda" a sea dike in the shape of the Garuda (a bird or bird-like creature), the national symbol of Indonesia.

There were, however, two other triggers that fostered the creation of the Great Garuda branding campaign. First, the Great Garuda project was economically appealing for the Dutch government and water sector. The complex project would be a prestigious addition to the portfolio of the Dutch water sector. The Dutch water sector has an excellent reputation for preventing flooding (OECD 2014), but risks using outdated knowledge because the country has not experienced a major flooding event since the 1950s. The future generation of engineers was born after the near flooding of the late twentieth century and has no personal experience with flooding. Therefore, the Dutch water sector is constantly looking for opportunities at home and abroad to create innovative projects that will update their knowledge and reputation and that can create economic opportunities for the Dutch water sector. Second, although Indonesia experienced a heavy flooding in 2007, flood management was not receiving much attention from Indonesian politicians. As such, it was not appealing to Indonesian politicians to address the flooding issue or place it on the political agenda, implying that the Dutch would miss a lucrative economic project to add to their water portfolio. Hence, the Dutch consortium realized that to increase policy support for the project, it should be made more attractive for politicians by, for example, using a branding campaign. *Or as Respondent 18 (a Dutch consultant) stated:* "…at this point flood management is not a very sexy topic, so you won't find any political champion out there that wants to finish building flood channels. That's not sexy. But if you work on the Giant Sea Wall or Garuda or some other magnificent things, there you might be able to find a political champion, because that is a bit more sexy."

A third trigger to create a brand was a decision of the national government to not invest in the construction of the sea dike. To generate sufficient revenue to construct this sea dike, the Dutch consortium therefore decided to combine the construction of the dam with large-scale urban development. These large-scale urban developments were envisioned on reclaimed land next to this dam. The Dutch consortium believed this strategic move would make it more interesting for the Indonesian private (real estate) sector to invest in the water project. The president of Indonesia approved the plans of the Dutch consortium but only on one condition: the strategic plan of the sea dike had to be integrated with existing policy plans for creating 17 artificial islands in Jakarta Bay. As such, in the promotion of the Great Garuda project (Fig. 2), one can notice visual images of "Hawaiian islands" in the hope to attract investments from the private sector. In sum, the necessity to place flooding on the political agenda, the need to attract investors to finance the project, and the

[3] http://ahok.org/berita/news/btp-soal-pulau-buatan-dan-sistem-giant-sea-wall/

Fig. 2 Cropped image of the "Great Garuda" from KuiperCompagnons website. https://www.kuipercompagnons.nl/nl/projecten/great_garuda_beschermt_de_kust/ (accessed: 2019-08-16)

possibility for Dutch companies to add a new project to their water portfolio resulted in the effort of the Dutch consortium to create the Great Garuda brand.

Energizing: Gaining Support for the Brand

According to Kerr (2009), "high-level support is vital for a government's branding campaign to succeed." Therefore, "brand managers need to understand and use the power base which exists within a city to avoid the risk of failure." In other words, they need to ensure "having 'the right people' on board of the branding campaign" (Kerr 2009, p. 107, emphasis added). In addition, Kerr (2009) argues the brand should align with the core beliefs of the target audience to resonate. To understand who the "right people" are in this case, it is important to provide a brief introduction to the power structure in Jakarta.

Indonesia is a unitary, democratic republic state, headed by a president. Indonesian ministries have large autonomy on their policy domain and huge "blockage power." It occurs, for example, that one ministry denies another ministry a permit because their project goes against the ministry's interests or own policy course. Furthermore, formally the central government exercises power over the lower levels of government. Nevertheless, in recent years, Indonesia has undergone several decentralization reforms in which power and authority are delegated to provincial and local government levels. So, what does that imply for the Great Garuda project? The project mainly concerned the city of Jakarta. In addition, the sea dike would stretch from the coast of Banten province to the coast of the province of West Java. The involvement of these three provinces justified interference of the national government because issues concerning multiple provinces or islands are transferred to the responsibility of the central government in Indonesia (Miller 2013). As a result, the Great Garuda project demanded support not only from the local government – in this case the city of Jakarta – but also from the relevant national ministries and the president of Indonesia.

Initially, the Dutch consortium focused on getting support for their Great Garuda project from the Indonesian government. After intensive talks and deliberations, the Dutch consortium even made the national Ministry of Economic Affairs (Perekonomian) co-author of the final report presenting the Great Garuda brand in April 2014. Simultaneously, Joko Widodo, then city-governor of the city of Jakarta, after intensive talks with the Dutch consortium, supported the Great Garuda project. This resulted in a collaboration between the city of Jakarta bureaucracy and the Dutch consortium to design extra flood measures for the city of Jakarta. Or as Respondent 2 (Dutch consultant) said, "the city of Jakarta clearly supported Garuda project. The city wanted an extra focus on on-shore measures for coastal safety, such as dikes, polder design, pumping stations and retention basins" (*personal correspondence*). In October 2014, Joko Widodo became president of Indonesia. In his new position, he adopted a more neutral position on the Great Garuda project, whereby he proclaimed himself to be neither a proponent nor an opponent of the project and the Great Garuda brand image.

Other political support came from the side of the Dutch government, which was highly committed to the Great Garuda project. The project's plans, which were several times presented by the Dutch consortium to the relevant Dutch ministries, appealed to the Dutch government for the central focus on structural flood infrastructure measures. In the proposal, the Dutch consortium clearly presented the Dutch as the "go to water engineers." Moreover, the consortium presented the water project in such a way that it fitted neatly in the portfolio of other prestigious Dutch flood projects. As a result, the Dutch government decided to provide the funds needed to initiate the follow-up stage of the project, in which Dutch consultants would create a detailed design for the Great Garuda plan. To this end, there was sufficient support from both the relevant (national and local) Indonesian political actors and the Dutch government officials to continue with the project, due to the efforts of the Dutch consortium to create "political energy" for the project.

Concepting: Adapting and Planning

The third managerial activity Kerr (2009) describes is adapting and concepting. This activity entails the (further) development of the brand concept, including a communication strategy and the creation of marketing materials (slogans, visualizations, etc.). As shown in Fig. 2, the Dutch consortium presented an offshore sea dike as a protector of the city of Jakarta against flooding. The Dutch consortium aimed to create more sympathy from the Indonesian people for the project by using the image of the Garuda, which is Indonesia's national symbol. Garuda is half human, half eagle and originates from Hindu mythology. Garuda takes a central position in Indonesia's coat of arms. The initial name of the project was National Capital Integrated Coastal *Defence* Strategy, hinting at the protective nature of both a dam and the Garuda. Also, the Garuda shape was used to emphasize that the dam would be there first and foremost to protect Indonesians. However, the brand consisted of other associations as well.

The brand also connected to what Emma Colven has termed the *allure of grand infrastructure* at the side of the Indonesian government, whereby the Indonesian government aimed to invest in, "large infrastructure projects, despite the challenges of realizing them" (Colven 2017, p. 251). This is illustrated by the phrasing of "a new cosmopolitan water-front city" in a promotional video[4] of the project. Over time, the project was officially re-named National Capital Integrated Coastal *Development*, because constructing the dam would bring along massive urban development as well. In abovementioned promotional video, the sea dike design was described in words like "iconic" and "spectacular." The Dutch consortium and Indonesian officials stated that the Great Garuda would "define the future" of Indonesia, so that "one day Jakarta will be able to compete with other major cities in the world" (Yani Sarwo Handayani, Deputy Governor of Spatial Planning of the city of Jakarta).

Concepting thus resulted in a vision of the future waterfront of Jakarta, whereby flood safety and urban development were integrated. The Great Garuda presented a vision of what Jakarta *could* look like in the future: a cosmopolitan water-front city. According to Kerr (2009, p. 113), place branding is often "aspirational" and thus "a statement of where it 'wanted to go'." Such visionary brands, then, might be far removed from the existing place identity (Kerr 2009). Resources and communication plans would, as such, be needed to influence society-induced change and create support for the policy plans. However, at the time of the launch of the Garuda plans and the first promotional videos of the sea dike design, the Dutch consortium did not have a concrete brand strategy, communication plan, place brand architecture, collaborative dialogue with non-public stakeholders (like societal organizations), or even brand manager to take the brand to the Indonesian society and beyond. It seemed the Garuda brand was strong enough by itself to bind people to the grand and futuristic policy plans; no communication strategy was necessary – and, if, in any event this would be the case, communications would be ad hoc, like presentations of the policy plans.

Expanding: Taking the Brand to the Community and Beyond

The Great Garuda brand was launched in 2014. Until this point, the brand existed only in the inner circles of the Dutch consortium members and involved national and local Indonesian policy-makers. However, from this moment onward, the Garuda brand also became visible to the general public. In the period that followed, the Great Garuda was occasionally given a place in public presentations and on the project's website. Because the design of the project included the reclamation of 17 islands in Jakarta Bay on the orders of the Indonesian president, people incrementally started to associate the project also with land reclamations. However, the issue

[4] https://www.youtube.com/watch?time_continue=292&v=l_YR6GydZAM, 16 August 2019.

of land reclamation in the Jakarta Bay had been a controversial topic in Indonesian politics from 1995 onward. Hence, for the first time, the Dutch consortium noticed a slight change in the *brand* message the general public received about the project. Nevertheless, despite little opposition for the land reclamation plans, the Dutch consortium did not change their branding approach for the Garuda project.

The Dutch consortium started to feel the need to communicate about the Garuda project more structurally to outsiders, for example, through media appearances. Most media appearances were generally done by Dutch consortium members. Indonesian government officials were reluctant to do media appearances. The reluctance of Indonesian government officials to communicate about the Great Garuda brand seems to originate from both the controversial issue of land reclamation and the lack of formal confirmation from powerful ministries and/or the president that Garuda project would actually be implemented. Also, the Great Garuda was by Indonesian government officials mostly viewed as a prestigious project that Dutch water companies could add to their portfolio, possibly leading to lucrative follow-up projects: "(…) then that would be a good spin-off or by-catch if this project yields something for the Dutch economy" (Respondent 39, Dutch Enterprise Agency). As such, the main message that followed from the media appearances was the Dutch consortium was involved in the "great" Garuda project and would, as such, protect all Indonesian people from future disasters.

Initially, the Great Garuda received acclaim and positive media coverage. However, criticism started to come from several fronts. While the Great Garuda plan places strong narrative emphasis on issues such as rapid land subsidence, climate changes, and sea level rise, and while it asserts a causal link between subsidence, uncontrolled groundwater extraction, and extreme limitations in piped water access, the plan does not directly solve these problems. Instead, the plan proposes a radical redesigning of the capital, wherein a new city rising from the waters of Jakarta Bay serves not only to protect the existing city from future marine inundation but to supply new avenues for economic development.

The area were the Garuda interventions would take place were mainly inhabited by poor fishermen. The fishermen, but also other members of the poor urban communities, were aware of the Garuda plans but did not actively participate in the decision-making process as they had other priorities and limited access to Indonesian policy-makers or the Dutch consortium. These urban poor spent most time of the day making a living for themselves. Most of them earn just enough to come by on local markets. From our anthropological observations and interviews, it became clear that the main concerns of the "poor" not so much had to do with the floods or the land reclamation but mostly with evictions (Hellman et al. 2018; van Voorst 2014).

Flood mitigation policies in Jakarta often come with forced resettlement of riverbank communities to low-cost housing. In fact, thousands of slum dwellers have become homeless due to forced evictions without being offered alternative, affordable housing (Hellman and van Voorst 2018; Padawangi 2018; van Voorst 2014). As such, many inhabitants in the poorer urban communities were afraid the "future city plans" would imply for them new forced evictions without compensation. As the Garuda plans initially did not discuss affordable relocation plans for the urban poor, most fishermen started to believe the city government was only acting out of self-interest and not thinking about its "poorer" inhabitants (Stepputat and van Voorst 2016).

Just like the poorer fishermen communities, also Indonesian non-governmental organizations (NGOs) started – with the increased media attention – to question the strategic plan. These Indonesian NGOs and other critics had four main concerns. First, the strategy itself (building a sea dike in the bay) was disputed by NGOs and sectoral ministries because without additional measures, garbage and sewerage water would freely flow into Jakarta Bay, transforming it into "a giant septic tank" (Bakker et al. 2017). Second, the Great Garuda plans did not address the root cause Jakarta's sinking (Bakker et al. 2017), meaning that the sea wall would only temporarily delay flooding by lowering the water level in the bay. Third, other critics agreed that the Great Garuda would only benefit the rich part of the population and be disastrous for the poor coastal (fishermen) communities. The envisioned urban development plans only included high-rise buildings and shopping malls, only available and affordable for the upper class. Concurrently, coastal communities would be cut off from their fishing grounds in open sea.

Moreover, the plan would entail that slum dwellers would be resettled to low-cost housing or would simply lose their house, a practice that characterizes Jakarta's history (Hellman and van Voorst 2018; Padawangi 2018; van Voorst 2014). Fourth, the Great Garuda was presented as a prestigious project for Indonesia, but in the media mainly Dutch consortium members flaunted with the design. As a result, several NGOs questioned the role and trustworthiness of the Dutch consortium, particularly because of their economic interests and presumed role in potential follow-up projects (Bakker et al. 2017; various interviews 2017–2018). Some even view it as a form of "new colonialism" by the Dutch by using economic strategies.

These criticisms resulted in the emergence of a counter-brand initiated by NGOs and the media. The associations created by the counter-brand were totally different from the original brand message. Opponents of the sea dike introduced the image of Jakarta Bay as "the biggest toilet in the world" (Bakker et al. 2017, p. 39) or a "septic lagoon" (The Guardian 2016). Although several respondents acknowledge the Indonesian government rather than the Dutch consultants was ultimately responsible for the project – for example, Respondent 20 from an Indonesian NGO clarifies: "They [consultants] are just doing their job" – they point out that despite the counter-branding campaign, the Dutch consortium members continued to describe the Garuda project as if it is inherently Dutch: "We are here working on the Garuda project, there will be a real Garuda built by Dutch consultants" (Respondent 23, Dutch government representative in 2018). Overall, this reluctance to adjust communications contributed to an image conveyed by the counter-branding campaign in which the Great Garuda was something "imposed" on Jakarta by "the Dutch."

A third element of the counter-brand consisted of framing the urban development as constructing high-rise buildings for the urban rich at the expense of local fishermen. "*If the fishing communities are further marginalized, the gap between rich and poor in Jakarta will grow even wider*" (Bakker et al. 2017, p. 39). Lastly, the counter-brand exploited doubts regarding the feasibility of the plans. The word "megalomaniac" had been used by both opponents and – more surprisingly – proponents of the project (various interviews, 2017–2018), whereby the first considered this as more as a funding challenge, while the latter translated this to "unrealistic" and "unwanted." In sum, the counter-brand introduced new and negative associations about the Great Garuda plans and portrayed the project as being something "foreign," "irresponsible," and "unhygienic." This counter-brand was widely adopted, and, at the time of interviewing (2017–2018), not only NGOs, media, and other actors skeptical of Garuda project used words like "megalomaniac" and "unrealistic" to describe the policy plans but even Jakarta policy officials and some of the Dutch consortium members (i.e., actors central in the brand community).

Re-energizing: Backfiring

The Dutch consortium did not have an immediate response to the counter-branding campaign. It can even be argued that during the whole trajectory, little attention was devoted by the Dutch experts how to "sell" the Great Garuda to the general public, as they viewed this the responsibility of their Indonesian counterparts. In consequence, a "communication vacuum" emerged that provided opportunity for opposing parties to keep spreading negative associations about the Great Garuda project. During this counter-branding campaign, NGOs made a convincing argument that the Great Garuda would not address the root problem of the land subsidence causing the flooding issue. As a result, government officials of the city of Jakarta started questioning the feasibility of the Great Garuda plan: "But to be honest, there's no study whether Jakarta has the carrying capacity, whether the carrying capacity or supporting capacity of Jakarta Bay can accommodate that kind of development" (Respondent 10).

In addition, several Indonesian ministries – especially those acting in the interest of fishermen and the environment – started to declare they would no longer support the existing project plans. The death blow for the Great Garuda brand came when Anies Baswedan promised to stop the project when being elected into office during his 2017's governor election campaign. "The previous governor, Mr. Ahok, he supported, I think more than 100% of the Garuda project (…) but the new governor he does not to support this project" (Respondent 15, national-level Indonesian government official).

After the election of the new governor, the Dutch consortium decided to "retire" the original brand. They rejected the Great Garuda concept and re-shaped the strategic plan. This updated plan still includes an offshore dam but no longer in the shape of a bird. The offshore dam is now presented as the "outer sea wall." In addition, the Dutch consortium stopped mentioning the Great Garuda in the media and reduced the amount media performances by Dutch actors in general. As such, the Dutch consortium decided to stop seeking active media attention for the project and started pushing Indonesian government officials to take over communications about the policy plans toward the general public. The counter-branding campaign thus successfully hijacked the attention for the Garuda project.

Illustrative for the success of the counter-branding campaign is a meeting at the Amsterdam International Water Week in 2017 (Fig. 3). The Dutch consultants started their presentation to a crowd of Dutch and Indonesian representatives from the private and public sector by stressing the Great Garuda was just one of the alternatives considered in the water project and the current preferred strategy no longer featured the Garuda. This was a final strategy the Dutch consortium used to try and change the negative narrative. Dutch consortium members still struggle with the failed attempt to use the Garuda symbol as a vehicle to gain more sympathy and support for the project. Behind the scenes, they face opposition of the local government of Jakarta and the other involved ministries. According to multiple respondents, some government officials prefer an "all-Indonesian" solution and wish to terminate collaborate with the Dutch consortium.

Conclusion

So far, empirical research on the management of a branding campaign has been scarce. Especially, few cases have been described where brand management faced a counter-branding campaign or a backfiring of their activities. Such cases are insightful to understand better the ways in which branding practices can bind or differentiate actors involved with a certain project. The brand management strategies of a Dutch consortium of water experts regarding the Great Garuda in Jakarta, Indonesia, proved a relevant and particularly telling case to add to our understanding of brand management. From our case analysis, it becomes clear that, despite the great potential branding has to bind actors to a policy (Eshuis and Klijn 2012), in reality, it takes a lot of managerial effort, time, and (incremental) adaptability of the branding campaign to make sure that a broad group of stakeholders (varying from politicians, civil society organizations, and citizens) start to see a similar image of a place, develop a shared brand identity, and, eventually, add or upload their own contributions (e.g., videos, short impressions of the area, events in the neighborhood, etc.) to the brand and policy proposal. Brand management, as such, can help to create awareness and support for a policy. At the same time, brand management can sometimes be a plain unworkable exercise.

We furthermore developed two causal pathways to theorize on potential explanations why a branding campaigned aimed to gain support for a grand water management project went astray. We used these causal pathways to cluster which managerial challenges were present in the empirical case and, subsequently, to see what management interventions, respectively, enhanced or impeded the branding campaign in Jakarta.

Overall, we believe that based on our case two important "brand" lessons stand out. One, the legitimacy of a branding campaign increases if a campaign is inclusive and considers the perspectives of all stakeholder groups that will be affected by the policy, for which the branding strategy is used. Two, branding can be used as a vehicle to mobilize a community of stakeholders, but at the same time, when the brand message does not create the right emotional "attachment" with the target audience, opposition for the campaign can be generated. Brands work largely through the visual, triggering emotions and trying to entice people instead of using words, argumentation, and regulation through rules or more or less rational incentives (see Eshuis and Klijn 2012). In this way, branding opens the symbolic, visual, and emotional side of governance processes, and a brand manager, therefore, also needs to be aware of the emotional consequences of its brand activities.

These findings have consequences for further theoretical development of branding. In our view, this should focus more on the creation of brand communities and the governance dimension of branding (e.g., combining brand theory with governance theory). Brand communities can represent a large group of stakeholders who feel they have a common identity by sharing symbols or feeling the same emotional attachment to the phenomenon that is branded (Muniz and O'Guinn 2001). Within these communities, brands can play a role in the identity-formation processes because individual stakeholders, like politicians, civil society organizations, or companies, can use brands to build and shape their organizational identities. In addition, building brand communities can lead to more resilience in society. That is to say, brand communities can ideally be used to mobilize stakeholders and sustain participation and collective action to tackle social problems. In that sense, brand communities can contribute to society's governance capacity to tackle severe social problems, like climate adaptation, urban renewal, or sustainability.

As such, possible future studies can address research questions such as: how can brand management help to build a brand community? Under what circumstances do individual stakeholders feel like they are a part of a brand community? The case described in this chapter clearly did not manage to attract this kind of support, as multiple categories of stakeholders felt the promoted solution harmed their interests. This raises the new question to what degree a brand community can serve as a vehicle to give account to those stakeholders that are directly affected by a government's effort to revitalize a place area? We see opportunities to answering this question by connecting the literature on brand management to the body of literature on governance. Especially governance literature in public administration already has a long history on this and thus is attractive to combine with the insights from branding.

This research naturally has certain limitations. In this chapter, we already addressed the small-N problem of a single case study and thereby the issue of context-dependent generalizations. Nevertheless, this case is illustrative of how a created brand failed to create positive associations by those directly affected by the plans that were branded. In response, those stakeholders' interests temporary aligned when they placed the disadvantages of these plans in the spotlight. As such, their efforts followed a conflictual strategy, i.e., aiming to preventing the proposed without proposing alternative policy directions. Future research could deduct whether the observed action-reaction pattern is indeed typical for branding of spatial strategies or whether other mechanisms may exist in backfiring brands as well.

Additionally, we only studied the management of a process in which the branding effort eventually backfired. We did not compare our findings to a situation where this was not the case. In consequence, the lessons that we suggest being most beneficial for branding projects can have a larger positive impact than we proclaim. Therefore, we propose that prospective studies examine the management of a successful branding campaign. Such an analysis would verify or falsify our research findings. In this way, we believe that the research niche of public branding, and in particular place branding, can further mature and thereby enrich the scholarly debates on how branding can spur participation and policy support in the public sector.

Funding and Acknowledgments This project has received funding from the Institute of Housing and Urban Development Studies (IHS), Deltares, the European Union's Horizon 2020 Research and Innovation Programme under grant agreement No. 726840 (TROPICO), the Vital Cities and Citizens program of the Erasmus University, and the VICI-NWO program No. 45314013.

References

Abidin, H. Z., Andreas, H., Gumilar, I., Fukuda, Y., Pohan, Y. E., & Deguchi, T. (2011). Land subsidence of Jakarta (Indonesia) and its relation with urban development. *Natural Hazards, 59*(3), 1753–1771. https://doi.org/10.1007/s11069-011-9866-9.

Adger, W. N., Arnell, N. W., & Tompkins, E. L. (2005). Successful adaptation to climate change across scales. *Global Environmental Change, 15*(2), 77–86. https://doi.org/10.1016/j.gloenvcha.2004.12.005.

Baker, B. (2012). Destination branding for small cities. In *Destination branding for small cities* (2nd ed., pp. 17–22). Portland: Creative Leap Books. https://doi.org/10.1057/bm.2008.38.

Bakker, M., Kishimoto, S., & Nooy, C. (2017). *Social justice at bay: The Dutch role in Jakarta's coastal defence*. Amsterdam: ENDS, SOMO and Transnational Institute (TNI).

Balmer, J. M. T., & Greyser, S. A. (2006). Corporate marketing: Integrating corporate identity, corporate branding, corporate communications, corporate image and corporate reputation. *European Journal of Marketing, 40*(7–8), 730–741. https://doi.org/10.1108/03090560610669964.

Bucx, T., van Driel, W., de Boer, H., Graas, S., Langenberg, V., Marchand, M., & van de Guchte, C. (2014). *Comparative assessment of the vulnerability and resilience of deltas – extended version with 14 deltas – synthesis report*. Delta Alliance report number 7. Delta Alliance International, Delft-Wageningen, The Netherlands.

Cai, L. A. (2002). Cooperative branding for rural destinations. *Annals of Tourism Research, 29*(3), 720–742. https://doi.org/10.1016/S0160-7383(01)00080-9.

Colven, E. (2017). Understanding the allure of big infrastructure: Jakarta's great garuda sea wall project. *Water Alternatives, 10*(2), 250–264.

Coulthard, S. (2008). Adapting to environmental change in artisanal fisheries-insights from a South Indian Lagoon. *Global Environmental Change, 18*(3), 479–489. https://doi.org/10.1016/j.gloenvcha.2008.04.003.

David, P. A. (2001). Tragedy of public knowledge "commons"? Global science, intellectual property and the digital technology boomerang. In *MERIT-Infonomics Research Memorandum series*. Retrieved from https://collections.unu.edu/eserv/UNU:1084/rm2001-003.pdf.

Eshuis, J., & Klijn, E.-H. (2012). *Branding in governance and public management* (S. P. Osborne, Ed.). New York: Routledge.

Eshuis, J., Klijn, E. H., & Braun, E. (2014). Place marketing and citizen participation: Branding as strategy to address the emotional dimension of policy making? *International Review of Administrative Sciences, 80*(1), 151–171. https://doi.org/10.1177/0020852313513872.

Gaggiotti, H., Low Kim Cheng, P., & Yunak, O. (2008). City brand management (CBM): The case of Kazakhstan. *Place Branding and Public Diplomacy, 4*(2), 115–123. https://doi.org/10.1057/palgrave.pb.6000081.

Hankinson, G. (2004). Relational network brands: Towards a conceptual model of place brands. *Journal of Vacation Marketing, 10*(2), 109–121. https://doi.org/10.1177/135676670401000202.

Hanna, S., & Rowley, J. (2011). Towards a strategic place brand-management model. *Journal of Marketing Management, 27*(5–6), 458–476. https://doi.org/10.1080/02672571003683797.

Hellman, J., & van Voorst, R. (2018). Claiming space in Jakarta: Megaprojects, city planning and incrementalism. In J. Hellman, M. Thynell, & R. Van Voorst (Eds.), *Jakarta: Claiming rights and space in the city* (pp. 157–172). London: Routledge.

Hellman, J., Thynell, M., & van Voorst, R. (2018). *Jakarta: Claiming spaces and rights in the city* (1st ed.). London: Taylor and Francis. https://doi.org/10.4324/9781315111919.

Hommes, C., & Wagener, F. (2008). Complex evolutionary systems in behavioral finance. In T. Hens & K. R. Schenk-Hoppé (Eds.), *Handbook of financial markets: Dynamics and evolution*. Amsterdam: North-Holland. Retrieved from http://hdl.handle.net/10419/86740.

Kavaratzis, M. (2004). From city marketing to city branding: Towards a theoretical framework for developing city brands. *Place Branding, 1*(1), 58–73. https://doi.org/10.1057/palgrave.pb.5990005.

Kerr, G. M. (2009). *The process of implementing aspirational place brands*. University of Wollongong.

Koppenjan, J., & Klijn, E. H. (2004). *Managing uncertainties in networks: A network approach to problem solving and decision making*. London: Routledge.

Laws, E., Scott, N., & Parfitt, N. (2002). Synergies in destination image management: A case study and conceptualisation. *International Journal of Tourism Research, 4*(1), 39–55. https://doi.org/10.1002/jtr.353.

Liotta, P. H. (2002). Boomerang effect: The convergence of national and human security. *Security Dialogue, 33*(4), 473–488. https://doi.org/10.1177/0967010602033004007.

Lucarelli, A., & Brorström, S. (2013). Problematising place branding research: A meta-theoretical analysis of the literature. *The Marketing Review, 13*(1), 65–81. https://doi.org/10.1362/146934713X13590250137826.

Lucarelli, A., & Olof Berg, P. (2011). City branding: A state-of-the-art review of the research domain. *Journal of Place Management and Development, 4*(1), 9–27. https://doi.org/10.1108/17538331111117133.

Martens, K., & Wolf, K. D. (2009). Boomerangs and Trojan horses: The unintended consequences of internationalising education policy through the EU and the OECD. In A. Amaral, G. Neave, C. Musselin, & P. Maassen (Eds.), *European integration and the governance of higher education and research* (pp. 81–107). Dordrecht: Springer.

Miller, M. A. (2013). Decentralizing Indonesian city spaces as new "centers". *International Journal of Urban and Regional Research, 37*(3), 834–848. https://doi.org/10.1111/j.1468-2427.2013.01209.x.

Muniz, A. M. J., & O'Guinn, T. C. (2001). Brand community. *Journal of Consumer Research, 27*(March), 412–432. https://doi.org/10.2139/ssrn.1444833.

Nielsen, J. Ø., & Reenberg, A. (2010). Cultural barriers to climate change adaptation: A case study from Northern Burkina Faso. *Global Environmental Change, 20*(1), 142–152. https://doi.org/10.1016/j.gloenvcha.2009.10.002.

OECD. (2014). *Water governance in the Netherlands*. https://doi.org/10.1787/9789264102637-en.

Padawangi, R. (2018). In search of alternative development in post-reformasi Jakarta. In *Jakarta: Claiming spaces and rights in the city* (pp. 173–194). London: Taylor and Francis. https://doi.org/10.4324/9781315111919.

Pielke, R., Prins, G., Rayner, S., & Sarewitz, D. (2007). Climate change 2007: Lifting the taboo on adaptation. *Nature, 445*(7128), 597–598. https://doi.org/10.1038/445597a.

Schmitt-Beck, R. (2015). Bandwagon effect. In G. Mazzoleni, K. G. Barnhurst, K. Ikeda, R. C. M. Maia, & H. Wessler (Eds.), *The international encyclopedia of political communication* (pp. 1–5). Hoboken/New York: Wiley-Blackwell.

Stepputat, F., & van Voorst, R. (2016). *Urban governance and sustainable development: Cities on the agenda*. Retrieved from https://www.prismaweb.org/nl/wp-content/uploads/2017/07/Cities-on-the-agenda-Urban-governance-and-sustainable-development | Finn-Stepputat-and-Roanne-van-Voorst | April-2016.pdf.

Stevens, V. (2019). Exponential Random Graph Modelling to analyze interactions in participatory practices of place branding. *The Marketing Review, 19*(1–2), 29–41. https://doi.org/10.1362/146934719X15633618140747.

SWECO. (2017). *National capital integrated coastal development*. Amsterdam, the Netherlands.

Syvitski, J. P. M., Kettner, A. J., Overeem, I., Hutton, E. W. H., Hannon, M. T., Brakenridge, G. R., et al. (2009). Sinking deltas due to human activities. *Nature Geoscience, 2*(10), 681–686. https://doi.org/10.1038/ngeo629.

The Guardian. (2016, November 22). *$40bn to save Jakarta: The story of the Great Garuda*. Philipp Sherwell. Retrieved from https://www.theguardian.com/cities/2016/nov/22/jakarta-great-garuda-seawall-sinking.

van Buuren, A., & Warner, J. (2014). From bypass to bathtub: Backfiring policy labels in Dutch water governance. *Environment and Planning C: Government and Policy, 32*(6), 1000–1016. https://doi.org/10.1068/c12212.

van Voorst, R. (2014). The right to aid: Perceptions and practices of justice in a flood-hazard context in Jakarta, Indonesia. *Asia Pacific Journal of Anthropology, 15*(4), 339–356. https://doi.org/10.1080/14442213.2014.916340.

Wesselink, A. (2016). Trends in flood risk management in deltas around the world: Are we going "soft"? *International Journal of Water Governance, 3*(4), 25–46. https://doi.org/10.7564/15-IJWG90.

How to Close the Gap? Branding as a Solution to Declining Youth Participation

Axelle Verdonck, Jan Boon, and Vidar Stevens

Introduction

In Europe, there seems to be an overall decline in traditional forms of political participation among young people. People are less and less involved in a political party, voter turnout keeps decreasing all around Europe, and young people are less interested in politics than ever before (Bhatti et al. 2018; Kavadias et al. 2016). Political participation, however, is a necessary element of today's Western democracies relying on citizen's votes to perform effectively. This chapter is interested in closing the gap between youth and politics.

Scientific research argues that young people are less involved in politics than their parents were (Calenda and Meijer 2009). In the research context of this chapter, studies show that only 19% of Flemish youth (northern part of Belgium) are engaged in a political party, trade union, or environmental or neighborhood community organization (Coppens and Weckx 2018). Political attitudes among young people have changed as well, as attitudes of disaffection and apathy toward politics are growing (Soler-i-Martí 2015).

This change has however been nuanced by some authors who claim young people are not necessarily less involved in politics, as it is rather a case of a different kind of involvement (Calenda and Meijer 2009). They claim young people are less interested in conventional institutionalized forms of political participation and focus more on unconventional ad hoc forms of participation such as street protests (Calenda and Meijer 2009). The attractiveness of these unconventional forms became clear in 2019 when action group Youth for Climate got 35,000 youngsters to march in the streets of

A. Verdonck · J. Boon (✉)
University of Antwerp, Antwerp, Belgium
e-mail: jan.boon@uantwerpen.be

V. Stevens
Mulier Institute, Utrecht, The Netherlands
e-mail: v.stevens@mulierinstituut.nl

S. M. Zavattaro (ed.), *Public Branding and Marketing*,
https://doi.org/10.1007/978-3-030-70505-3_13

Brussels to demand a climate change policy. On the one hand, these forms of political participation can lead to the inclusion of formerly excluded citizens (Marien et al. 2010). On the other hand, issues exist pertaining to the representativeness and democratic legitimacy of forms of unconventional participation. In addition, their lack of institutionalization often leads to effects that are not enduring or stable.

To fill this void, there is a need for an effective way to mobilize young people, not just through unconventional ad hoc means of participation but also through institutionalized participatory governance. In this study, we focus on youth councils as a case of conventional youth political participation. We ask whether branding strategies can be used to engage young people to participate in youth councils (to different extents of activeness). Branding has shown its effects by using symbols and message strategies to mobilize people to participate in politics and affect people's attitude toward the government. Branding strategies have shown their significance in increasing political engagement, social inclusion, trust in policy, and positive economic effects (Ashworth and Kavaratzis 2018; Karens et al. 2016). To achieve these effects, a variety of message strategies can be used.

The use of a visual strategy, e.g., a logo, can enhance trust in policy or public services. A transformational strategy in which communicative mechanisms are used to link the text message with the reader's emotions can affect citizen's attitudes (Piotrowski et al. 2019). What has yet to be determined, however, is whether these strategies can be used to fill in the void between youth and politics. We contribute to the literature on branding in several respects. Research in the branding literature often focuses on what a branding campaign should look like but does not examine the result. As such, testing the attitudinal effect a branding campaign has on respondents allows the research field to gain more knowledge of the conditions under which branding has an impact. We address the following research question: How can brand message strategies be applied to enhance youth political participation? We rely on the experimental method to address this question.

In the remainder of this chapter, we present a brief literature overview of branding in the public sector. A theoretical framework is developed based on the message strategies used to affect citizen attitudes. We then outline the experimental research design, and afterward, we present and discuss the results.

Literature Overview: Branding in the Public Sector

Governments worldwide face complex challenges, leading to more and more forms of horizontal governance such as public-private partnerships and stakeholder involvement (Eshuis and Klijn 2012). These governance strategies often use branding to bind actors, influence public opinion and stakeholder perceptions, and manage the public perceptions of persons, places, organizations, projects, and physical objects (Eshuis and Klijn 2012). Branding is the process in which meaning is added to a product by using symbols. Through this mechanism, ideas and attitudes are deliberately influenced by making positive psychological and emotional associations with the product or service (Eshuis and Edwards 2013).

A brand consists of four elements (Eshuis and Klijn 2012). First, a brand gives meaning to something by making positive associations. The second element is that a brand adds value to a product or service. Third, adding a brand to a product differentiates it from other similar products and makes it identifiable. Lastly, a brand is a visible representation of a design, a name, or a sign. A brand is thus used to distinguish the branded product from other similar products, using visible aspects such as signs or symbols, to link positive associations to the product or service. Brands build associations to influence the process through which perceptions are built. Used in the private sector, brands enhance trust in products or organizations. When implemented into the public sector, it could enhance trust in policy and the government (Karens et al. 2016).

While branding originated in the private sector, the public sector increasingly relies on branding as a governance strategy. Branding can be applied to places, processes, persons, and organizations. First, place branding has been shown to attract tourists, inhabitants, investments, and development funds (Henninger et al. 2016) and to increase social inclusion, political engagement, political participation, and trust in policy and the government (Ashworth and Kavaratzis 2018; Karens et al. 2016). Second, process branding tries to link positive emotions to a policy, institution, or service (Eshuis and Klijn 2012). Some changes to services, for example, can cause a lot of inconvenience to citizens. By branding policies and stressing positive attributes, citizens can be less aggravated when being confronted with potential inconveniences. For instance, the city of Antwerp (Belgium) has done this in its Smart to Antwerp (Dutch: "Slim Naar Antwerpen") campaign. By branding a campaign about construction works, the city tried to convince the public of positive alternatives to come to the city (Antwerpen Mobiel 2015).

Third, personal branding is used by political leaders and public professionals (Eshuis and Klijn 2012). By using branding, they try to associate slogans or logos with the person they are referring to. The "Hope" campaign of Barack Obama in 2008 was a mix of slogans and images to convince voters of his vision. The branding process tried to link these slogans and images to the person of Obama himself. Fourth, organizational branding is a form of branding mostly used in the private sector, although it is also used in the public sector. Universities often try to brand themselves as an organization with a specific identity that provides specific services (Eshuis and Klijn 2012). They try to make the association between the positive outcomes of these services and their organization. In the current study, we focus on how emotions can be governed in a context of youth participation councils, which is a form of organizational branding.

Branding Message Strategies

The branding process works using different kinds of message strategies. In this study, we build on two established dimensions to categorize branding message strategies: visual (vs. non-visual) strategy and informational (vs. transformational) strategy.

First, we distinguish visual strategies from non-visual strategies. Visual strategies typically involve symbolic elements such as logos, figures, colors, etc. entangled in public communications and strategically designed to evoke associations (Alon-Barkat and Gilad 2017). Social psychologists demonstrated how such symbols could change attitudes through associative transfer of affect (De Houwer 2012). Karens et al. (2016), for instance, found that using the European Union (EU) brand in policy proposals increases citizens' trust in those policies. Similarly, Alon-Barkat and Gilad (2017) found using a familiar symbol increases trust and can even compensate for poor performance of the product. Building on these insights, our first hypothesis reads:

H1: A visual branding strategy will positively influence youth political participation

Second, we build on the message strategy classification of Laskey et al. (1989) who divided message strategies used in advertising into two basic categories: informational and transformational advertising. This division is already used in public sector research as well (Piotrowski et al. 2019). An informational strategy relies on presenting factual information to citizens. Facts and figures are presented without explicitly persuading citizens. A transformational strategy emphasizes how certain factual information relates to one's personal experiences and tends to be more involving. Piotrowski et al. (2019) further divide the transformational strategy into the "user image" strategy and the "brand image" strategy. The user image strategy focuses on the users' expectations, needs, or lifestyles. The brand image strategy promotes the "brand identity" and the quality, prestige, and status that are linked to the usage of the product or service.

The transformational strategy used in the current experiment is measured using a user image strategy. This strategy has proven successful in increasing identification with the branded product, as shown in the research of Balaji et al. (2016). They found giving factual information about a certain university to future students increases their identification with the university. This leads to more participation in university activities. As already stated, Piotrowski et al. (2019) used the division between user and brand image strategies as well. They expected an increase in citizens' trust and satisfaction after using one of these two transformational message strategies. Although they did not find any significant differences between the user and brand image strategies, their hypothesis has partially inspired us to form our second hypothesis:

H2: A transformational branding strategy will positively influence youth political participation

Methods

Empirical Context

This study focuses on youth councils as a typical instrument of youth political participation. A first "raison d'être" for youth councils is their advisory function. In Belgium, the empirical setting for this study, most youth councils exist at the local level (cities or municipalities) to advise the local government. Youth councils are typically composed of young individuals who represent a diverse set of youth organizations (though these councils can also include engaged young people without organizational ties). As such, youth councils aspire to include a broad range of voices and interests. A second function of youth councils relates to stimulating youth participation in politics and, ultimately, contributing to the formation of a sense of citizenship among young people. Youth participation can be achieved in diverse ways: creating awareness, sharing information, organizing opinion polls, holding demonstrations, etc. Youth councils, therefore, represent a wide range of participatory activities and instruments.

While large variety exists in how youth councils function, their existence and composition is legally determined. As of 2003, Belgian municipalities are obliged to organize a youth council with a certain diversity in their members' composition. Youth councils are a form of conventional government-initiated political participation. While arguably citizen-initiated unconventional political participation (e.g., climate protests) are increasing, we see several reasons why the study of conventional participation remains important.

Unconventional participation occurs outside traditional institutions of representative democracy. Next to its benefits, the non-institutionalized nature of unconventional participations risks several disadvantages, pertaining to their lack of enduring effects and robustness and their lack of representativeness. In addition, conventional political participation has been shown to be a determinant of political interest and engagement. Participation "regimes," therefore, may benefit from a combination of unconventional participation and conventional government-initiated participation. We therefore argue that governments and citizens gain from (youth) participation in institutions such as advisory (youth) councils. The question, however, is how to achieve such participation, which ultimately rests on the engagement of (young) individuals in a current climate of distrust in government and politics.

Data Collection

Our study uses an experimental method to test the effect of branding strategies on youth political participation. Because we are interested in the effect of different branding strategies across groups, rather than in the longitudinal effect of branding strategies on the same respondents, we use a between-subjects design. Experiments are commonly acknowledged as highly suited to test causal assumptions (Piotrowski et al. 2019). This study investigates the causal relation between using a particular branding strategy and youth political participation.

The experiment consists of a 2×2 design according to the two theorized dimensions of branding strategies: visual vs. non-visual strategy and informational vs. transformational strategy. Respondents were randomly allocated to one of four experimental groups. The experimental design was pretested by the authors and piloted among students at the faculty of social sciences of the University of Antwerp. The pilot showed no flaws in the experimental design.

The experiment involved the presentation of a fictional local youth council ("Youth Council"). We choose a fictional youth council – as opposed to, say, the local youth council of the city of Antwerp – to account for the risk that students might not be interested in participating because they do not identify with the presented region rather than with the concept of a youth council. Manipulations were introduced in terms of how the Youth Council was presented: using a logo or not (visual branding) and including psychological components related to the user image effect or not (transformational branding).

The survey experiment was filled in by bachelor-level students at the University of Antwerp (aged between 18 and 24). Classes in different faculties were attended by members of the research team. The classes represented sufficient variation in terms of student profiles (faculty of social sciences, faculty of economics, and faculty of linguistics). Classes were attended in the first or second year of university to ensure that respondents were in the age category which would make them candidates for youth councils (those who were not were later excluded from analyses). The Qualtrics software was used to collect the data. The presence of members of the research team ensured respondents received a consistent introduction to the survey. Respondents were guaranteed anonymity, and 404 respondents filled in the survey. After removing respondents who were not in the age category of 18–24, a final sample of 367 respondents remained.

In accordance with standards of practice of experiments (Baekgaard et al. 2015), the survey separated a series of pre-test questions (which inquired into the social and demographic background of respondents), from the treatment (explained below), the items that tapped into the dependent variable (youth political participation), and, finally, the manipulation check.

Operationalization

The dependent variable "youth political participation" consists of several items tapping into the diversity of participatory actions and instruments a youth council can stimulate and has the following: (1) Follow Youth Council on social media; (2) Sharing a political message of the Youth Council on social media; (3) Sign a petition of the Youth Council; (4) Visit lectures organized by the Youth Council; (5) Join a demonstration organized by the Youth Council; (6) Join monthly meetings of the Youth Council; and (7) Organize activities for the Youth Council. These items were formulated to distinguish online from offline participation, as well as more active from more passive forms of youth political participation. In the analyses, the distinct impact of branding strategies on each form of youth political participation is examined in separate analyses (with each form as dependent variable).

The independent variables "branding message strategies" were measured in the four experimental groups tapping into the branding message strategies youth councils have at their disposal to stimulate youth political participation (absence/

Table 1 Operationalization branding message strategies

Visual strategy: logo	Message strategy: informational vs. transformational
DRUK JE STEMPEL OP DE TOEKOMST (in English: "Leave your mark on the future")	Informational strategy: The Youth Committee is an organization where young people get the opportunity to present their viewpoint and influence policy-making. In this committee, young people from your neighborhood are represented. They will convene on a monthly basis to advise local government. In addition, the Youth Committee organizes other activities such as lectures, debates, and demonstrations
	Transformational strategy: Hey, you! Do you also think that politicians take young peoples' opinions not into account? Should young people have a stronger influence on their own futures? If the answer on these questions is "YES," you are the sort of person we are looking for in the Youth Committee! The Youth Committee is an organization for and by young people. We strive for change in our neighborhood. You'll have the opportunity to give your opinion and influence policy. How do we go about this? We organize activities, such as lectures, debates, and demonstrations, and have monthly get-togethers to discuss and change our neighborhood. Are you young and opinionated? Join us! Leave your mark!

presence of visual strategy; absence/presence of transformational strategy). The presence of a visual branding strategy was operationalized by inclusion of a logo, which is not uncommon in branding studies (Alon-Barkat and Gilad 2017; Karens et al. 2016). Following Piotrowski et al. (2019), the presence of a transformational strategy was operationalized by introducing psychological components into the message (as opposed to mere factual information). Whereas informational strategies are intended to provide factual information, transformational strategies are deliberately intended to influence the perceptions of message receivers (Piotrowski et al. 2019). Here are the texts we used in the informational vs. transformational groups (see Table 1).

Results

The sample was checked for homogeneity among some demographic variables. As a balance check, we used an ANOVA to identify whether the differences between our control and treatment groups concerning political preference (left-right continuum) and political interest were significant. An ANOVA is a statistical technique used to compare means of more than two populations. These turned out not to be significant, implying that our groups are well balanced, so we did not need to control for these individual-level variables (Nielsen and Baekgaard 2015). More specifically, the results of the ANOVA were the following: political interest ($F = 1.752$, $p > 0.5$) and political party vote ($F = 0.571$, $p > 0.5$).

We then checked whether the participants perceived the experimental treatment (i.e., the branding strategy) in the way intended. This manipulation check was performed by comparing the perceived strategy in the groups. As a manipulation check, we included two questions after our vignettes aimed at identifying whether the treatment had affected the respondents. The first question – "did you see the logo of the Youth Council?" – was used as a check to see whether the treatment affected the respondents. By running the ANOVA, this turned out to be true ($F = 32, 741, p < 0.01$). The second question – "to what extent do you agree with the proposition that government officials do not take into account the considerations of young people?" – was only mentioned in the transformational message strategy groups and also turned out to affect the respondents (on the basis of independent sample T-tests, $F1 = 3.941, p < 0.05, F2 = 7.062, p < 0.01$).

Overall, what stands out from the analyses is the lack of significant results. Within the more active forms of political participation (sign a petition, visit lectures, join a demonstration, join monthly meetings, organize activities), no significant effects are retrieved. Within one of the more passive offline forms of political participation ("Sharing a political message of the Youth Council on social media," cf. Table 2), we find one significant effect: respondents exposed to an informational strategy with a logo were more likely to share political messages of the Youth Council on social media compared to respondents exposed to a transformational strategy without a logo.

We conclude that the use of an informational strategy may have a small effect on more passive forms of political participation. Overall, however, our findings raise some concerns concerning the ability of branding strategies to close the youth-politics gap. This finding, however, should be interpreted considering the experimental setup of this study. Because we wanted to avoid tapping into existing emotions of students related to the existing local youth council, we presented them with a fictional example and logo. The logo in this study (megaphone) was developed to visually exploit dissatisfaction among young people of not being heard. Yet being a fictional logo, no familiarity existed among respondents with the logo. The minimal effect of the use of a logo, therefore, can be explained by the lack of familiarity of respondents with the proposed fictional Youth Council and logo.

Somewhat surprising, the use of a transformational strategy does not seem to influence any of the dependent variables. This finding is somewhat harder to interpret. One explanation might be that the theorized mechanism (i.e. tapping into feelings of not being heard) does not matter for stimulating activity in youth councils. Another explanation might relate to the length and language used in the transformational text. In comparison to the informational strategy, this text was significantly longer which could have caused fatigue or negative attitudes resulting in a more negative evaluation of the Youth Council. In addition, in trying to appeal to young people, the language used in the transformational strategy may have had an unintended effect: its informal tone might conflict with the pre-existing image of what respondents have of youth councils and what they should be about.

Table 2 Multiple comparisons for dependent variable "Sharing a political message of the Youth Council on social media"

	Reference group	Comparative groups	Mean difference (I–J)	Std. error	Sig.	95% confidence interval	
						Lower bound	Upper bound
Tukey HSD	Informational strategy – no logo	Informational strategy – logo	−.231	.165	.501	−.66	.20
		Transformational strategy – no logo	.245	.175	.502	−.21	.70
		Transformational strategy – logo	−.056	.169	.987	−.49	.38
	Informational strategy – logo	Informational strategy – no logo	.231	.165	.501	−.20	.66
		Transformational strategy – no logo	.476[a]	.170	.028	.04	.92
		Transformational strategy – logo	.174	.164	.712	−.25	.60
	Transformational strategy – no logo	Informational strategy – no logo	−.245	.175	.502	−.70	.21
		Informational strategy – logo	−.476[a]	.170	.028	−.92	−.04
		Transformational strategy – logo	−.302	.174	.310	−.75	.15
	Transformational strategy – logo	Informational strategy – no logo	.056	.169	.987	−.38	.49
		Informational strategy – logo	−.174	.164	.712	−.60	.25
		Transformational strategy – no logo	.302	.174	.310	−.15	.75

[a]The mean difference is significant at the 0.05 level

Conclusion

In this study, we examined whether branding could be a solution for the decline in youth political participation. The legitimacy of democratic systems hinges on citizens' participation, be it through voting or more active means of political participation. In this sense, the observation that young people seem to turn away from conventional institutionalized forms of political participation – such as youth councils – is problematic. This chapter explored the potential of branding strategies to mobilize young people to participate in youth councils, effectively closing the gap between youth and politics.

Branding has proven to be effective in other contexts to increase political engagement, social inclusion, trust in policy, and positive economic effects (Ashworth and Kavaratzis 2018; Karens et al. 2016), enhance trust in policy or public services, and affect citizen's attitudes (Piotrowski et al. 2019). Two forms of message strategies were theorized and tested to have an effect on different forms of youth political participation (varying in their activeness) in youth councils: a visual (vs. non-visual) strategy and a transformational (vs. informational) strategy (Laskey et al. 1989; Karens et al. 2016; Piotrowski et al. 2019).

An experiment was conducted in which the effects of a visual and a transformational branding strategy were tested. Overall, the scarcity of significant effects stood out as most hypotheses were neither confirmed nor rejected. The more active forms of participation (attending lectures/debates, attending demonstrations, attending council meetings, organizing council activities, or signing a petition) were not affected using any branding message strategy. Among the more passive participation forms, sharing a political message on social media was positively affected using an informational message strategy with a visual element (logo). This significant effect, however, was only found in comparison to the use of a transformational strategy without a logo.

This study, therefore, has several practical implications and concrete avenues for further research. Overall, our study suggests that "getting the message right" – in terms of both visual familiarity and content – is a challenging endeavor in building a connection with target groups (here: young people) and a brand (here: youth council).

First, an important take-away for branding scholars and practitioners concerns the debate on the importance of design/aesthetics vs. familiarity in branding strategies, both of which are considered to be important elements by social psychologists that contribute to the effectiveness of visual elements (such as a logo). Contrasting our findings, which point at the absence of an effect of a fictional logo that was designed to represent existing feelings of being unheard, with previous research that used visual elements that tapped into existing emotions of respondents had stronger effects (Alon-Barkat 2019; Alon-Barkat and Gilad 2017; Karens et al. 2016), suggests the importance of familiarity with existing logos. When using a new logo, the familiarity between the brand and the logo does not exist. Familiarity is an important part of image-forming that can lead to certain affections toward a brand. It promotes more efficient processing, leading to the demand of fewer attentive resources than with new logo images (Buttle and Westoby 2006).

Second, our finding on the absence of an effect of transformational message strategies points at the importance – and challenge – of tapping into the right emotions among target audiences. While we hoped to trigger identification among youngsters with the fictional youth council by introducing psychological components into the message, the absence of an effect might suggest that either no or the wrong emotions were triggered. We suggest investing sufficient time and resources on getting to know your audience. Within the context of limited resources in which this study had to operate, we had to base our expectation on what emotions would motivate young people to participate politically on desk research. We recommend practitioners to use qualitative techniques such as focus groups to see which emotions are triggered by each strategy.

References

Alon-Barkat, S. (2019). Can government public communications elicit undue trust? Exploring the interaction between symbols and substantive information in communications. *Journal of Public Administration Research and Theory, 30*(1), 77–95. https://doi.org/10.1093/jopart/muz013.

Alon-Barkat, S., & Gilad, S. (2017). Compensating for poor performance with promotional symbols: Evidence from a survey experiment. *Journal of Public Administration Research and Theory, 27*(4), 661–675.

Antwerpen Mobiel. (2015). *Communicatiecampagne "Slim naar Antwerpen" gelanceerd.* Retrieved from https://www.oosterweelverbinding.be/nieuws/algemeen/communicatiecampagne-slim-naar-antwerpen-gelanceerd.

Ashworth, G. J., & Kavaratzis, M. (2018). The roles of branding in public administration and place management: Possibilities and pitfalls. In E. Ongaro & S. Van Thiel (Eds.), *The Palgrave handbook of public administration and management in Europe* (pp. 425–439). London: Palgrave Macmillan UK.

Balaji, M. S., Roy, S. K., & Sadeque, S. (2016). Antecedents and consequences of university brand identification. *Journal of Business Research, 69*(8), 3023–3032. https://doi.org/10.1016/j.jbusres.2016.01.017.

Baekgaard, M., Baethge, C., Blom-Hansen, J., Dunlop, C. A., Esteve, M., Jakobsen, M., Kisida, B., Marvel, J., Moseley, A., Serritzlew, S., Stewart, P., Thomsen, M. K., & Wolf, P. J. (2015). Conducting Experiments in Public Management Research: A Practical Guide. *International Public Management Journal, 18*(2), 323–342.

Bhatti, Y., Dahlgaard, J. O., Hansen, J. H., & Hansen, K. M. (2018). Can governments use Get Out The Vote letters to solve Europe's turnout crisis? Evidence from a field experiment. *West European Politics, 41*(1), 240–260.

Buttle, H., & Westoby, N. (2006). Brand logo and name association: It's all in the name. *Applied Cognitive Psychology, 20*(9), 1181–1194.

Calenda, D., & Meijer, A. (2009). Young people, the Internet and political participation: Findings of a web survey in Italy, Spain and the Netherlands. *Information, Communication & Society, 12*(6), 879–898. https://doi.org/10.1080/13691180802158508.

Coppens, T., & Weckx, L. (2018). *First voters: Jongeren en de gemeenteraadsverkiezingen van 2018* [Powerpoint]. Retrieved from https://www.vrt.be/vrtnws/nl/2018/10/01/first-voters/.

De Houwer, J. (2012). Evaluative conditioning. In N. M. Seel (Ed.), *Encyclopedia of the sciences of learning* (pp. 1179–1181). New York: Springer.

Eshuis, J., & Edwards, A. (2013). Branding the city: The democratic legitimacy of a new mode of governance. *Urban Studies, 50*(5), 1066–1082. https://doi.org/10.1177/0042098012459581.

Eshuis, J., & Klijn, E.-H. (2012). *Branding in governance and public management.* New York: Routledge.

Henninger, C. E., Foster, C., Alevizou, P. J., & Frohlich, C. (2016). Stakeholder engagement in the city branding process. *Place Branding and Public Diplomacy, 12*(4), 285–298. https://doi.org/10.1057/s41254016-0002-1.

Karens, R., Eshuis, J., Klijn, E. H., & Voets, J. (2016). The impact of public branding: An experimental study on the effects of branding policy on citizen trust. *Public Administration Review, 76*(3), 486–U424. https://doi.org/10.1111/puar.12501.

Kavadias, D., Siongers, J., & Stevens, F. (2016). *Jongeren en politiek Onderzoeksgroep TOR.* Vakgroep Sociologie, VUB.

Laskey, H. A., Day, E., & Crask, M. R. (1989). Typology of main message strategies for television commercials. *Journal of Advertising, 18*(1), 36–41.

Marien, S., Hooghe, M., & Quintelier, E. (2010). Unconventional participation and the problem of inequality: A comparative analysis. In *New forms of citizen participation. Normative implications* (pp. 131–146). Baden-Baden: Nomos Verlag.

Nielsen, P. A., & Baekgaard, M. (2015). Performance Information, Blame Avoidance, and Politicians' Attitudes to Spending and Reform: Evidence from an Experiment. *Journal of Public Administration Research and Theory, 25*(2), 545–569.

Piotrowski, S., Grimmelikhuijsen, S., & Deat, F. (2019). Numbers over narratives? How government message strategies affect citizens' attitudes. *Public Performance & Management Review, 42*(5), 1005–1028.

Soler-i-Martí, R. (2015). Youth political involvement update: Measuring the role of cause-oriented political interest in young people's activism. *Journal of Youth Studies, 18*(3), 396–416.

From a Dog's Breakfast to a King's Table: Branding and Marketing a Major Research University

Julia Cummings and Ken McConnellogue

The Visual Disconnect

Many a college and university communications and marketing professional has had to convene the conference table meeting. You gather top administrators. You find the largest conference table in the biggest conference room at the institution and lay out dozens (or, in our case, hundreds) of brochures, magazines, business cards, viewbooks, laptops displaying web pages, and much more of the plethora of print and electronic publications and images representing the organization. The mission is to dramatically show just how tangled and confusing the institution's visual identity – and by extension its brand – has become. There is probably an app for it these days, but when we did it at the University of Colorado (CU) just more than a decade ago, it was a decidedly large conference table.

One administrator at CU called the materials on display "a dog's breakfast." Few knew quite what that meant, but we all understood it to have many unappetizing ingredients. And that was the image the university was projecting. It was not an image befitting a major educational and research institution that is among the top public universities in the country graduating some 15,000 students annually and ranking among the top 10 nationally in research prowess.

The original version of this chapter was revised. The correction to this chapter is available at https://doi.org/10.1007/978-3-030-70505-3_17

J. Cummings (✉)
CU Denver School of Education & Human Development, Denver, CO, USA
e-mail: julia.cummings@ucdenver.edu

K. McConnellogue
University of Colorado, Office of the President, Denver, CO, USA
e-mail: ken.mcconnellogue@cu.edu

S. M. Zavattaro (ed.), *Public Branding and Marketing*,
https://doi.org/10.1007/978-3-030-70505-3_14

CU is a large and complex institution, a collection of diverse people, programs, and places. It has four campuses around Colorado: the flagship in Boulder (nearly 36,000 students), an urban campus in Denver (almost 15,000 students), a regional campus in Colorado Springs (just over 12,000 students), and an acclaimed medical campus in Aurora (4,300 students). Its fully online options are housed within CU Online. It boasts 5 Nobel laureates, 10 MacArthur "Genius" fellows, and some 20 alumni-astronauts, who have spanned the US space program, from the early Mercury missions to the International Space Station today. It is Colorado's third-largest employer and has an annual budget approaching $5 billion and an annual economic impact of $14.5 billion.

When it comes to branding, marketing, and visual identity, many public universities such as CU are not command-and-control organizations. The leader cannot say "Here is our logo. Use it." Instead, a centralized-decentralized approach means units (CU's departments and entities) act with a fair amount of independence on each campus and within its schools and colleges, much less within the larger organization. The result was obvious on the conference room table. The print collateral and digital resources were sorely mismatched. Fonts, colors schemes, and photography styles were inconsistent. Hand-sketched logos and graphic attempts with clipart were outliers but definitely in the mix. This disjointed mosaic did not reflect our mission-driven university, one of Colorado's most important assets.

Why Brand?

We had to do something about it. The events that followed our conference table meeting proved often to be painful in the short run and productive in the long run. And the outcome resonates more than ever today.

We had a personal and professional stake in the brand strategy and outcome. We are not only professionals whose duties include serving as keepers and promoters of the institution's brand, but we ourselves are CU alumni. We have also sent our children to CU. We care deeply for the place that continues to influence our life trajectories and aims to positively transform the lives of thousands of others. We work in different parts of the CU system and therefore offer macro and micro perspectives.

It is important to note the difference between branding and marketing. Branding is the promise of who you are as an organization, using key imagery, narratives, and culture to permeate the place. It is what you stand for and how your institution drives emotional connection. Marketing is how you promote that brand and build strategic relationships, ranging from target audience segmentation and customer journey mapping to orchestrating authentic marketing mix touchpoints, such as traditional and digital advertising and interactive websites.

It was time to embrace the lessons of the conference table exercise. We needed a renewed focus on brand management, strategy, and execution, including a deeper focus on digital engagement and research. Fortunately, at CU, we had a president who recognized the problem. As an experienced business and civic leader, he understood the necessity of effective branding and marketing in private and public spheres of influence.

He set us on the sometimes painful but ultimately productive path of getting our arms around the university's brand and marketing efforts in ways that were networked, consistent, and visually appealing. His mantra was that the hundreds of disjointed messages and images confused our stakeholders, and there were many to confuse.

Like any large research university, CU's audiences include prospective students, current students, faculty, staff, parents, alumni, donors, elected officials, media, the business community, community organizations, research partners, and more. The dog's breakfast approach was inefficient (particularly in the use of funding). But, worse, it was ineffective because CU exists in a crowded higher education marketplace where competition – for students, for donations, for state funding, for faculty, for research dollars, and for attention – was and is extreme. We needed to create alignment and greater cooperative networks and consistency across campuses to line up under a master brand umbrella.

When we started our project, state support for colleges and universities, particularly in Colorado (which ranked 48th nationally in state support for higher education and is now 47th), was in a downward spiral accelerated by the recession of 2008–2009. A crop of online institutions poached students and faculty. Worthy charities (especially during the economic downturn) competed for philanthropic dollars. Additionally, about one-third of people in our state (less nationally) have a college degree. This means nearly two-thirds of people do not have an affinity with the university outside shirttail connections, such as my cousin or grandchild went there, or I like the football team.

Perhaps most important, the flood of images and messages from ever more sophisticated and technologically savvy corporations, organizations, and causes made effective branding and marketing imperative to CU and its colleague colleges and universities.

Navigating Branding Speed Bumps

Branding and marketing have not always played well inside higher education. A common refrain is a version of "We are *educators*, imparting knowledge and conducting important research, not grubby businesses making and selling widgets. Be away with your branding and marketing!" Yet the market realities facing higher education skewed that somewhat true but antiquated notion.

To help make the case, CU engaged a nationally known branding and marketing firm. We warily eyed a few speed bumps: a history of the university studiously avoiding even the appearance of anything so pedestrian as branding and marketing, a skeptical internal audience perfectly comfortable with not being told what to do, and a public ready with criticism of "wasting taxpayer dollars." Yet, we waded into the fray.

We will not paint a rosy picture. Getting the wider university community on board with branding was part persuasion and part trench warfare. We found that the conference table exercise was the tip of the iceberg. In an inventory of logos alone, we uncovered more than 500 across the CU system. Since the 1980s, CU's visual identity has been anchored by an interlocking version of the letters *C* and *U*. The more successful units still used it. Some lesser offenders corrupted it with various colors, orientations, and additions. Most just created an entirely new logo that seemingly had no connection with the university whatsoever. When we asked about its origin, we often heard a version of "the director's niece designed it" or the all-too-common (and proudly delivered) "our students created it!"

Another problem was the athletics logo on our Boulder campus, known as the Ralphie logo. Ralphie is the live buffalo mascot that leads the CU football team onto the field before games. The logo is a canted interlocking *CU* inside the outline of a buffalo. It is wildly popular. The Boulder campus administration had been waging its own campaign to limit it to athletics and spirit uses. It is trademarked and licensed, so unauthorized use of it costs the athletics department lost revenue. Additionally, there was a contingent on the campus who felt an athletics logo, no matter how compelling, should not represent an academic institution.

Whether using Ralphie or a niece's or student's design, units fiercely held their positions. "We are competing against the University of X!" they cried. "We have to differentiate ourselves from Y," they said. Our favorite was those who felt compelled to differentiate themselves from similar schools or colleges on CU's other campuses. In other words, we were competing against ourselves. "We are special. We are different. You do not understand!" There is a certain irony in a communications and marketing professional hearing from a Philosophy Department chairman that we just do not get it.

The first step was to get a sense of the magnitude of the problem through market research and a competitive assessment. The university needed to ascertain what key audiences thought of us – not what we thought of ourselves. We surveyed our national competitors. We sent out 40,000 electronic surveys and received an impressive 25% response rate. We held focus groups with alumni, donors, current and prospective students, parents, and business and community leaders. We presented at endless faculty meetings and staff meetings. The market research played out in similar fashion across the university, on a broad scale, and in colleges and departments.

Our committed colleagues in communications and marketing across our system were allies in the cause and invaluable to the effort, including employees in central communications offices and communicators at college and department levels. They were on the frontlines every day, making connections, listening, sympathizing, refuting, and weaving inspiring narratives. They always held the line and steered the more recalcitrant back into the fold. During the branding initiative, we convened several in-person brand summits, attended by hundreds of communications professions, to get everyone on the same page.

Central university communications offices at all four campuses developed brand promises, elevator speeches, messaging platforms, branding websites, advertising, and more. All cascaded from higher-level messaging that leadership and communications and marketing professionals collaborated on and agreed to. Within months, they engaged school and college deans, communicators, and selected faculty to collaborate on unit-specific messaging platforms, creative strategies, and marketing mix elements. Everything they created laddered up to the campus and CU system's branding and positioning.

Nobody likes having things taken away from them, especially things they hold dear. However misguided about the value of their micro approach, the partisans were loath to give up on their logos and visual identities, no matter how unattractive or ineffective. Nor did it help that the people asking them to do so were from the distant, faceless central administration or the communication's office: "What do they know of us?!"

Our centers and smallest subunits with more independent funding streams were most likely to oppose giving up decades-old logos. Their directors said things like: "Our partners at X university and Y community organization are not going to work with us anymore because that color scheme does not mesh with theirs." Another saying we heard was: "People know us as a national center and it's better to stick with the old logo because the new CU logo may hurt our funding prospects."

In our estimation, they substantially overestimated the impact of their marks and underestimated the broader university's brand equity. Because CU is a national research leader, an identity forged with CU is highly likely to bring in more partnerships and more funds. We witnessed some negotiations and personalized guidance take place in these arenas. For instance, communicators took the extra time to deliver newly written and designed websites paired with professionally refreshed print collateral in exchange for brand compliance. This is not to say that the squeaky wheel always gets the grease. Some branding negotiations remain ongoing in these subunits. The brand also needs continual policing because people new to the organization often dive right into their own brand redesigns, unaware of the larger university effort.

Launching Our Efforts

The launch of the new branding solution was multipronged. It started with a sustained internal campaign that had the backing of the university president and each campus chancellor. We laid out a compelling case that described the problem and its consequences and solutions. We used third-party market research to make the case that no, the units were not as special as they imagined. Their logo was just an ingredient in a dog's breakfast that was indistinguishable from many other ingredients. A unique logo would not help differentiate subunits. They only confused stakeholders. Worse, many outside observers did not know that some units were even affiliated with CU. They were like free agents floating around in the market.

University culture reveres research. When our constituents saw the breadth and depth of our quantitative and qualitative market research efforts, detailed earlier, they were impressed. The epiphany was obvious to our stakeholders. Our brand equity rested with the University of Colorado, not its component parts. If people had a touchpoint with the university or even if they did not, they knew CU, what it did, and what it stood for. They did not necessarily know the Philosophy Department or the Parking Services unit or the College of Arts and Sciences specifically, but they knew CU. This gave us a toehold to advance in the trenches.

The headlines and messages that would soon hit the market were previewed with internal audiences during the launch. While they proved popular, we discovered many covert art critics with decided opinions. There was also a chunk of people who just would not buy it, no matter how compelling. But the vast majority did. Our key to headline and messaging success was engaging and carefully listening to stakeholders early and often. Research with our audiences helped articulate our brand personality, illustrating the university as a person with a unique personality. We convened endless meetings to inform the university's common mission and vision of who we are, what we do, and what we aspire to become.

Along the way, people at the university had lots of opinions, and few of them could be distilled into concise sentences or statements, which is the lifeblood of branding and marketing. But through continual engagement, feedback loops, input and refinement, lots of listening, and some imploring, we got there. Our branding teams developed messaging around four of the university's fundamental activities (expanding each to a two-word idea): learning and teaching, discovery and innovation, health and wellness, and community and culture. The broad messaging buckets encompassed virtually everything the university does and allowed campuses, schools, colleges, and units to tailor their specific messages to their key audiences. Most everyone went home happy.

Campuses created beautiful marketing campaigns with refreshed messaging, which were consistent with the umbrella brand. For instance, CU Boulder's "Be Boulder" campaign showcased diverse students and famous alumni, often pictured with dramatically beautiful mountain images and well-known campus buildings on what many view as one of the most beautiful campuses in America. CU Denver launched with a "Learn with Purpose," which morphed into a "CU in the City" campaign. These campaigns featured highly diverse students and faculty faces, paired with urban Denver photography scenes. These campaigns were blessed at top levels of the university. They graced airport terminals, viewbooks, city billboards, banners on city and campus light poles, and more.

The Importance of Graphic Expertise, Branding Toolkits, and Brand Boards

How do you replace hundreds of logos with a simple yet powerful version of the interlocking *CU*? The secret ingredient was talented graphic designers across our campuses, who took great pride in the work. This was nobody's niece. These were highly skilled professionals who knew about color and line spacing and fonts and images – all those seemingly small details that are critical to successful imagery.

There was also a bit of a carrot and stick involved. The carrot was the wonderful visual identity that made it easy for campus units to adopt. Branding toolkits were made readily available, tailored to units. These key resources made units feel empowered. They were connected to the larger university but also appreciated that their department or unit had a certain prominence within it.

We created a board on each campus responsible for brand identity standards, along with a systemwide board comprised of representatives from all the campuses plus the president's office. The systemwide board members are the arbiters of all things brand: mediating disputes, fielding requests, and evangelizing. This group created a brand identity manual that clearly laid out the rules of the road, with lots of visual examples.

The president and chancellors clearly and consistently communicated why we were engaged in the work, what the outcome would be, and what it meant for campuses, schools, and departments. We also codified the work in university policy, the amorphous guide to all things at a university. CU's Board of Regents, its governing board, approved the policies and let it be known that they expected compliance for the good of the university.

Still, it was not all sweetness and light at the end of the darkness. There were naysayers who clung to the old ways. The loneliest job at the university was that of brand cop, those poor souls dispatched to the far reaches of the campuses to engage would-be brand offenders and convince them to join the fold. For instance, some units asked to create T-shirts and giveaways with older logos or off-brand marks. Others wanted to place independently created department taglines on merchandise. They might say things like "My husband owns a clothing design company. Can't we employ his graphic design services to make this happen?" The answer was a resounding no. Campus units needed to purchase university giveaways, clothing, and merchandise from one of our seven licensed vendors, who had all been officially trained on university branding standards. Our vendors cooperate with internal clients, communicators, and branding experts to get signoffs on design before printing. As a result, we no longer have stretched or independent logos, or unapproved taglines, on university giveaways.

As another example, "brand cops" worked to calm faculty members asking for uniquely designed business cards and office stationery. We needed to stick to university policy and say no. All orders for CU business cards and stationery use a standard CU template and are printed at our university print shop for consistency. The only variation is multilingual elements added to faculty and staff business cards, if applicable to their campus role and responsibilities.

A marketing and communications leader at another university, upon hearing about our project, said, "The woman who led that effort at our university was the most hated person on campus for 2 years; now, she's beloved."

Branding Success: Measurements, Outcomes, and Feedback Loops

Success came gradually and then suddenly. We narrowed 500 logos down to a half-dozen. Soon, the furor around the project subsided almost entirely. New materials looked good. They reflected a serious and valuable research university. People suddenly understood that the brand and visual identity were important and that it was no longer the Wild West where units did what they wanted. There would be the occasional brushfire, but nothing that could not be doused.

Still, there were aberrations. The entire Colorado Springs campus negotiated with the president to use UCCS in its typography because that is what it is commonly known as. The compromise was that it would be used alongside the interlocking *CU*. Not ideal, by any means, but not horrible. Ironically, in market research the campus conducted nearly a decade after the branding project started, fully one-third of respondents did not know the campus was affiliated with the University of Colorado. Images and words matter.

The healthcare system affiliated with the university was another aberration. Just as the branding project was wrapping up, the University of Colorado Hospital teamed up with other health systems and hospitals to create UCHealth. It was confusing because the University of Colorado Hospital is a separate entity from the university. Although all if its doctors are faculty in the CU School of Medicine, in the early 1990s, the Colorado Legislature made it an authority separate from the university. When the hospital merged with other hospitals and healthcare systems, it was governed by its own independent board. We at the university certainly tried to bring them into the branding fold, but they determined they needed brand autonomy to deal with their own market imperatives and competitive pressures. It was probably a lost opportunity (for both entities, we believe), but one we could not control.

We use a variety of metrics to track success of our branding efforts. Those relating to lead generation and our student enrollment pipelines are tracked via a sophisticated data visualization software tool. These measures include application numbers and enrollment yield information. Administrators may view these numbers at a high level and then dive deep to see them play out at the unit and major levels across time. With this software, we see trends relating to student diversity, student success, and much more. We track advertising, social media, and PR with metrics from impressions, reach, clicks, likes, and engagement from stakeholders. Our mass emails now feature branding templates. We track open rates, clicks, registrations, and the like.

For the university overall, the outcomes have been positive. CU set annual fundraising records for almost the entire decade after the project was implemented. Enrollment at the four campuses hit record highs. Research funding topped $1 billion annually for the first time. While it may be a stretch to attribute all that success to branding and marketing, they certainly contributed.

They also laid the groundwork for CU's successful marketing campaign of recent years with the tagline "All Four: One." It aimed to show that the university had four campuses and offered something for every kind of learner, in addition to being an asset to Colorado and its communities. It can be seen in a marketing sponsorship of the high-speed train to Denver International Airport, on television, on social media, in print, and elsewhere. Annual research assessing the project shows a rising needle on favorable impressions.

Conclusion

The university we chose as undergrads so many years ago is still our continuing passion. The institution's public mission, paired with its accomplished and diverse students, faculty, staff, and alumni, propel it forward. While it was a long and winding brand journey, all campuses and units are now living under the same iconic University of Colorado umbrella. We project consistent, coordinated images and narratives. Our brand value is conveyed in strategic, thoughtful, clear, and compelling ways. It is good for us to have the perspective of nearly a decade to look back to see where we came from and where we are today. The success makes us proud as communications and marketing professionals. But it makes us even more proud that it is our alma mater.

Who Learns from Who in Participatory Practices of Place Brand Co-creation? An Exponential Random Graph Modelling Analysis on the Determinants of Learning Interactions in a Belgian Branding Process

Vidar Stevens and Peter Nafzger

Introduction

This chapter examines the conditions under which individual stakeholders are likely to engage with other participants in learning activities during processes of place brand co-creation. *Learning* is in this study understood as a behavioural activity of a single stakeholder to get a better understanding of another stakeholder's point of view, whereas *place brand co-creation* is interpreted as, 'participatory processes in which multiple stakeholders develop and realize place brands – i.e. signs that identify places, and evoke associations that imbue places with a certain meaning – for geographical locations with the aim to trigger positive associations about this area and to distinguish the place from other places' (Aitken and Campelo 2011; Eshuis et al. 2014: 5).

Especially, in areas that are challenged by population decrease or a lack of commercial activities, developing and forging new emotional and psychological associations with a place (like associations of cosmopolitan, inclusiveness or innovation) can stimulate new residents to settle in the area, or companies and end-users to get their goods and services from manufactures of a particular place (Eshuis and Klijn 2012; Kavaratzis and Ashworth 2005). This, in turn, contributes to the revitalization of this specific location. As such, place branding, as an activity to create new views and positive associations of a place's uniqueness, can boost certain areas for revenue.

Creating a place brand is, however, not an easy exercise. The biggest challenge for governments when developing place brands is to avoid creating a brand which is 'alien to the place' (Kavaratzis and Hatch 2013: 7). That is to say, developing a place brand without an identity, or at least an identity that is not recognized or appreciated by stakeholders that make use of the area. Therefore, governments are

V. Stevens (✉) · P. Nafzger
Mulier Institute, Utrecht, The Netherlands
e-mail: v.stevens@mulierinstituut.nl; p.nafzger@mulierinstituut.nl

S. M. Zavattaro (ed.), *Public Branding and Marketing*,
https://doi.org/10.1007/978-3-030-70505-3_15

advised to organize participatory processes, when creating a place brand – so-called processes of *brand co-creation* – to ensure that everybody who gives meaning to a place feels they have the possibility to add to the development of a place brand and, in the long run, are more likely to embrace the place brand in their individual marketing activities (Hatch and Schultz 2010; Ind and Bjerke 2007; Kavaratzis 2009; Kavaratzis and Hatch 2013).

This participatory view of branding and brand co-creation has featured surprisingly little in place branding studies as yet (but see Aitken and Campelo 2011 and Warnaby 2009 for exceptions). What particularly lacks in the literature is a micro-level understanding of the interactive dynamics of stakeholder involvement in participatory processes of place brand co-creation (Hana and Rowley 2008; Hankinson 2001; Lucarelli and Berg 2011; Niedomysl and Jonasson 2012; Syssner 2010). Hence, by looking at the conditions under which individual stakeholders are likely to learn from one another, we aim to get a better understanding of how through dialogue and learning place brands are co-created by a multitude of stakeholders who are active in a certain place area. Specifically, we examine the learning interactions between 12 stakeholders within a Flemish case of place brand co-creation in the coastal area.

In order to make inferences about the learning interactions among these stakeholders, the statistical network methodology of Exponential Random Graph Modelling (ERGM) is used. Up and until now, ERGMs have not been applied in place branding studies. The methodology has a longer tradition in the fields of conflict management and peace studies (Cranmer et al. 2012), disease studies (Rolls et al. 2013), and neurosciences (Teleford et al. 2011). ERGM is a methodology that aims to explain tie-formation (Goodreau 2007). This means that the methodology is capable of drawing inferential conclusions about why individuals have a tendency to connect (e.g. learning) with some people and not with others in collaborations. This outcome variable (i.e. tendency to connect or not), and thereby the overall purpose of the methodological tool, thus makes the ERGM methodology well-equipped for examining the learning dynamics between individual stakeholders in the Flemish process of place brand co-creation.

This chapter advances as follows. First, we elaborate on the phenomenon of place brand co-creation. Then it is discussed how the concept of learning is perceived in this study, and what factors possibly hinder of foster the learning dynamics among participants in processes of place brand co-creation. Subsequently, the case, data and chosen methodology are reviewed. In the results section, the findings of the ERGM analyses are presented. This chapter ends by reflecting on the main observations that can be drawn from this study regarding learning activities between individuals in participatory processes of place brand co-creation. In addition, in the end the value of the ERGM methodology as tool to analyse the interactive dynamics of place brand co-creation process is assessed.

The Phenomenon of Participatory Place Brand Co-creation

In today's society, towns, cities and regions are in competition with each other for visitors, labor, financial investments, residency and tourists (Connie et al. 2013). Especially in areas that are challenged by population decrease or a lack of commercial production, local governments go the extra mile to boost certain areas for revenue. Different strategies are used by local governments to revive areas; for example, Bilbao embarked on a process of economic and cultural restructuring by employing architecture as a catalyst for urban renewal (Klingmann 2007). Another public management strategy that is frequently used by municipalities to revitalize city areas, and which is also the locus of this study, is place branding (Eshuis et al. 2014; Eshuis and Klijn 2012; Giovanardi et al. 2013; Lucarelli and Berg 2011).

Place branding refers[1] to the development of brands for geographical locations with the aim to trigger positive associations about this area and to distinguish the place from other places (Eshuis et al. 2014). Brands are signs that identify places and evoke associations that imbue places with a certain meaning. For example, the city of Amsterdam used the place brand 'I *am*sterdam' to imbue the city with meanings such as 'inclusion' and 'exciting.' These associations make the city valuable to residents, visitors or other target groups both psychologically (e.g. it offers a sense that the city welcomes everybody) and socially (e.g. it offers opportunities for living out a cosmopolitan lifestyle). As such, the public management strategy of place branding influences people's ideas by forging particular emotional and psychological associations with a place. When done properly, these emotional and psychological associations can stimulate tourists to visit the area, or companies and end-users to get their goods and services from manufactures of a particular area – which, in turn, contributes to the revitalization of this specific area or location.

The use of place branding as a tool to revitalize city areas has gained scholarly attention. Lucarelli and Berg's (2011) meta-analysis identified 217 journal articles dealing with place branding in the period 1988–2009, and also in recent years new studies have been conducted on the use and value of place branding as public management tool (Andersson 2014; Braun 2012; Braun et al. 2014; Eshuis and Klijn 2012). One prominent research puzzle in the field of place branding which is, however, in need for more scholarly attention is the dynamics of the so-called participatory (place) branding approach (Kavaratzis and Hatch 2013: 4). Scholars have recognized that local governments are not the only 'producers' of a place brand because they are not the only stakeholders active in, and which give meaning to, a particular area that is in need for revitalization. If these stakeholders (i.e. companies, factories, residents and civil society organizations) are not sufficiently included

[1] There are also other definitions of the concept of place branding. Zenker and Braun (2010: 5), for instance, define place branding as 'a network of associations in the consumers' mind based on the visual, verbal, and behavioral expression of a place, which is embodied through the aims, communication, values and the general culture of the place's stakeholders and the overall place design' (Kavaratzis and Hatch 2013: 2).

in a participatory design process of a place brand, the branding efforts of local governments will only lead to a *brand alien* to the place (Houghton and Stevens 2011; Kavaratzis and Hatch 2013: 7; Therkelsen et al. 2010); that is, the efforts of a local government to brand a place will most likely run aground because the users of the area do not recognize or appreciate the place brand. Therefore, different scholars have promoted the idea of participatory branding processes, where, through *learning* between stakeholders who are active in the place area, a place brand is created which symbolically fits and represents the area (Bergvall 2012; Berthon et al. 2007).

The few studies that have been conducted on the topic of participatory processes of place branding merely indicated that a proper coordination of the participatory process regarding rights, roles, responsibilities and relationships (i.e. the four Rs of participatory place branding processes), eventually increases the understanding of a sense of place, fosters brand ownership among a multitude of stakeholders and assures brand authenticity (e.g. Aitken and Campelo 2011: 927). What is absent in the literature, however, is a clear micro-level understanding of how individual stakeholders *learn* in participatory processes of place brand co-creation, why individual stakeholders *learn* in such a way and how emergent learning interaction patterns between individual stakeholders impact the development of a place brand. Hence, we aim to explain in this chapter the patterns of learning behaviour of individual stakeholders in participatory brand co-creation processes.

The central argument that we make in this chapter is that despite the highflying expectations of scholars in the capacities of participatory processes of brand co-creation, the extent to which collaborations can actually contribute to the development of a place brand, which is appreciated and supported by a multitude of stakeholders, depends on the different intentions and forms of learning behaviour of individual stakeholders in the co-creation process. Participants of collaborations can, for example, hold different perceptions of a place, may be reluctant to collaborate and learn from one another or may paralyze the brand co-creation process for strategic reasons. Therefore, we argue that a clear micro-level understanding of the stakeholders' behaviours is necessary to get more knowledge of how participants of collaborative brand creation processes can be 'managed' to start moving together in the same direction. As such, this chapter makes a contribution to the place brand literature, in its emphasis on the learning behaviour of individual stakeholders in participatory place branding processes, and the factors that influence these micro-level interactive dynamics between stakeholders in these sort of co-creation processes. Before we focus on the hypotheses that are drafted to test the learning interactions between individuals in our empirical case, the next section specifies how learning is perceived is this research.

Learning Behaviour in Processes of Place Brand Co-creation

For this study, learning is understood as a specific form of actor behaviour. This definition enables us to examine how actors behave differently in relation to one another. Analytically, actor learning behaviour is regarded as distinctive from other, more self-referential, forms of actor behaviour, such as avoidance, defiance or manipulation (Oliver 1991). We anticipate 'learning actors' have a more positive attitude towards a trust-based circulation and cross-fertilization of ideas, views and brand stories across organizational boundaries (Sørensen and Torfing 2011), and to be more open to the formation of joint ownership and responsibility for the selection and implementation of a place brand.

Van den Bossche et al. (2011: 298) developed, tested and validated a measurement scale that captures 'learning behaviour' in collaborative work environments as a distinctive analytical term. In their article, Van den Bossche et al. (2011) ascribe nine different behavioural manifestations, clustered into three components, to the concept of 'learning behaviour'. First of all, they argue that 'learning actors', '(1) listen carefully to the contributions of others in discussions and deliberations, and (2) ask questions of clarification if something is unclear about the frames of reference of other participants in the collaboration.' They regard this cluster of behavioural manifestations as the socio-cognitive behaviour[2] of 'construction' (2011: 287). This sort of socio-cognitive behaviour helps individuals in collaborations to get an understanding of each other's ways of looking at a problem situation, or in our case the different visions of what the peculiarities, stories or history, of the place are.

The second cluster of behavioural learning manifestations is considered as the socio-cognitive behaviour of 'co-construction.' This entails that learning actors, '(3) draw conclusions from the ideas that are being discussed, (4) elaborate on each other's information and ideas, and (5) complement their initial frame of reference with information and ideas that follow from the interactions with the other actors in the collaborative arrangement.'

However, if a group of actors wants to move beyond conventional wisdoms and generate a story about a place that can be appreciated by a multitude of relevant stakeholders, then it is not enough for the actors to gain and refine understandings of each other's perspectives on a place based on dialogue and reflection alone. According to Van den Bossche et al. (2011: 287), discrepancies in understanding are only likely to be overcome if the divergence in meaning leads to deep-level processing and contestation of the diverse information and viewpoints among participants in a collaboration.

---

[2] Socio-cognitive behaviour is in the cognitive psychology literature understood as a form of human interaction and information process (Gioia and Sims 1986), whereby participants through dialogues and discussions with other people acquire more knowledge about the problem situation which, in turn, allows them to adapt or refine their original interpretation of the problem addressed.

This requires that 'learning actors', '(6) share the relevant information and ideas they have, (7) handle differences of opinions by addressing them directly, (8) verify opinions and ideas of other stakeholders by asking them critical questions, and (9) act upon comments of other stakeholders on their own perceptions and ideas.' These behaviour manifestations combined are by Van den Bossche et al. (2011: 298) understood as the socio-cognitive behaviour of 'constructive conflict', and account for the third cluster of behavioural manifestations that makes the concept of 'learning behaviour' distinctive from other forms of actor behaviour.

In this study, an adapted version (see Table 3) of the scale of Van den Bossche et al. (2011) is used to ask stakeholders who participated in the co-creation process for the development of a place brand for the Flemish coast about who of the other stakeholders in the collaboration behaved in the interactions with them as 'learning actors'. In this way, we are able to gain a notion from who (i.e. all 'other' possible stakeholders *j*) each of the stakeholders *i* tried to learn (Fig. 1). The stakeholders were not asked to indicate in relation to whom they showed learning behaviour, to avoid that a 'positive' self-reporting would error the results. In the next sections, we discuss theoretical arguments which are tested with the ERGM methodology to explain why a certain stakeholder *i* is more likely to show (signs of) learning behaviour in relation to some stakeholders *j* than others in participatory processes of place brand co-creation.

Fig. 1 Learning activities of stakeholder *i* with some possible stakeholders *j* (j^2 and j^4), but not with others (j^1 and j^3)

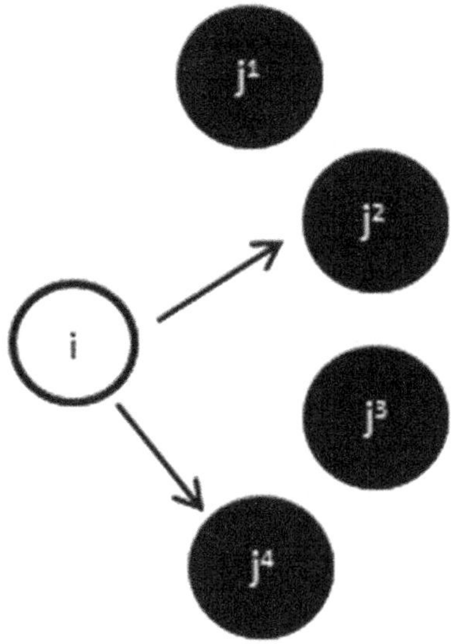

Theoretical Expectations

When looking at learning interactions between individuals in processes of place brand co-creation, various mechanisms can be at play that determine why there are differences in propensity of stakeholders to engage with one another in learning activities. Individuals will not have the same reasons to *decide* whether they connect with a network partner or not. To understand the richness of the interactive dynamics, it is therefore important to focus on a wide range of mechanisms to try and explain the variances in learning activities. Table 1 provides an overview of the nine hypotheses that will be analysed with the ERGM methodology, as well as the expectations in terms of causality. The hypotheses are derived from different theoretical streams, as we believe that different theories help explain individual learning behaviours place brand co-creation processes.

The Advocacy Coalition Framework (ACF) teaches us that individuals are cognitively limited in processing all information they encounter in decision-making processes. Due to these cognitive limitations, individuals use shortcuts, or heuristics, to simplify and filter the incoming information. According to Calanni et al. (2014: 903), 'they do this by filtering the incoming information based on how it lines up with their own beliefs and precognitions, so as to reduce the mental discomfort (also referred to as "cognitive dissonance" [Festinger, 1957]) new and often discordant information create'. Therefore, the ACF asserts that individuals have the tendency to focus on opinions that confirm their pre-existing expectations and

Table 1 Hypotheses for explaining learning interactions in place brand co-creation processes

Concept:	Hypothesis: *a stakeholder will show (more) learning behaviour towards a network alter, if...*	Theoretical roots
H1: Belief homophily	…this network alter has similar stories and views about the identity of the place area	Advocacy Coalition Framework
H2: Benevolence-based trust	…this network alter goes out of his/her way to bring the place brand co-creation process to a good end	Social Capital Theory
H3: Ingenuity	…this network alter is capable of combining different views and stories that are presented in the place brand co-creation process	Social Capital Theory
H4: Earlier bilateral contact	…this network alter had earlier bilateral contact with the network ego	Game Theory
H5: Chairman	…this stakeholder is the chairman of the place brand co-creation process	Collaborative Leadership Theory
H6: Societal value	…the network ego believes that this collaboration makes it easier for society at large to cope with the addressed issues	Policy Alienation Framework
H7: Outdegree	…the network ego intrinsically (i.e. by nature) connects more easily with other stakeholders	Social Network Analysis Literature
H8: Reciprocity	…the network alter has shown the intention to learn from the network ego	Social Network Analysis Literature
H9: Transitivity	…the value of this network alter's views has been scrutinized by another stakeholder in the network	Social Network Analysis Literature

beliefs rather than insights that contradict these (Weible and Sabatier 2005). Hence, we expect that a stakeholder is more likely to show learning behaviour towards a network partner in a place brand co-creation process, if the network alter has similar views and stories about the identity of the place.

The Social Capital Theory (SCT) identifies 'trust' as an influential determinant for starting, maintaining or terminating (micro-level) relationships with other individuals in collaborations (Agranoff and McGuire 2001; Lubell 2007). Mayer and Davis (1999), *inter alia*, distinguish between benevolence-based trust (i.e. trusting a person because this individual goes out of his/her way to bring the collaboration to a good end) and ingenuity (i.e. trusting a person based on his/her ability to combine different insights in collaborations to create a collaborative advantage). For that reason, we include both concepts as theoretical expectations in our explanatory model.

The 'shadow of the future' is a concept that stems from the game theory literature (Heide and Miner 1992). Essentially, it expresses the idea that people behave differently when they expect to interact with someone repeatedly over time. The common take on this is that the shadow of future is good for cooperation, as it allows individuals to escape prisoners' dilemma situations by using conditional retaliation strategies. Hence, if two individuals have had earlier bilateral contact, we expect that they are more likely to show learning behaviour to one another.

Tummers et al. (2009) have shown public stakeholders can feel estranged from programmes or other activities they develop, implement or organize. Such a feeling of estrangement is by them called 'policy alienation.' This is of concern, as for satisfactory development and implementation of a programme (like a place branding process), some identification with the programme is required. One of the conditions to measure the amount of 'policy alienation' a stakeholder experiences is 'societal meaninglessness,' that is, professional's perception that a programme is not actually dealing with specific societal problems, or with the provision of desirable public goods and services. We use a positive reading of this concept as a hypothesis for our analyses; expecting that stakeholders who believe that the place branding process is of benefit for society at large to cope with the targeted issues, are more likely to show learning behaviour towards another stakeholder.

The Collaborative Leadership Theory asserts that the core task of a manager of a collaborative process is to 'facilitate' stakeholders to make things happen (Ansell and Gash 2008, 2012). Ansell and Gash argue that there are three different roles of managers to facilitate actors make things happen in collaborative processes: as a *steward*, *mediator* or *catalyst*. As mediator, managers have to serve as broker between different involved stakeholders (2012: 11). Because stakeholders hold diverse perspectives and interests, they often do not see eye-to-eye. Therefore, managers are called upon to facilitate positive exchanges between different stakeholders through adjudication of conflict, to arbitrate between different positions, to stabilize the conditions for positive exchange and to promote trust-building. Against this background, we expect that a manager will show much learning behaviour towards network alters in the participatory process of place brand co-creation to take on the role of mediator as good as possible.

In collaborative settings, individual behaviour is also influenced by the presence or absence of relationships among other participants, and their behaviours, in the arrangement (Lubell et al. 2012). These influences, which are structural effects inherent to the collaboration itself, are in the social network analysis literature understood as 'tie-interdependence effects'. In this study, three tie-interdependence effects are included, namely, outdegree, reciprocity and transitivity.

We include the tie-interdependence effect of *outdegree* in the ERGM analysis, as it represents the basic tendency of individuals to have ties at all, and in a decision-theoretic approach, its parameter could be regarded as the balance of the benefits and costs of an arbitrary[3] tie (Snijders et al. 2010: 10).

The concept of *reciprocity* determines whether individuals forge dyadic learning relationships (Lee et al. 2012). Reciprocity entails individuals who *receive* information, advice or knowledge through (learning) activities return the favour to the sender in equivalent form. The expectation is that when an individual decides to learn from another participant, most of the time the learning activity will (over time) turn into a dyadic activity. There is one particular reason for this. Increased discussions on each other's views, stories and interests allow individuals to become more acquainted with both their intentions in the collaboration (Koppenjan and Klijn 2004). Such an increased awareness thus makes it easier to develop a modus vivendi for dealing with their differences in discussions on a place brand.

The third tie-interdependence term is the *transitivity* effect. Transitivity is within the social network analysis literature better known as the 'friends of friends become friends effect' or in graph-theoretic terminology: two-paths tend to be, or to become, closed. Within dialogue and learning situations, the concept of transitivity can best be interpreted as the tendency of an individual to learn from persons whose 'value' to learn from has been scrutinized by others from who the individual in question already is learning. In the case of Fig. 2, this would imply that person i forges a learning tie with person k, because person j, from whom person i is already learning, has positively scrutinized the learning value of person k.

There are various reasons why an individual would forge a transitive learning relationship. First of all, it allows an individual to increase the amount of knowledge it (already) obtained through dialogues with other participants in the collaboration. Second, and perhaps most important, transitive learning relationships enable the formation of 'learning alliances'. With 'learning alliances', we mean a social clustering of actors that together engage in creative search processes to develop a comprehensive place brand and ad campaign applications. In this sense, the transitivity effect can be regarded as the most basic level of a tightly clustered network (Lee et al. 2012). Hence, for the analyses, it is expected that when a stakeholder is motivated, and sees the value of, setting up a branding campaign with other stakeholders (and their organizations) involved, it will forge transitive learning relationships.

[3] Arbitrary meaning here that the other actor j ('receiver') has no characteristics or tie pattern making him/her especially attractive for actor i ('sender').

Fig. 2 Transitive learning relationship of actor *i*

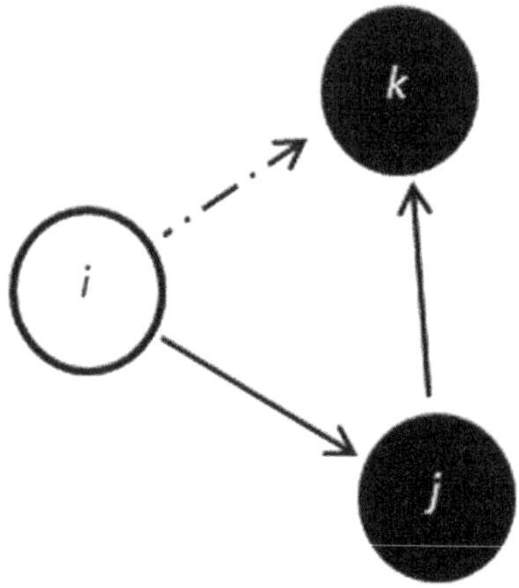

Introducing the Case

We now enter the empirical case examining the learning interactions in the Flemish process of place brand co-creation for the coastal area. Within the place brand co-creation process 12 different organizations participated. The co-creation process started in 2015. For this study, the specific focus is on an 'episode' between December 2015 and September 2016. During this episode, the stakeholders deliberately tried to develop a place brand to capture the identity of the coastal region of Flanders. This identity had to be a unique identifying mark, and ultimately deliver the intended benefits, like attracting new tourists, companies and residents to the area, while at the same time make residents more aware of urgent issues of climate change (e.g. coastal erosion, navigation, flood *protection* and hydrologic *issues*). As such, the aim of the group of stakeholders was not only to develop a slogan and logo (i.e. 'the Flemish Coast as a Metropolitan Area'), but also move beyond ad campaign applications and think of policy actions that could as a spill-over effect be implemented for the benefit of the co-branding campaign (e.g. new architectural buildings, ecological test hubs for climate adaption, better parking- and shuttle services around the beaches).

In total, the group of stakeholders met eight times during the episode we analyse. The outcome of the brand co-creation process was that under the brand umbrella of 'Metropolitan Coastal Landscape,' a new working group would be established that could give more substance to the vision of a city-like appearance of the Flemish coast including transitions in climate adaptation, innovation and coastal protection. We asked the involved stakeholders in the survey to 'evaluate' the episode we study (more information about the survey in the methodology section). The results of this evaluation are presented in Table 2. This table shows that the participatory brand co-creation process on the whole did not run smoothly. There was clearly quite some difference in opinion between stakeholders. However, eventually most respondents agree that the developed place brand is advantageous for the area in the long

Table 2 Evaluation of the dynamics in the place brand co-creation process

Evaluation criteria: *(On a scale from 1 to 10)*	Δ	σ	Min.	Max.
Do you think there was much difference in opinion between stakeholders?	5.3	3.4	3	8
Do you think the process ran smooth without any blockades or stagnations due to differences of opinion between stakeholders?	3.9	3.9	2	9
Do you think that much innovative branding ideas are developed in this project, compared to other projects	5.9	3.1	3	9
Do you think that the aims of the different stakeholders have been connected sufficiently?	6.7	1.9	3	8
Do you think in general that all involved actors have delivered a recognizable contribution to the development of the place brand?	6.0	2.7	3	8
Do you think the place brand is advantageous for the area in the long term?	6.4	3.9	2	9
Do you think conflicts between stakeholders were resolved effectively?	6.3	4.4	3	9
Do you think that substantive conflicts eventually lead to more mutual understanding between stakeholders?	6.9	3.6	3	9
Do you think that the results from the network collaboration can expect the continuous support of all the home-organizations of the representatives	6.8	2.7	4	9
Do you think – in general – that benefits exceed the costs of participatory co-branding process?	8.0	2.5	5	10
Do you think that new durable relationships have been developed between involved actors during the brand co-creation process?	7.5	1.8	4	9
Do you think that the extent to which actors trust each other has increased during the collaborative process?	7.4	2.6	4	10

term, that all involved actors delivered a recognizable contribution to the development of the place brand, and that much innovative branding ideas were developed in this project, compared to other projects. In addition, the stakeholders are very positive about the benefits that arise from collaborating, the new durable relationships that were developed during the process and the increase in trust among the stakeholders involved. Hence, this chapter sheds light on how, in a relatively 'successful' case of place brand co-creation, among a multitude of stakeholders (with competing needs, opinions, branding stories and ideas concerning the role of branding) interactions emerged, of which some were learning interactions, with the goal to create a unique and advantageous identity for the coastal region to attract new residents, tourists and (start-up) companies.

Data Collection and Analysis

The data collection proceeded in two steps. The involved stakeholders were asked to fill out a survey with standardized questions about the interactive dynamics within the participatory process of place brand co-creation for the Flemish coast.

Second, the respondents were interviewed to ask follow-up questions or questions of clarification about their responses to the standardized survey questions (Dawson 2002). All 12 stakeholders were surveyed and interviewed.

Table 3 provides an overview of the standardized questions (with the possible answer categories) of the survey. In several of the questions, the phrase 'list up to five stakeholders' is used. Such a phrase makes the interview question a so-called name-generator network question (Lubell et al. 2014). Name-generating survey questions help to minimize the burden for respondents to give answers to questions about (the behaviour of) their network alters (Marin and Hampton 2007). To elucidate, in the survey, respondents were not only asked about the learning behaviour of

Table 3 Standardized questions survey

Concept:	Standardized question:	Inspired by the work of:
DV: Individual learning behaviour (three dimensions) 1. Asking question of clarification: 2. Elaborating on information of others: 3. Sharing of information *On the basis of the responses to these questions the predictor variables of outdegree, reciprocity and transitivity were also modelled*	*Could you please list up to five stakeholders who most frequently* *... asked questions of clarification if something you said was unclear to them* *... elaborated on the information and ideas you exchanged* *... shared the relevant information and ideas they have*[a]*?*	Stevens (2018a, b; Van den Bossche et al. (2011))
Belief homophily	*Could you please list the stakeholders who had rather similar stories and views about the identity of the place area?*	Calanni et al. (2014: 910)
Benevolence-based trust	*Could you please list up to five stakeholders who did go out of their way to bring the collaboration to a good end?*	Mayer and Davis (1999)
Ingenuity	*Could you please list up to five stakeholders who were during the collaboration the most ingenious in combining stories, from the discussions, about the coastal area?*	Mayer and Davis (1999)
Earlier bilateral contact	*Could you please list the stakeholders with whom you have had earlier bilateral contact?*	
Chairman	*Who was the chairman of the collaboration?*	
Societal value	*On a scale from 1 ('not so much') to 10 ('very much'), could you please indicate to what extent you expect that this collaboration makes it easier for society at large to cope with issues surrounding the Flemish coastal area?*	Tummers (2012)

[a]For the analyses, the adjacency matrix of the answers to these three questions was transposed, to ensure that the rows of the adjacency matrix represent the behavioural action of the sender and the columns indicate who received the behavioural activity

network alters in relation to them (i.e. dependent variable), but also about other aspects of the collaboration and the contribution of every network alter therein (e.g. issues of trust and belief homophily). Hence, a survey about the composition, dynamics and interrelations among stakeholders in a collaborative setting can easily become a very time-consuming activity. As such, a name-generating question, which compels an interviewee to only name a maximum number of network alters (though, the respondent is allowed to name less network alters), helps to reduce the survey time and thereby creates 'extra time' to ask more survey questions (Marin and Hampton 2007: 5).

Yet, apart from 'minimizing the response-burden reasons,' there is another, more deliberate, reason – especially, for the dependent variable – why for this study it was decided to use such a name-generating technique for collecting data. Namely, the learning literature suggests that it is extremely difficult for both scholars and respondents to observe and pinpoint 'learning' or 'learning behaviour' in practice (*therefore* Van den Bossche et al. 2011 inter alia focus on 'learning behaviour' instead of the cognitive process of learning). A person can, for example, ask questions of clarification, but this does not have to entail that this person really tries to 'learn' from the other person he or she engages with.

Therefore, to ensure that the identified 'tie-formations' in the analyses really represent behavioural learning manifestations instead of ordinary interactions, the respondents were asked to indicate their most *substantive* learning ties (i.e. up to five network alters), and distinguish these from other, more 'weak', (learning) ties (Granovetter 1973). Hence, a tie in this research is specifically understood as 'a *substantive* behavioural effort of a stakeholder to learn from another stakeholder by showing socio-cognitive behavioural signs of construction, co-construction and/or constructive conflict in its interactions towards this other stakeholder.' Eventually, the survey and additional information from the follow-up interviews were coded and stored the data in a CSV-Excel file. Subsequently, the CSV-Excel file was used in the statistical program of *R* to perform the Exponential Random Graph Modelling (ERGM[4]) analyses.

To analyse which predictor variables influence the distinct dimensions of the dependent variable (i.e. the specific forms of individual learning behaviour), we followed an analysis strategy that is commonly used in ERGM studies (Goodreau 2007). For each dimension, we started with a so-called Bernoulli/Erdös-Rényi model, which with only a single term (i.e. edges) captures the density of the network under study (Goodreau 2007: 239). Second, we included the selected endogenous or tie-interdependence effects in the analysis. Within ERGM studies, this is considered as a 'reduced homogeneous realization-dependent model' (idem: 240). By analysing the reduced homogeneous realization-dependent model, a researcher starts to get an idea of which properties inherent to the participatory place branding process itself influence the interactive dynamics between stakeholders.

[4] This chapter has a compendium article in the journal *The Marketing Review* in which the value of ERGM models (in place branding) is explained compared to traditional methods like case study research and regression analysis (Stevens 2019).

Third, we added the exogenous nodal covariates (i.e. chairman and societal value) to the model with the tie-interdependence terms. Fourth, we analysed a model that includes the endogenous terms as well as the exogenous dyadic covariate effects (i.e. belief homophily, trust, earlier bilateral contact and ingenuity). If necessary, we tested a fifth model – a more optimized model of the dyadic covariate effects model – to ensure a better fit of the (final) explanatory model. Hence, for each dimension of the dependent variable, we searched for four, or sometimes five, 'best fitting models' – and from this sample of models, we selected the model that was most capable in explaining the interaction patterns in the empirical data.

Inspired by the work of Goodreau (2007), we used three general approaches to examine the goodness of fit of the models. First, we checked for degeneracy and model convergence. A minimum requirement for a model to fit well is for estimation of parameters to converge on finite parameter values. It must also be non-degenerate, that is, not place all of its probability mass on a few networks entirely unlike the observed network, such as a full or empty network.

Second, we compared the Akaike Information Criterion (AIC) between models. Models that exhibit dyadic independence can be fit with standard logistic regressions, which yield a likelihood measure for the model. Those models that are dyad dependent must be fit with MCMC, which also yields an estimate of the likelihood. We used the given likelihood to calculate AIC in order to compare models, with a lower AIC implying a significant increase in model fit.

Third, we viewed the goodness of fit by plotting the model against higher order statistics. This approach is described in detail by Hunter et al. (2008a, b). The logic entails generating new networks according to the probability distribution implied by the fit model. Because the normalizing constant is still present in the fit models, this must be done using the same MCMC approach employed during the estimation procedure. A statistic of interest is then calculated on the original network and on the set of networks generated from the model, and these are plotted for comparison. If the original network is inconsistent with the networks generated from the model, this suggests that the structure of the network differs from those predicted by the model, and the model is not well fit. Multiple statistics can be compared visually to provide detailed information about the systematic ways in which the data and the model predictions differ.

For this third approach, the higher order network statistics we specifically used to compare the selected models to include the following: in-degree, out-degree, edge-wise shared partners, dyad-wise shared partners and the minimum geodesic distance. Each of these higher order network statistics was plotted on the log-odds scale for the sake of greater visibility across the covered range of values. For a more detailed account of the analyses, research decisions and model selections for this study, as well as the code for performing the ERGM analysis and the replication data, please go to this chapter's Harvard Dataverse: https://doi.org/10.7910/DVN/HSCUNB. What follows are the results that emerged from our ERGM analyses. We start with the model that was the best fit for the learning interactions in the place branding process in terms of 'asking question of clarification'.

Results

Factors Influencing the Learning Behaviour of Asking Questions of Clarification

Table 4 presents the parameter estimates, standard errors and *P* values of the model that best *fits* the data regarding the learning behaviour of asking questions of clarification to another representative in the participatory place branding process. It turns out that this form of tie-formation is (well) explained by the following significant factors: chairman, earlier bilateral contact and (perceived) societal value. The parameter estimates (i.e. the thetas) of the model are presented in log-odds. On the basis of these parameter estimates, the odds ratio for each of the significant variables can be calculated by using the formula $e^{(log\text{-}odds)} = odds\ ratio$.

The odds ratio of the variable of 'chairman' equals $e^{(4.1057)} = 60.69$. In the analysis, the base of this variable was set at 1 – which means that the 'chairman' was used a reference category to compare the individual learning behaviour of ordinary stakeholders in the network to. Hence, the odds ratio shows that the odds of a chairman asking questions of clarification towards another representative in the participatory co-branding process compared to not asking questions of clarification are $1/e^{(4.1057)} = 0.0165$ times more than for an ordinary stakeholder in the collaboration. Or stated differently, the chairman used a very *hands-off* managing style to foster learning interactions (in terms of asking questions of clarification) in the collaborative arrangement.

The variable of societal value proves to be a significant parameter. It has an odds ratio of $e^{0.8856} = 2.42$. The odds ratio indicates that the more a stakeholder believes that the collaboration is beneficial for society at large, the higher the likelihood in odds that he or she will ask questions of clarification to network alters in the participatory co-branding process.

A final significant parameter is the variable of earlier bilateral contact; it has an odds ratio of $e^{0.0181} = 2.72$. As this value is greater than 1, this odds ratio shows that within participatory co-branding process, stakeholders were eager to asking questions of clarification to network alters with whom they already had collaborated (or met) in an earlier collaborative process.

Factors Influencing the Learning Behaviour of Elaborating on Information

Table 5 presents the parameter estimates, standard errors and *P* values of the model that best *fits* the data regarding the emergent learning interaction patterns of elaborating of information of others. Two significant parameters seem to explain for why a stakeholder in the participatory co-branding process elaborates more on the

Table 4 Factors influencing the learning behaviour of asking questions of clarification

	Reduced homogeneous realization-dependent model			Nodal covariate effects model			Dyadic covariate effects model			Optimized dyadic covariate effects model		
	Θ	(se)	*P* value	*θ*	(se)	*P* value	*θ*	(se)	*P* value	*θ*	(se)	*P* value
Endogenous factors												
Edges	−0.1911	0.7562	0.801	−11.0708	3.8739	0.00499**	−12.99244	4.73635	0.00701**	−13.4330	4.5002	0.00342**
Mutual	0.1824	0.5857	0.756	–	–	–	−0.66292	0.69000	0.338859	−0.5711	0.6619	0.38993
Gwodgree ($\alpha = 0.7$)	1.5949	0.3382	<1e-04***	0.8551	2.9556	0.77281	0.86305	2.99790	0.77393	0.7430	4.5158	0.86958
Ttriple	–	–	–	−0.1255	0.1802	0.48753	−0.15737	0.21255	0.46048	−0.1460	0.2052	0.47808
Nodal covariates												
Chairman				3.4005	1.2115	0.00580**	3.85899	1.51864	0.01232*	4.1057	1.4377	0.00504**
Societal value (perception)				0.7374	0.2756	0.00845**	0.85496	0.34357	0.01419*	0.8856	0.3273	0.00776**
Dyadic covariates												
Belief homophily							−0.09635	0.53287	0.85682			
Earlier bilateral contact							0.97102	0.45716	0.03570*	1.0018	0.4498	0.02773*
Ingenuity							0.02230	0.61130	0.97096			
Trust							0.57204	0.52393	0.27708			
Goodness of fit	LL = −80.17243, df = 4, AIC = 168.3, BIC = 179.9			LL = −69.6057, df = 6, AIC = 151.2, BIC = 168.5			LL = −66.13461, df = 11, AIC = 154.3, BIC = 186			LL = −66.78044, df = 8, AIC = 149.6, BIC = 172.6		

Table 5 Factors influencing the learning behaviour of elaborating on information

	Reduced homogeneous realization-dependent model			Nodal covariate effects model			Dyadic covariate effects model			Optimized dyadic covariate effects model		
	Θ	(se)	*P* value	*θ*	(se)	*P* value	*θ*	(se)	*P* value	*θ*	(se)	*P* value
Endogenous factors												
Edges	−0.9613	0.2571	0.000276***	−11.07.08	3.8739	0.00499**	−9.9613	4.0130	0.0144*	−10.4343	3.7153	0.00579**
Mutual	0.1787	0.5848	0.760367	–	–	–	−0.6716	0.7638	0.3810	−0.6438	0.7531	0.39431
Gwodgree ($\alpha = 0.7$)	–	–	–	0.8551	2.9556	0.77281	0.6815	6.1229	0.9116	–	–	–
Ttriple	–	–	–	−0.1255	0.1802	0.48753	−0.4202	0.3147	0.1843	−0.3411	0.2994	0.25686
Nodal covariates												
Chairman				3.4005	1.2115	0.00580**	2.3067	1.1231	0.0421*	2.3536	1.1545	0.04361*
Societal value (perception)				0.7374	0.2756	0.00845**	0.6888	0.2672	0.0111*	0.6928	0.2725	0.01225*
Dyadic covariates												
Belief homophily							−0.1043	0.5527	0.8506	–	–	–
Earlier bilateral contact							0.6518	0.4461	0.1466	0.6512	0.4504	0.15077
Ingenuity							−0.5653	0.6856	0.4113	−0.5489	0.6646	0.41043
Trust							−0.5970	0.5812	0.3063	−0.6864	0.5367	0.20332
Goodness of fit	LL = −79.18852, df = 2, AIC = 162.4, BIC = 168.1			LL = −69.60567, df = 6, AIC = 151.2, BIC = 168.5			LL = −65.59517, df = 11, AIC = 153.2, BIC = 184.9			LL = −66.01051, df = 8, AIC = 148, BIC = 171.1		

information and ideas of some network alters, and not, or to a lesser extent, on the information and ideas of others.

First of all, the parameter of chairman is again a significant parameter. The odds ratio equals $e^{(2.3536)} = 10.52$. The base of this variable was set at 1 – which means that the 'chairman' was used a reference category to compare the individual learning behaviour of ordinary stakeholders in the network. Hence, the odds ratio shows that the odds of a chairman elaborating on the information and ideas of other stakeholders in the participatory co-branding process compared to not elaborating on the information and ideas are $1/e^{(2.3536)} = 0.5002$ times more than for an ordinary stakeholder in the collaboration. This means that *the* laissez-faire approach of the chairman was helpful to stimulate learning interactions (in terms of elaborating on information and ideas of network alters) among the stakeholders in the participatory co-branding process.

Second, the variable of societal value proves to be a significant parameter. It has an odds ratio of $e^{0.6928} = 1.9993$. The odds ratio thus indicates that the more a stakeholder believes that the collaboration is beneficial for society at large, the higher the likelihood in odds that he or she will elaborate on the information and ideas of network alters in the participatory co-branding process.

Factors Influencing the Learning Behaviour of Sharing of Information

The final ERGM analyses focused on the emergent learning interaction patterns, in terms of information-sharing, of the participatory co-branding process. The respondents were asked in the survey to indicate who of the other representatives in the participatory process of place branding were most open in giving information and ideas to them. Table 6, *inter alia*, presents the parameter estimates, standard errors and *P* values of the model (i.e. optimized dyadic covariate effects model) which best *fits* this relational data.

The parameter of outdegree turns out to be significant. The odds ratio equals $e^{1.4634} = 4.32$. Hence, the odds ratio indicates that stakeholders who intrinsically (i.e. by nature) connect more easily with other stakeholders are more likely to share information and ideas with network alters.

The significant variable of transitivity has a negative theta ($\Theta = -0.6026$). This means that the variable has an odds ratio of $e^{-0.6026} = 0.55$. As this value is smaller than 1, this odds ratio shows that within the participatory co-branding process, stakeholders were not eager to form strong (learning) alliances – and thus did not share information and ideas among an 'inner-group of stakeholders'. As the variable of reciprocity is not a significant factor, we can infer that most sharing of information and ideas went at random without a specific preference for network alters.

Another significant parameter is the variable of earlier bilateral contact. It has an odds ratio of $e^{1.7760} = 5.91$, which means a stakeholder in the participatory

Table 6 Factors influencing the learning behaviour of sharing of information

	Reduced homogeneous realization-dependent model			Nodal covariate effects model			Dyadic covariate effects model			Optimized dyadic covariate effects model		
	Θ	(se)	*P* value	θ	(se)	*P* value	θ	(se)	*P* value	θ	(se)	*P* value
Endogeneous factors												
Edges	1.0449	1.2104	0.38962	−2.9852	3.0894	0.3358	−1.3908	3.5693	0.697473	0.3281	1.2646	0.795747
Mutual	−0.3472	0.6689	0.60463	−0.4685	0.6745	0.4886	−0.7795	0.7002	0.267812	−0.5721	0.6773	0.399958
Gwodgree ($\alpha = 0.7$)	0.4379	0.4404	0.00141**	1.0361	1.1341	0.3627	1.1070	0.8770	0.209279	1.4634	0.5193	0.005634**
Ttriple	−0.5673	0.2693	0.03714*	−0.6206	0.2593	0.0182*	−0.6370	0.2566	0.014421*	−0.6026	0.2576	0.020951*
Nodal covariates												
Chairman				1.2674	0.9516	0.1853	1.0027	1.0980	0.362947	0.5908	0.7101	0.407036
Societal value (perception)				0.2662	0.2120	0.2116	0.1161	0.2414	0.631365	–	–	–
Dyadic covariates												
Belief homophily							0.4126	0.5634	0.465324			
Earlier bilateral contact							1.7607	0.4432	0.000121***	1.7760	0.4434	0.000106***
Ingenuity							1.4067	0.6817	0.041203*	1.5154	0.6459	0.020554*
Trust							−0.9770	0.6044	0.108622	−0.8569	0.5660	0.132570
Goodness of fit	LL = −75.24069, df = 5, AIC = 160.5, BIC = 174.9			LL = −73.44748, df = 7, AIC = 160.9, BIC = 181.1			LL = −62.64306, df = 11, AIC = 147.3, BIC = 179			LL = −63.75145, df = 9, AIC = 145.5, BIC = 171.4		

co-branding process was more likely to share information and ideas with a network alter, if the network alter already collaborated in an earlier project with the network ego.

The last significant effect of this analysis is the variable of ingenuity. The theta has a value of 1.5154, which means that the odds ratio equals 4.55. Thus, a stakeholder was in the participatory co-branding process more likely to share information and ideas with a network alter, if the network alter was regarded by the stakeholder as 'very ingenious' in combining different branding stories, from the discussions, about the coastal area.

Discussion and Conclusion

By using the ERGM methodology, we were capable of generating several insights in the social processes and tie-formation mechanisms which generated the emergent learning activities between stakeholders in the Flemish place brand co-creation process. Based on these ERGM analyses, we can draw four inferences about the decision of a single network ego to learn from a network alter in participatory processes of place brand co-creation.

First of all, the ERGM analyses reveal that relational history matters. That is to say, the factor of earlier bilateral contact came up a significant predictor variable for the socio-cognitive learning behaviours of construction and constructive conflict: meaning that network egos were more likely to ask questions of clarification and share ideas and information, if they knew the network alters from earlier collaborations. Based on these outcomes, we can conclude that as stakeholders collaborate repeatedly in different and earlier branding projects, and thereby define mutual norms of equity, they place greater reliance on each other not to act opportunistically – resulting in a more stable modus vivendi. Better place brands (in terms of unique identity and support) are thus more likely to be developed, if the co-creation process happens between stakeholders who already 'know each other' and have a record of collaborating.

Second, the ERGM analyses show that learning activities between stakeholders in participatory processes of brand co-creation are influenced by a person's perception of the societal relevance of the place branding process. If an individual stakeholder believes it is meaningful to use a place brand to revitalize an area, he or she is more likely to show learning behaviour towards another stakeholder in the participatory process of place brand co-creation. At the operational level, this means that a 'socially engaged' stakeholder will be consciously assessing during the process of place brand co-creation whether the branding activities, and also the activities of the group of stakeholders, contribute to solving specific societal problems in the targeted area, like unemployment, housing, etc. As such, it can be concluded that more in-depth learning activities between stakeholders in participatory processes of place brand co-creation occur, when stakeholders believe and experience that their actions have an immediate impact on handling specific societal problems or delivering public goods.

Third, it turns out that less managerial involvement (see, the variable of chairman) increases the amount of learning interactions between stakeholders in participatory processes of place brand co-creation. Fourth, the factor of ingenuity came up as significant predictor variable. This means that stakeholders who were capable of combining different branding stories into one coherent vision – and, as such, catalyze the participatory process of place brand co-creation – were more likely to receive different learning ties from network alters. To some extent, these 'ingenious' stakeholders can be regarded as brokers, that is, individuals who play an intermediary role in ameliorating differences in perceptions and opinions between network partners (Stevens and Agger 2017). From our data, it is hard to tell whether these 'brokers' took on these roles for their own gain or for the benefit of the collective. What we can conclude, however, is that the role the 'ingenious' stakeholders took on in the participatory process of place brand co-creation, ensured that they sat in positions where their involvement accrued a form of power within the collective, that is, the power to receive more information and ideas from other network partners about what the brand message should entail.

Of course, there can be many other tie-formation effects at play in learning activities during participatory processes of place brand co-creation. For this empirical, we just used a few hypotheses from the available scientific literature to study the learning dynamics within the participatory place branding process in the Flemish governmental context. A downside to the analysis, definitely, is that it does not allow to get a gist of how much the stakeholders as a group learned and how much each individual stakeholder learned. That being said, with the help of the ERGM methodology we got a sense of under what circumstances individuals are likely to interact with each other in participatory processes of place branding in the public sector.

Of course, the method of Exponential Random Graph Modelling has its limitations. The method still undergoes a lot of changes to become a more mature methodology for scholarly use. Nevertheless, for the field of place branding we see a huge potential for gaining a better understanding in what Pasquinelli (2011: 234) and Gummesson (2002) have called the 'relationscape of a place brand', that is, the nano-relationships between 'many-headed' stakeholders that behind a brand are potentially capable of building communities through collective dialogue, exchange, and in some cases conflict among the stakeholders. Specifically, the ERGM allows to gain more insights in the social processes and tie-formation mechanisms which generate the emergent interaction patterns in participatory processes of place branding. The potential of the ERGM is in our opinion optimally used as an addition to case study research – where the ERGM methodology provides an overview of the 'bones' of the interactive structures and the case study method provides more 'flesh', and thus substance, to the ERGM observations. In the end, we hope that this chapter contributes to the exploration of new venues in the field of place branding, and encourages scholars to add new equipment to their methodological toolbox by making use of relatively new research methods, like Exponential Random Graph Modelling.

Acknowledgements This project has received funding from the Vital Cities and Citizens programme of the Erasmus University Rotterdam.

References

Agranoff, R. & McGuire, M. (2001). Big questions in public network management research. *Journal of Public Administration Research & Theory, 11*(3), 295–326.

Aitken, R., & Campelo, A. (2011). The four Rs of place branding. *Journal of Marketing Management, 27*(9–10), 913–933.

Andersson, I. (2014). Placing place branding: An analysis of an emerging research field in human geography. *Geografisk Tidsskrift-Danish Journal of Geography, 114*(2), 143–155.

Ansell, C. & Gash, A. (2008). Collaborative governance in theory and practice. *Journal of Public Administration Research & Theory, 18*(4), 543–571.

Bergvall, S. (2012). Brand ecosystems. In *Brand Culture* (J. Schroeder & M.S. Morling, Eds.). Routledge.

Berthon, P. R., Pitt, L. F., McCarthy, I., & Kates, S. M. (2007). When customers get clever: Managerial approaches to dealing with creative consumers. *Business Horizons, 50*(1), 39–47.

Braun, E. (2012). Putting city branding into practice. *Journal of Brand Management, 19*(4), 257–267.

Braun, E., Eshuis, J., & Klijn, E. H. (2014). The effectiveness of place brand communication. *Cities, 41*(1), 64–70.

Calanni, J. C., Siddiki, S. N., Weible, C. M., & Leach, W. D. (2014). Explaining coordination in collaborative partnerships and clarifying the scope of the belief homophily hypothesis. *Journal of Public Administration Research and Theory, 25*(1), 901–927.

Connie, S., Jonas, L., Michael H., & Jørgen, O. B. (2013). Experiencing spatial design, Chapters. In J. Sundbo & F. Sørensen (Ed.), *Handbook on the Experience Economy* (chapter 16, pp. 310–324). Edward Elgar Publishing.

Cranmer, S. J., Desmarais, B. A., & Menninga, E. J. (2012). Complex dependencies in the alliance network. *Conflict Management and Peace Science, 29*(3), 279–313.

Dawson, C. (2002). *Practical research methods. A user-friendly guide to mastering research techniques and projects.* Trowbridge: Cromwell Press.

Eshuis, J., & Klijn, E. H. (2012). *Branding in governance and public management.* New York: Routledge.

Eshuis, J., Klijn, E. H., & Braun, E. (2014). Place marketing and citizen participation: Branding as strategy to address the emotional dimension of policy making? *International Review of Administrative Sciences, 80*(1), 151–171.

Giovanardi, M. et al (2013). Towards brand ecology: An analytical semiotic framework for interpreting the emergence of place brands. *Marketing Theory, 13*(3), 365–383.

Gioia, D.A. & Sims, H.P. (1986). Cognition-behavior connections: Attribution and verbal behavior in leader-subordinate interactions. *Organizational Behavior and Human Decision Processes, 37*(2), 197–229.

Goodreau, S. M. (2007). Advances in exponential random graph (p*) models applied to a large social network. *Social Networks, 29*(1), 231–248.

Granovetter, M. S. (1973). The strength of weak ties. *American Journal of Sociology, 78*(6), 1360–1380.

Gummesson, E. (2002). Relationship marketing in the new economy. *Journal of Relationship Marketing, 1*(1), 37–57.

Hatch, M. J., & Schultz, M. (2010). Toward a theory of brand co-creation with implications for brand governance. *Journal of Brand Management, 17*(8), 590–604.

Hanna, S. & Rowley, J. (2008). An analysis of terminology use in place branding. *Place Branding and Public Diplomacy, 4*, 61–75.
Hankinson, G. (2001). Location branding: A study of the branding practices of 12 English cities. *Journal of Brand Management, 9*, 127–142.
Heide, J.B. & Miner, A.S. (1992). The Shadow Of The Future: Effects Of Anticipated Interaction And Frequency Of Contact On Buyer-Seller Cooperation. *Academy of Management Journal, 35*(2), 265–291.
Houghton J.P., Stevens A. (2011) *City Branding and Stakeholder Engagement*. In: Dinnie K. (eds) City Branding. Palgrave Macmillan, London. https://doi.org/10.1057/9780230294790_6.
Hunter, D. H., Goodreau, S. M., & Handcock, M. S. (2008a). Goodness of fit of social network models. *Journal of the American Statistical Association, 103*(481), 248–258.
Hunter, D. H., Handcock, M. S., Butts, C. T., Goodreau, S. M., & Morris, M. (2008b). ERGM: A package to fit, simulate and diagnose exponential-family models for networks. *Journal of Statistical Software, 24*(3), 1–29.
Ind, N. & Bjerke, R. (2007). *Branding governance: A participatory approach to the brand building process*. John Wiley & Sons.
Kavaratzis, M., & Ashworth, G. J. (2005). City branding: An effective assertion of identity or a transitory marketing trick? *Tijdschrift voor Economische en Sociale Geografie, 96*(5), 506–514.
Kavaratzis, M. (2009). Cities and their brands: Lessons from corporate branding. *Place Branding and Public Diplomacy, 5*, 26–37.
Klingmann, A. (2007). *Brandscapes: Architecture in the experience economy*. MIT Press.
Koopenjan, J. & Klijn, E. (2004). *Managing uncertainties in networks*. Routledge.
Kavaratzis, M., & Hatch, M. J. (2013). The dynamics of place brands: An identity-based approach to place branding theory. *Marketing Theory, 13*(1), 69–86.
Lee, Y., Lee, I. W., & Feiock, R. C. (2012). Inter-organizational collaboration networks in economic development policy: An exponential random graph model analysis. *Policy Studies Journal, 40*(3), 547–573.
Lubell, M. (2007). Familiarity breeds trust: Collective action in a policy domain. *The Journal of Politics, 69*(1), 237–250.
Lubell, M., Scholz, J., Berardo, R., & Robins, G. (2012). Testing policy theory with statistical models of networks. *Policy Studies Journal, 40*(3), 351–374.
Lubell, M., Robins, G., & Wang, P. (2014). Network structure and institutional complexity in an ecology of water management games. *Ecology and Society, 19*(4), 1–14.
Lucarelli, A., & Berg, O. P. (2011). City branding: A state-of-the-art review of the research domain. *Journal of Place Management and Development, 4*(1), 9–27.
Marin, A., & Hampton, K. N. (2007). Simplifying the personal network name generator: Alternatives to traditional multiple and single name generators. *Field Methods, 19*(2), 163–193.
Mayer, R. C., & Davis, J. H. (1999). The effect of the performance appraisal system on trust for management: A field quasi-experiment. *Journal of Applied Psychology, 84*(1), 123–136.
Niedomysl, T. & Jonasson, M. (2012). Towards a theory of place marketing. *Journal of Place Management and Development, 5*(3), 223–230.
Oliver, C. (1991). Strategic responses to institutional processes. *Academy of Management Review, 16*(1), 145–179.
Pasquinelli, C. (2011). Place branding and cooperation: Can a network of places be a brand? In A. Pike (Ed.), *Brands and branding geographies* (pp. 230–247). Cheltenham: Edward Elgar Publishing Ltd.
Rolls, D.A., Sacks-Davis, R., Jenkinson, R., McBryde, E., Pattison, P., Robins, G., & Hellard, M. (2013). Hepatitis C Transmission and Treatment in Contact Networks of People Who Inject Drugs. PLOS, https://doi.org/10.1371/journal.pone.0078286.
Snijders, T. A. B., Van de Bunt, G. G., & Steglich, C. E. G. (2010). Introduction to stochastic actor-based models for network dynamics. *Social Networks, 32*(1), 44–60.
Sørensen, E. & Torfing, J. (2011). Enhancing collaborative innovation in the public sector. *Administration & Society, 43*(8), 842–868.

Stevens, V. (2018a). Individual learning behaviour in collaborative networks. In C. Dunlop, C. Radaelli, & P. Trein (Eds.), *Learning in public policy: Analysis, modes and outcomes* (pp. 97–122). Basingstoke: Palgrave-MacMillan.

Stevens, V. (2018b). Replication data for: Exponential Random Graph Modelling to analyze interactions in participatory practices of place branding. *Harvard Dataverse, V2*. https://doi.org/10.7910/DVN/HSCUNB.

Stevens, V., & Agger, A. (2017). Managing collaborative innovation networks – practical lessons from a Belgian spatial planning initiative. *Journal of Public Administration and Governance, 7*(3), 154–173.

Stevens, V. (2019). A concept in search of a stronger foundation: Three ideal types of place branding architectures. *The Marketing Review, 19*(3–4), 279–292.

Syssner, J. (2010). Place branding from a multi-level perspective. *Place Brandin and Public Diplomacy 6*, 36–48 (2010). https://doi.org/10.1057/pb.2010.1.

Teleford, Q.T., Simpson, S.L., Burdette, J.H., Hayasaka, S., & Laurienti, P. (2011). The Brain as a Complex System: Using Network Science as a Tool for Understanding the Brain. *Brain Connectivity, 1*(4), 295–308.

Therkelsen, A., Halkier, H., & Jensen, O. B. (2010). Branding Aalborg: Building community or selling place? In G. Ashworth & M. Kavaratzis (Eds.), *Towards effective place brand management* (pp. 136–155). Cheltenham: Edward Elgar Publishing Ltd.

Tummers, L. et al (2009). Policy alienation of public professionals. *Public Management Review, 11*(5), 685–706.

Tummers, L. (2012). Policy alienation of public professionals: The construct and its measurement. *Public Administration Review, 72*(4), 516–525.

Van den Bossche, P., Gijselaers, W., Segers, M., Woltjer, G., & Kirschner, P. (2011). Team learning: Building shared mental models. *Instructional Science, 39*(3), 283–301.

Warnaby, G. (2009). Toward a service-dominant place marketing logic. *Marketing Theory, 9*(4), 403–423.

Weible, C. & Sabatier, P. (2005). Comparing policy networks: Marine protected areas in California. *Policy Studies Journal, 33*(2), 181–201.

Zenker, S. & Braun, E. (2010). Branding a city: A conceptual approach for place branding and place brand management. Paper presented at the 39th EMAC Annual Conference 2010, Frederiksberg, Denmark.

Conclusion

Staci M. Zavattaro

I will end as the book began – with a nod to the rapid changes in our world affecting communities globally. The COVID-19 pandemic continues to shake our sense of place and urgency. Many people thought something like this could not happen again in an advanced society. Epidemiologists, though, warned about a virus-spreading crisis repeatedly (Cox et al. 2003). "A pandemic is considered likely if the novel virus is readily transmissible from person to person and causes disease and if there are large populations worldwide that lack immunity to the virus" (Cox et al. 2003, p. 1801).

Some popular press has covered how to brand during a pandemic, focusing largely on the corporate sector. According to Michelson (2020) in her Forbes article, "how companies and leaders respond in this crisis may well define leadership – and their brands – for many years to come" (para. 4). As she points out, many organizations had to pivot (perhaps another word to define 2020) when pandemic conditions shut down normal operations. Some restaurants turned their empty dining rooms into spaces for local makers to sell goods. Some companies shifted production totally to producing vital personal protective equipment for frontline workers. Those brand strategies could build trust and brand citizenship (Michelson 2020).

Places and destinations are of course not immune from this change and associated need to pivot. For instance, Hawaii announced a temporary residence program called Movers and Shakas to revitalize the local economy by offering free roundtrip flights to people who can move to the state for at least 30 days to work remotely (Elassar 2020). Additionally, the World Travel and Tourism Council released its Safe Travels guidelines and stamp of approval to signal places and organizations complying with their recommended safety protocols (WTTC 2020).

While many of the chapters in this book were written before COVID-19, their lessons still apply as the pandemic continues in many countries at the time of this

S. M. Zavattaro (✉)
School of Public Administration, University of Central Florida, Orlando, FL, USA
e-mail: Staci.Zavattaro@ucf.edu

S. M. Zavattaro (ed.), *Public Branding and Marketing*,
https://doi.org/10.1007/978-3-030-70505-3_16

writing. While New Zealand declared itself virus free in late December 2020 (Ministry of Health 2020), the United States still sets records with its daily death tolls. Stories like those will decidedly affect each nation's brand image – how people perceive each nation. The same can be said for schools taking measures to keep students and teachers safe or not. Internal workplaces will have to take precautions to keep people safe on property. Just as the economic recovery will take years, the same can be said for reputation and image recovery damaged during the crisis.

References

Cox, N. J., Tamblyn, S. E., & Tam, T. (2003). Influenza pandemic planning. *Vaccine, 21*, 1801–1803.

Elassar, A. (2020). *Hawaii is offering free round trips to remote workers who want to live there temporarily*. Retrieved from: https://www.cnn.com/2020/12/05/business/hawaii-free-trip-remote-workers-trnd/index.html.

Michelson, J. (2020). *Branding in this pandemic crisis – what works and what doesn't*. Retrieved from: https://www.forbes.com/sites/joanmichelson2/2020/03/31/brand-citizenship-in-this-pandemic-crisis%2D%2Dwhat-works-what-doesnt/?sh=5f6990911d40.

Ministry of Health. (2020). *No new cases of COVID-19*. Retrieved from: https://www.health.govt.nz/news-media/media-releases/no-new-cases-covid-19-55.

WTTC. (2020). *Safe Travels: Global protocols & stamp for the new normal*. Retrieved from: https://wttc.org/COVID-19/Safe-Travels-Global-Protocols-Stamp.

Correction to: From a Dog's Breakfast to a King's Table: Branding and Marketing a Major Research University

Julia Cummings and Ken McConnellogue

Correction to: Chapter 14 in: S. M. Zavattaro (ed.), *Public Branding and Marketing*, https://doi.org/10.1007/978-3-030-70505-3_14

The text "*We think this should this be "propel" since it's about mission + our people. That's the way we had it in the original draft*" and "*We submitted some short bios to Staci. Will those go after the conclusion?*" has been accidently included in the chapter. This has now been corrected throughout the chapter.

The updated online version of this chapter can be found at
https://doi.org/10.1007/978-3-030-70505-3_14

S. M. Zavattaro (ed.), *Public Branding and Marketing*,
https://doi.org/10.1007/978-3-030-70505-3_17

Index

S. M. Zavattaro (ed.), *Public Branding and Marketing*,
https://doi.org/10.1007/978-3-030-70505-3

The manufacturer's authorised representative in the EU is Springer Nature Customer Service Centre GmbH, Europaplatz 3, 69115 Heidelberg, Germany. If you have any concerns regarding our products, please contact ProductSafety@springernature.com

Printed and bound by CPI Group (UK) Ltd, Croydon, CR0 4YY
30/09/2026
02225497-0013